Automated Highway Systems

Automated Highway Systems

Edited by

Petros A. Ioannou

Center for Advanced Transportation Technologies
University of Southern California
Los Angeles, California

PLENUM PRESS • NEW YORK AND LONDON

Library of Congress Cataloging in Publication Data

Automated highway systems / edited by Petros A. Ioannou.
p. cm.
Includes bibliographical references and index.
ISBN 0-306-45469-6
1. Intelligent Vehicle—Highway Systems—United States. 2. Electronics in transportation. 3. Motor vehicles—Automatic control. I. Ioannou, Petros A.
TE228.3.A93 1996
388.3'12—dc21 96-48638
CIP

Information on a software package called SPACING for calculating safe intervehicle spacing, capacity and studying the effects of vehicle characteristics on vehicle collisions can be obtained by contacting the editor Petros Ioannou at ioannou@bode.usc.edu [tel. (213) 740 4452].

ISBN 0-306-45469-6

A Division of Plenum Publishing Corporation
233 Spring Street, New York, N. Y. 10013

http://www.plenum.com

10 9 8 7 6 5 4 3 2 1

Printed in the United States of America

CONTRIBUTORS

MATTHEW J. BARTH • College of Engineering, Center for Environmental Research and Technology, University of California, Riverside, California 92521

DAVID A. BLANCETT • Advanced Technology and Development Center, Northrop Grumman Corporation, Pico Rivera, California 90660-0158

ANNE BRESNOCK • California State Polytechnic University, Pomona, California 91766

F. BROWAND • Department of Aerospace Engineering, University of Southern California, Los Angeles, California 90089

GREGORY H. DAVIS • Advanced Technology and Development Center, Northrop Grumman Corporation, Pico Rivera, California 90660-0158

RANDOLPH W. HALL • Department of Industrial and Systems Engineering, University of Southern California, Los Angeles, California 90089-0193

PAUL HELLMAN • California PATH Program, University of California at Berkeley, California 94804

FU-SHENG HO • Center for Advanced Transportation Technologies, University of Southern California, Los Angeles, California 90089-2562

PETROS IOANNOU • Center for Advanced Transportation Technologies, University of Southern California, Los Angeles, California 90089-2562

ALEXANDER KANARIS • Center for Advanced Transportation Technologies, University of Southern California, Los Angeles, California 90089-2562.

IOANNIS KANELLAKOPOULOS • Department of Electrical Engineering, University of California, Los Angeles, California 90089

EDWARD H. LECHNER • Science Applications International Corporation, Systems Control Technology Group, Los Altos, California 94022

MARK A. MILLER • California PATH Program, University of California, Berkeley, California 94804

PROKOPIOS PANAGIOTOU • Communication Sciences Institute, University of Southern California, Los Angeles, California 90089-2565

SHERYL A. PAYNE • Advanced Technology and Development Center, Northrop Grumman Corporation, Pico Rivera, California 90660-0158

ANDREAS POLYDOROS • Communication Sciences Institute, University of Southern California, Los Angeles, California 90089-2565

MOHAMMAD SHARAFSALEH • California PATH Program, University of California at Berkeley, California 94804

STEVEN E. SHLADOVER • California PATH Program, University of California, Berkeley, California 94804

WILLIAM B. STEVENS • National AHS Consortium, Bethesda, Maryland 20817

CHARLES E. TAYLOR • Advanced Technology and Development Center, Northrop Grumman Corporation, Pico Rivera, California 90660-0158

P. TOKUMARU • Department of Aerospace Engineering, University of Southern California, Los Angeles, California 90089

MASAYOSHI TOMIZUKA • Department of Mechanical Engineering, University of California, Berkeley, California 94804

H.-S. JACOB TSAO • California PATH Program, University of California, Berkeley, California 94804

JERRY D. WARD • Ward Consulting, El Cajon, California 92019.

YOUNGBIN YIM • California PATH Program, University of California, Berkeley, California 94804

M. ZABAT • Department of Aerospace Engineering, University of Southern California, Los Angeles, California 90089-1191

FOREWORD

Today's vehicle highway system operates with remarkable safety and efficiency, given the complex factors that are at work in even the most mundane trips. In a typical U.S. freeway environment, the weather, the driver, the vehicle, other vehicles, other drivers, ramp signaling systems, pavement quality, and the roadway configuration and markings smoothly interact millions of times a day, to serve our travel needs and move the goods upon which we thrive. Even so, the system is far from perfect, with congestion and accidents all too common. Clearly, the system is being stressed, a result of ever-increasing travel demands as our population grows and shifts, and as our life-styles depend increasingly on mobility. The promise of automating our highways, once considered farfetched, is seen by increasing numbers of transportation professionals and public decision makers as a viable, even essential, component of tomorrow's transportation system.

The vehicle highway system in the United States will continue as our primary mode of transportation for the foreseeable future. This system presently carries 89% of all passenger miles of travel and 32% of the nation's ton miles of freight. It is a primary and essential link in our national economy, with a massive in-place infrastructure. The vehicle highway system plays a key societal role as well, offering mobility unparalleled worldwide. The desire for an enriching quality of life calls us to seek green mobility, that is, more efficient transportation with proper care for the environment, land use, and community values. Our society has moved beyond the paradigm of more roads, more concrete. At the same time, portions of our highway transportation system are rapidly reaching the limit of their ability to handle additional travel demand and serve the growing transportation needs of the public and commerce, as is vividly reflected in the increasing daily congestion on our highways. And, the public rightly demands a higher level of safety from the vehicles and roads they depend on.

Where is the greatest leverage for improvement? The current system can be divided into four components: highway design, highway operation, vehicle design, and vehicle operation. Let's look at each of these in turn. Our current interstate highway system is the result of decades of continual improvement in highway designs; this field is highly evolved. Improvements in highway operations provided the kernel for what is now the Intelligent Transportation Systems (ITS) initiative; advanced traffic management systems will become commonplace in the coming years, lessening congestion, increasing safety, and streamlining transportation management activities. Vehicle design is highly evolved in the fields of crashworthiness, fuel efficiency, and emissions controls, and continues to see innovation in the area of driver assistance systems.

But what about vehicle operations? It is the orphan child, the neglected fundamental term in the equation. While today's drivers are generally quite skilled, there is nevertheless a huge variability and inconsistency in vehicle operation that is clearly responsible for most accidents and plays a major role in congestion. The dramatic improvements in each of the other components remain limited by a system of manually controlled vehicles. Automation of the driving task, through vehicle control integrated into an intelligent highway infrastructure, offers dramatic increases in system performance. Here is the vital leverage. In fact, Automated Highway Systems (AHS) are capable of providing a level of performance and service that is a generation beyond other ITS services. An AHS, deployed on existing highway infrastructure, can double or triple the throughput of today's most congested highway lanes while significantly increasing safety and trip quality. AHS capability offers benefits to all highway users, opening new opportunities for transit bus operations, enhancing the safety and productivity of heavy trucks, and offering improved convenience and dependability to the traveling public. Its efficiency can help reduce both fuel consumption and individual vehicle emissions, at the same time ensuring maximum use of our existing highway infrastructure investment.

Current research in highway automation began in 1992 as part of the U. S. Department of Transportation's (DOT) ITS program. The program focused on the prudent application of technology and systems engineering to provide information to travelers, operate our roadways with greater safety and efficiency, optimize freight and transit operations, and design smarter automobiles, through partnerships in both the public and private sectors. U. S. DOT's ITS activities are responsive to guidance contained in the Intermodal Surface Transportation Efficiency Act of 1991 (ISTEA). Specific to AHS, the act calls for development of an automated highway and vehicle prototype, setting a goal of 1997 for initial automated capability on a test track.

A long-term strategic program and partnership is key to the eventual deployment of highway automation. The federal government role is essential in this endeavor, in supporting long-range, high-risk research and development that industry or state governments cannot undertake alone. Since the passage of ISTEA, the Federal Highway Administration (FHWA) has taken the lead for AHS research within U. S. DOT by initiating substantial research addressing the full range of AHS issues. These research efforts, called Precursor Systems Analyses of Automated Highway Systems, examined topics such as human factors, roadway configurations for AHS entry and exit, approaches to lateral and longitudinal control of vehicles, safety considerations, automation of alternative propulsion vehicles, evolutionary deployment, roadway operations and maintenance for an AHS, modeling of AHS traffic flows, societal and institutional issues, and preliminary cost/benefit factors. An overarching conclusion was that, while the challenges are many, AHS indeed appears feasible across all of these dimensions, with strong potential to deliver the envisioned benefits.

With this solid foundation of basic research knowledge, the FHWA has moved into the current system definition phase of the program, by establishing a government–industry–academia collaboration known as the National Automated Highway System Consortium (NAHSC). The nation's major vehicle/highway system stakeholders, industry, state and local governments, user representatives who will eventually build, own, operate, and use a future AHS, are represented within the NAHSC. This partnership is vital; neither the public nor the private sector can implement highway automation alone or provide all of the needed expertise. The vehicle manufacturers must cooperate with the highway builders and operators to develop vehicle and highway instrumentation that complement each other and gain the full benefits of automation. In this context, the NAHSC is charged with both the necessary R&D to develop and demonstrate highway automation capability, as well as the development of a national consensus of stakeholders as to the optimum approach for deployment of highway automation early in the next century.

A key underlying aspect of this national consensus is full awareness and understanding of the complex issues surrounding implementation of AHS within the technical community. In this respect, I am pleased to see this broad treatment of recent research brought together in this volume. The breadth and depth of the research is evident here, as is the high quality of academic researchers and industry professionals who have ably contributed to highway automation research to date. These efforts, combined with current and future R&D that lies ahead, will lead us to implementation of the next generation of highway transportation.

Richard Bishop, Jr.
Program Manager for Automated Highway Systems
Federal Highway Administration
U. S. Department of Transportation

PREFACE

The current surface transportation system in most major cities in the United States and abroad is overburdened and its operation is inefficient most of the time. Mobility is declining, congestion is rising, and safety remains a serious problem. Building additional highways in areas that are already overbuilt and crowded is no longer a viable solution. Intelligent Transportation Systems (ITS), formerly called Intelligent Vehicle–Highway Systems (IVHS), provide the tools that are necessary to improve the efficiency of the current system and free us from the necessity of building additional roadways. ITS is a broad area that includes near-term as well as long-term, more ambitious applications of advanced technologies. Automated Highway Systems (AHS) is an area under ITS that promises to dramatically improve the efficiency of the current highway system. AHS is looking further ahead in the future and will introduce dramatic changes in the way the current highway system operates both from the vehicle and infrastructure point of view. As a result, a considerable number of issues and uncertainties need to be faced and resolved before AHS can be implemented.

The purpose of this book is to present some of the ideas and discuss some of the main issues and uncertainties associated with the design and deployment of AHS as seen and understood by numerous experts in the field. The book covers new AHS concepts, technology issues and designs, benefits, institutional and societal issues, and other types, and provides good exposure to the ideas, concepts, and complexity of AHS.

Chapter 1 introduces AHS and gives a brief description of AHS activities in the United States, Europe, and Japan. In Chapter 2, Shladover explains the concept of "platooning" and the benefits and controversies associated with that. The feasibility of integrating AHS into existing California freeways are addressed in Chapter 3 by Mark Miller's group at PATH. In Chapters 4–7, experts from academia and industry discuss the evolutionary scenarios and deployments that will lead to the implementation of AHS. Choices of the intervehicle spacing for different AHS concepts and the resulting capacity values are discussed in Chapter 8. Tables with numerical values for spacing and capacity are presented that allow a quantitative comparison of the capacity benefits for different AHS concepts. In Chapter 9, Polydoros and Panagiotou discuss the possible communication technologies that can accommodate the communication requirements of AHS. The control and sensor issues and requirements in AHS are discussed in Chapter 10. The use of proper sensors and control techniques is very crucial for the success of AHS. The chapter presents a brief overview of some of the most obvious issues and requirements. In Chapter 11, Kanellakopoulos and Tomizuka discuss the application of Automated Vehicle Control Systems (AVCS) to commercial trucks and buses. Chapters 12 and 13 deal with the possible effects of AHS on fuel economy and emissions. The possible effect of AHS on mobility and congestion is addressed in Chapter 14. The deployment of a large-scale AHS in the Los Angeles region for the year 2025 is used as a case study. Chapter 15–17 deal with the institutional and societal aspects of AHS.

ACKNOWLEDGMENTS. We would first like to thank the authors of the chapters for their valued contributions. Almost all of them have been part of the family called AHS activity, and their ideas, interactions, and discussions were very valuable.

We are especially grateful to Mr. Dick Bishop for providing the Foreword to the book and for motivating and supporting AHS.

We thank Steve Shladover, Pravin Varaiya, Karl Hedrick, Masayoshi Tomizuka, and all of the other PATH researchers for many fruitful discussions and interactions on AHS and AVCS.

We thank Mike Shulman, Steve Eckert of Ford Motor Co., and their research groups for their support with the Precursor Systems Analyses (PSA) AHS effort.

We are also grateful to all researchers of the Raytheon PSA team for interactions and discussions on AHS. The interaction with a long list of researchers who worked on the PSA effort and those who are currently working with the National Automated Highway System Consortium is also appreciated.

Special thanks are due to Bill Stevens and his group at The MITRE Corporation for many useful discussions during the PSA effort.

Finally, we acknowledge the support of several organizations including Caltrans, PATH, and FHWA, whose support of AVCS and AHS made the development of this book possible.

Petros Ioannou
Center for Advanced Transportation Technologies
University of Southern California

CONTENTS

CHAPTER 1

INTRODUCTION

Petros Ioannou

CHAPTER 2

REASONS FOR OPERATING AHS VEHICLES IN PLATOONS

Steven E. Shladover

CHAPTER 3

INTEGRATION OF AUTOMATED HIGHWAY SYSTEMS INTO EXISTING CALIFORNIA FREEWAYS

Youngbin Yim, Mark A. Miller, Paul Hellman, and Mohammad Sharafsaleh

CHAPTER 4

SYSTEM CONFIGURATIONS: EVOLUTIONARY DEPLOYMENT CONSIDERATIONS

Randolph W. Hall

CHAPTER 5

STEP BY STEP TO AN AUTOMATED HIGHWAY SYSTEM—AND BEYOND

Jerry D. Ward

CHAPTER 6

THE EVOLUTION OF AHS AND CURRENT VEHICLE TRENDS IN LIGHT OF AEROSPACE SYSTEMS EVOLUTION

David A. Blancett, Gregory H. Davis, Sheryl A. Payne, and Charles E. Taylor

CHAPTER 7

EVOLUTION TO AN AUTOMATED HIGHWAY SYSTEM

William B. Stevens

CHAPTER 8

SPACING AND CAPACITY EVALUATIONS FOR DIFFERENT AHS CONCEPTS

Alexander Kanaris, Petros Ioannou, and Fu-Sheng Ho

CHAPTER 9

COMMUNICATION TECHNOLOGIES FOR AHS

Andreas Polydoros and Prokopios Panagiotou

CHAPTER 10

CONTROL AND SENSOR REQUIREMENTS AND ISSUES IN AHS

Petros Ioannou

CHAPTER 11

COMMERCIAL TRUCKS AND BUSES IN AUTOMATED HIGHWAY SYSTEMS

Ioannis Kanellakopoulos and Masayoshi Tomizuka

CHAPTER 12

AERODYNAMIC BENEFITS FROM CLOSE-FOLLOWING

F. Browand, M. Zabat, and P. Tokumaru

CHAPTER 13

THE EFFECTS OF AHS ON THE ENVIRONMENT

Matthew J. Barth

CHAPTER 14

REGIONAL MOBILITY IMPACTS ASSESSMENT OF HIGHWAY AUTOMATION

Mark A. Miller, Anne Bresnock, Steven E. Shladover, and Edward H. Lechner

CHAPTER 15

INSTITUTIONAL AND SOCIETAL ISSUES ASSOCIATED WITH AUTOMATED HIGHWAY SYSTEMS: AN ENVIRONMENTAL PERSPECTIVE

Mark A. Miller

CHAPTER 16

AUTOMATED HIGHWAY SYSTEM DEPLOYMENT: A PRELIMINARY ASSESSMENT OF UNCERTAINTIES

Randolph W. Hall and H.-S. Jacob Tsao

CHAPTER 1

INTRODUCTION

PETROS IOANNOU

The United States had managed to develop one of the biggest and most efficient roadway systems in the world that provided mobility, allowed the expansion of cities, and changed the life-style of people and the way they conduct business. The demand on the roadway system, however, kept increasing to the point that the construction of new highways is not as feasible as it used to be because of increasing land and labor costs, environmental considerations, and the long time required for construction. The lack of additional highways to meet the demand for traffic capacity led to congestion and inefficient usage of the current system. Congestion causes travel delays and inefficient movement of vehicles that reduces productivity, wastes energy, increases pollution, and adversely affects the quality of life. With more people driving today than ever before, traffic accidents are taking their toll in terms of congestion, productivity, and loss of life.

The solution to the current overburdened highway system is to make it more efficient by using advanced technologies. Advanced technologies could be used to automate some of the basic elements of the system, such as the vehicle and the infrastructure decision-making tasks, in order to eliminate sluggish, unpredictable, and nonresponsive behaviors that have negative effects on efficiency.

PETROS IOANNOU • Center for Advanced Transportation Technologies, University of Southern California, Los Angeles, California 90089-2562.

Automated Highway Systems, edited by Ioannou. Plenum Press, New York, 1997.

During the last two decades the rapid advances in electronics, computers, communications, controls, and sensor technologies have been followed by a simultaneous reduction in cost that makes advanced technologies feasible for widespread applications. Many of these technologies have already found their way into industry, defense, consumer products, and so forth. What appeared to be science fiction a few years ago, has turned out to be a reality and part of daily life. Unfortunately, surface transportation benefitted very little from this explosion in the development and implementation of advanced technologies. The current highway system operates as an open-loop system, which makes it very sensitive to disturbances such as those caused by accidents and driver behavior. The lack of automatic feedback in the system, the random behavior of drivers, and the lack of effective sensing techniques for improving the characteristics of traffic lead to the sluggish, unpredictable, and inefficient traffic flows observed in today's highway systems.

The Intelligent Vehicle–Highway Systems (IVHS) or as more recently called Intelligent Transportation Systems (ITS) programs around the world aim to improve the efficiency of the current transportation system by using new technologies that allow the implementation of feedback and more sophisticated decision-making techniques. Some of the main areas of ITS are:

- *Advanced Traffic Management Systems (ATMS).* This area covers freeway ramp metering, arterial signal control, traffic congestion management, dynamic traffic control systems, alternative routing instructions, and so forth.
- *Advanced Traveler Information Systems (ATIS).* This provides information that assists travelers in reaching their desired destination. It also includes on-board navigation systems that may accept inputs from ATMS and help travelers plan their trips. Information from ATMS may be accepted by other devices that are located at home, at the office, or are even portable and can be used for pretrip planning.
- *Advanced Vehicle Control Systems (AVCS).* AVCS includes a broad range of concepts and products that will enhance the driver's control of the vehicle. These products include collision warning systems, collision avoidance using automatic braking and/or automatic steering, blind spot detection, and so forth. The long-term AVCS includes fully automated vehicles operating on special highway lanes with support from the infrastructure. This long-term fully automated system is referred to as the Automated Highway System (AHS).
- *Commercial Vehicle Operations (CVO).* This area includes the application of ITS technologies to commercial vehicles such as trucks, buses, vans, taxis, and emergency vehicles. Technologies such as Automated Vehicle Identification (AVI) systems and on-board navigation devices have already started to appear in commercial vehicles.

- *Advanced Public Transportation Systems (APTS)*. This area uses ATMS, ATIS, and AVCS technologies to improve the operation of high-occupancy vehicles, such as transit buses, van- and car pools.

Each of the above areas is expected to introduce some near-term technologies that will improve the efficiency of the current transportation system. The area, however, that is expected to introduce dramatic changes that will lead to dramatic improvements in efficiency is AHS. In AHS, AVCS, ATMS, and ATIS technologies will be integrated to produce a highway system where fully automated vehicles are guided to their destinations and the flow of traffic is controlled and optimized for maximum efficiency and safety. The design of AHS is a challenging one and the issues involved are enormous from the technological, human factors, socioeconomic, legal, institutional, and environmental points of view. The purpose of this book is to present some of these issues as addressed by various experts in the field.

In the following subsections we present a brief description of the AHS activities that have occurred in the United States, Europe, and Japan over the last three decades. Further information on AHS and ITS activities in general can be found in several review papers[1–7] and publications of the Intelligent Transportation Society, formerly the Intelligent Vehicle and Highway Society, of America[8–11] and the U. S. Department of Transportation.[12]

1. AHS ACTIVITIES: UNITED STATES

The concept of AHS has its roots at the General Motors Pavilion at the 1939–40 World's Fair where the notion of vehicles that drive themselves while their drivers relax in the back seat was introduced to the public as a vision of the future of automobiles.[1,2] During the late 1950's, General Motors, in cooperation with the Radio Corporation of America (RCA) demonstrated at a test track, automatic lateral and longitudinal control of automobiles.[13]

In the late 1960s, the Bureau of Public Roads (BPR), the predecessor to the Federal Highway Administration (FHWA), initiated new research efforts whose objective was the improvement of safety and efficiency of highway travel through the use of advanced communication and control concepts. The BPR initiative had several projects related to ATMS and ATIS[2] as known today, but its most significant part was the AHS program. The AHS program involved system analysis studies conducted by CALSPAN[14] and General Motors[15] and vehicle network control experiments performed by Ohio State University (OSU)[16] under the sponsorship of BPR (which became FHWA) and the Ohio Department of Highways. The OSU program was a long-term research effort that started in 1965 and ended in 1980. In this program both automatic steering and longitudinal control were demonstrated on a test track.[16] Other studies that focused on automated ground transportation during the 1960s and 1970s include the MIT Project METRAN[17] and "Glideway" study,[18] a Federal Railroad Administration Northeast Corridor Study,[19] and a dual mode program supported by the Urban Mass Transportation Administration.[20–22] The dual mode transportation system was essentially a type of AHS configuration where a driver would operate normally on most streets but would release control to an automatic system once on a restricted guideway.

In 1981 there was a dramatic downturn in research in traffic systems. The new U. S. administration introduced new policies and brought new appointees whose priorities did not include advanced research in transportation systems. Because of this negative climate, attempts to revitalize some programs or develop new ones were not successful.

While in the early 1980s there was no significant effort in IVHS in the United States, the IVHS efforts in Europe and Japan, initiated in the mid-1960s, were continuing. During the same period, advances in computers and electronics that took place during the previous two decades started finding their way into consumer products leading to a growing realization that advanced technologies are no longer part of science fiction but part of daily life. The emergence of advanced technologies together with the deterioration of the efficiency of the highway system characterized by high levels of congestion in many U. S. cities led to the realization that an aggressive research program in IVHS must be pursued. By the mid-1980s urban traffic congestion became one of the major concerns of FHWA, state transportation departments, and related agencies that led to serious discussions and proposals for major research programs to deal with the problem.

In October 1986 the California Department of Transportation (Caltrans) sponsored a 3-day conference to consider the role of advanced vehicle–highway technologies in meeting the increasing demand for highway capacity and efficiency. The event set the stage for new IVHS research efforts that will in the future lead to a major AHS effort to be funded by FHWA. Caltrans in collaboration with the Institute of Transportation Studies of the University of California went ahead and initiated a small effort to study the use of automation in vehicle–highway systems. This effort led to the development of a California statewide program named Program on Advanced Technology for the Highway (PATH) in 1986, which in 1992 was renamed Program for Advanced Transit and Highways (PATH) in order to reflect the broadened scope of the program. PATH, with the support of Caltrans, started a very aggressive research program on vehicle and highway automation and laid the foundations for new AHS efforts.[3,4] Caltrans and PATH, however, realized that without federal support and the widespread support of private industry and transportation experts across the country, a sound AHS effort would not be possible.

Subsequent workshops that took place in November 1987 and March 1988 at an FHWA facility and at the University of California, Berkeley, managed to bring together interested parties from around the country to discuss the use of advanced technologies to deal with the transportation problem. These activities led to the formation of Mobility 2000 in 1988. Mobility 2000 was a group of transportation experts whose task was to suggest research and development that involved the use of advanced technologies to improve surface transportation. The name Intelligent Vehicle–Highway Systems initially used by Drs. Kan Chen and Robert Ervin of the Transportation Research Institute of the University of Michigan was adopted by Mobility 2000 to describe a wide range of activities that were grouped into four broad areas: Advanced Traffic Management Systems (ATMS), Advanced Driver Information Systems (ADIS), Commercial Vehicle Operations (CVO), and Advanced Vehicle Control Systems (AVCS). Later on ADIS became Advanced Traveler Information Systems (ATIS) and two other groups were added: the Advanced Public Transportation Systems (APTS) and the Advanced Rural Transportation Systems (ARTS).

The work of Mobility 2000[22] led to the development of a national IVHS program soon after. In 1990, the Intelligent Vehicle and Highway Society (IVHS) of America was formed as a national organization. A few years later, the name was changed to Intelligent Transportation Society (ITS) of America in order to reflect the broad scope of the organization. During the same time, the U. S. Department of Transportation (DOT) established a formal IVHS program office and recognized IVHS America as a utilized Federal Advisory Committee. In 1991, Congress passed the Intermodal Surface Transportation Efficiency Act (ISTEA). Part B, Section 6054(b) of the act states:

> The Secretary (of Transportation) shall develop an automated highway and vehicle prototype from which future fully automated intelligent vehicle–highway systems can be developed. Such development shall include research in human factors to ensure the success of the man–machine relationship. The goal of this program is to have the first fully automated roadway or an automated test track in operation by 1997. This system shall accommodate installation of equipment in new and existing motor vehicles.

In order to manage the AHS effort, U. S. DOT established an AHS program within FHWA under the guidance of the IVHS program office. The AHS program[23,48] is coordinated with the National Highway Traffic Safety Administration (NHTSA) programs such as the one on collision avoidance systems.

The AHS effort is divided into three phases:

1. *Analysis Phase.* This phase provides the analytical foundation for further efforts on AHS. It includes the Precursor Systems Analyses (PSA) which was a 1-year effort that ended in 1994. The PSA effort involved 15 contracts to different consortia across the country that studied more than 16 AHS topics using different approaches and philosophies.[24] It also included the AHS human factors study contracted to a consortium led by Honeywell to develop the "AHS human factors handbook for AHS Designers" and "Driver-based AHS Human Factors System and Development Guidelines" using driving simulator tests and other experiments.[12] The study started in October 1992 and is expected to be completed by March 1996. The NHTSA sponsored collision avoidance systems for warning that may someday evolve into AHS may also be considered as part of this phase.[12]
2. *Systems Definition Phase.* This is a multiyear effort that includes (a) the development of AHS performance and design objectives, (b) the evaluation of alternative AHS system concepts, (c) the demonstration of AHS concepts that involved fully automated vehicles in a controlled test track

environment, (d) the selection and testing of a preferred system approach, (e) the demonstration, testing, and evaluation of a prototype configuration of the preferred system approach, and (g) the documentation of the preferred AHS system configuration.

3. *Operational Evaluation Phase.* This phase includes one or more implementations of the preferred system at selected U. S. locations in order to evaluate its operation.

In 1994, two national consortia competed for a contract to pursue the systems definition phase of AHS effort. The selected consortium referred to as the National Automated Highway System Consortium (NAHSC) consists of General Motors, Hughes Aircraft, PATH, Bechtel, Caltrans, Delco Electronics, Martin-Marietta, Parsons-Brinckerhoff/Farradyne Systems, and Carnegie Mellon University as the core participants. In addition to the core members, the consortium also has a long list of associated members that include universities, companies, research institutes, and federal and state organizations from the United States and abroad. Further information on NAHSC can be obtained from: NAHSC Program Office, Suite 500, 3001 West Big Beaver Road, Troy, MI 48084.

More information on the ITS projects supported by the U. S. DOT in the United States can be found in Ref. 12. In addition to the efforts described above, almost all of the major automobile companies in the Unites States are engaged in several near-term AVCS technologies that include intelligent cruise control and collision warning devices.

2. AHS ACTIVITIES: EUROPE

Automated Vehicle Control Systems and IVHS-type programs similar to those in the United States were in place in Europe from the mid-1960s onward.[25] Automatic steering control experiments were conducted in England by the Road Research Laboratory in the late 1960s[26] whereas longitudinal control system developments took place in Germany[27] and France[28] during the 1970s.

The most significant efforts, however, in the area of IVHS, started in the mid-1980s with the DRIVE (Dedicated Road Infrastructure for Vehicle Safety in Europe) project funded by the European Community (EC). The DRIVE project was followed by DRIVE II and III and is ongoing. The DRIVE effort has been concentrating mainly on infrastructure-oriented IVHS research by focusing on pilot projects and demonstrations of the ATMS and ATIS type with some efforts on traffic safety that involve collision avoidance systems.[7,11] A program closer to AHS was initiated by the motor vehicle industry. It is called the PROMETHEUS (Program for European Traffic and Highest Efficiency and Unprecedented Safety) program and is funded by the EC's EUREKA program and the governments of several EC countries. The EUREKA program is designed to stimulate cooperation between industries and governments in Europe under the direction of the European Research Coordination Agency.

Most of the European research efforts concentrated on near-term AVCS implementations that include intelligent cruise control, cooperative driving, and driver assist systems.[29-33] An automated guided system using machine vision has been implemented and tested by Daimler Benz.[34] The equipped vehicle is an autonomous one and is capable of lane and vehicle following and forward and side obstacle detection.

As in the United States, every major automobile company in Europe is experimenting with intelligent cruise control systems, collision avoidance warning systems, and other driver assist devices. Volkswagen AG in association with the PROMETHEUS program went further to demonstrate automated driving in a convoy or a platoon, with short intervehicle spacing.[35] The details about these developments are either confidential or are published.

The equivalent of ITS America in Europe has been ERTICO (European Road Transport Telematics Implementation Coordination Organization) established in 1991 to promote and assist with the implementation of IVHS in Europe. In 1996, the supplementary title ITS Europe was adopted by the now-well-established ERTICO, which will be maintained as the legal name of the organization and as its corporate identity. Despite the extensive and well-funded IVHS activities in Europe, the AHS activities are focused on near-term implementations and they are small compared with the AHS program and activities in the United States.

3. AHS ACTIVITIES: JAPAN

The AVCS efforts in the United States and Europe during the 1960's were paralleled in Japan as well. The Mechanical Engineering Laboratory of the Ministry of International Trade and Industry (MITI) used an inductive cable as lane reference to demonstrate automatic driving.[36] Later on, under MITI's Intelligent Vehicle Systems (IVS) program, automatic steering control was demonstrated using machine vision at 30 km/h[37].

Currently most of the research efforts on vehicle automation are under MITI. These efforts include the Personal Vehicle System (PVS) and Super-Smart Vehicle System (SSVS) projects.[38,39] A PVS vehicle equipped with video cameras, sonar, and laser radar sensors has been developed by Nissan and other member companies of MITI's Mechanical Society Systems Foundation and used to demonstrate automatic driving[40] at speeds as high as 37 mph.

The SSVS project includes a variety of topics ranging from accident recognition and avoidance to driver assist devices.[41,42] Most of these topics are very similar to the AVCS topics of the U. S. IVHS program.

In addition to MITI, the Ministry of Transport focused on traffic safety and developed the concept of the "Advanced Safety Vehicle" (ASV) emphasizing collision avoidance and problems with aging drivers. The Ministry of Construction considers the automation of freight movements in tunnels that are to be exclusively used for goods distribution beneath the streets of major cities in Japan.[37] The same ministry has also been considering the concept of the Road–Vehicle Cooperative

Driving System.[39] This concept is emphasizing roadway to vehicle communications for safety warnings, collision avoidance, and relevant topics that may eventually lead to automated driving.

Some of the automobile companies in Japan have been pursuing their own internal research on automated vehicles in addition to near-term products such as intelligent cruise control and collision warning devices. Mazda developed fully automated vehicles that have been tested on test tracks.[43,44] Mitsubishi[45] and Toyota[46,47] also performed experiments with fully automated vehicles on test tracks.

Even though some of the IVHS efforts in Japan are classified as AHS activities, there is no AHS program in Japan of the same magnitude as that in the United States.

REFERENCES

1. Special Issue on Intelligent Vehicle Highway Systems, *IEEE Trans. Veh. Technol.* **40**, No. 1 (1991).
2. L. SAXTON, Mobility 2000 and the Roots of IVHS, *IVHS Rev.* pp. 11–27 (Spring 1993).
3. S. SHLADOVER, R. BUSHEY, AND R. PARSONS, California and the Roots of IVHS *IVHS Rev.* (Spring 1993).
4. S. SHLADOVER, Review of the State of Development of Advanced Vehicle Control Systems (AVCS) *Int. J. Veh. Mech. Mobil. Veh. Syst. Dyn.* **24**, 551–595 (1995).
5. R. FENTON, IVHS/AHS: Driving into the Future, *IEEE Control Syst.* **14**, No. 6 (1994).
6. R. FRENCH, Intelligent Vehicle/Highway Systems in Action, *ITE J.* 23–31 (Nov. 1990).
7. R. CASE, R. L. FRENCH, Y. NOGUCHI, C. QUEREE, K. SAKOMOTO, AND O. SRIDEN, IVHS Overview of U. S., Europe and Japan *IEEE Veh. Technol. Soc. News* **42**, No. 1 (1995).
8. *ITS Quarterly,* A Publication of the ITS of America.
9. National ITS Program Plan, ITS America, March 1995, Vol. 1, part II, edited by G. Euler and H. D. Robertson.
10. 1995 Who's Who in ITS Japan, ITS America (1995).
11. A comparison of IVHS Progress in the United States, Japan & Europe Through 1993, prepared by R. French & Associates for IVHS America (March 31, 1994).
12. Intelligent Transportation Systems (ITS) Projects, U. S. Department of Transportation, Federal Highway Administration, Publ. No. FHWA-JPO-95-001 (Jan. 1995).
13. K. GARDELS, Automatic Car Controls for Electronic Highways, General Motors Research Labs, Warren, MI, GMR276 (June 1960).
14. Practicality of Automated Highway System, CALSPAN Corp., Buffalo, NY, Final Report DOT-FH-11-8903 (Nov. 1977).
15. J. G. BENDER, An Overview of Systems Studies of Automated Highway Systems, *IEEE Trans. Veh. Technol.* **40**, 82–99 (1991).
16. R. E. FENTON AND R. J. MAYHAN, Automated Highway Studies at The Ohio State University—An Overview, *IEEE Trans. Veh. Technol.* **40**, 100–113 (1991).
17. M. E. HANSON (Ed.), *Project METRAN: An Integrated, Evolutionary Transportation System for Urban Areas* (MIT Press, Cambridge, MA, 1966).
18. *The Glideway System—A High-Speed Ground Transportation System in the Northeast Corridor of the United States* (MIT Press, Cambridge, MA, 1965).
19. Automated Highway Systems, TRW Systems Group, Redondo Beach, CA, (Dec. 1969) (available from NTIS as Reference PB-191696).
20. Dual Mode Transit System, Rohr Industries Inc.) (Sept. 1974) (available from NTIS as Reference PB 237 724).
21. Tomorrow's Transportation: New Systems for the Urban Future, U. S. Department of Housing and Urban Development (1968).

22. Dual Mode Transportation. Proc., conference conducted by Transportation Research Board, May 29–31, 1974, TRB Special Report 170 (1976).
23. L. Saxton and J. R. Bishop, The United States DOT automated highway system (AHS) program. *Proc. First World Congress on Applications of Transport Telematics and Intelligent Vehicle–Highway Systems,* Paris, 1994, pp. 1952-1958.
24. J. R. Bishop and W. Stevens, Results of precursor systems analyses of automated highway systems, *Proc. First World Congress on Applications of Transport Telematics and Intelligent Vehicle–Highway Systems,* Paris, 1994, pp. 1959-1966.
25. C. G. Giles and J. A. Martin, Cable Installation for Vehicle Guidance Investigations in the New Research Track at Crowthorne, Road Research Lab, Crowthorne, England, Report RN/4057/CGG (1961).
26. K. H. F. Cardew, The Automatic Steering of Vehicles—An Experimental System Fitted to a Citroen Car, Road Research Laboratory, RL 340, Great Britain (1970).
27. R. Hesse, Cabinentaxi—A personal public transportation system. Transp. **1**(3), 321–329.
28. J. P. Maury, The ARAMIS PRT system, SAE Paper 740143 (1974).
29. F. Broqua, G. Lerner, V. Mauro, and E. Morello, Cooperative driving: Basic concepts and a first assessment of 'intelligent cruise control' strategies. *Proc. Advanced Telematics in Road Transport,* Vol. II, DRIVE Conference, Brussels, 1991, pp. 908–929.
30. M. De Saint Blancard, B. Dubuisson, and M. Rombaut, PRO-LAB2: A vehicle with driving assistance, *Proc. First World Congress on Applications of Transport Telematics and Intelligent Vehicle–Highway Systems,* Paris, 1994, pp. 1871–1878.
31. S. Becker, M. Bork, H. T. Dorissen, G. Geduld, O. Hofmann, K. Naab, G. Nöcker, P. Rieth, and J. Sonntag, Summary of experience with autonomous intelligent cruise control (AICC). Part 2: Results and Conclusions, *Proc. First World Congress on Applications of Transport Telematics and Intelligent–Highway Systems,* Paris, 1994, pp. 1836–1843.
32. G. Reichart and K. Naab, Systems for lateral and longitudinal vehicle guidance, *Proc. First World Congress on Applications of Transport Telematics and Intelligent Vehicle–Highway Systems,* Paris, 1994, pp. 2126–2133.
33. M. Andrews, S. Jacobi, U. Reuter, and V. Salzer, Porche's contribution to safer and more comfortable driving in the future, *Proc. First World Congress on Applications of Transport Telematics and Intelligent Vehicle–Highway Systems,* Paris, 1994, pp. 1879–1886.
34. B. Ulmer, Autonomous automated driving in real traffic. *Proc. First World Congress on Applications of Transport Telematics and Intelligent Vehicle–Highway Systems,* Paris, 1994, pp. 2118–2125.
35. Tomorrow's Road Traffic, Obstacle Warning System/Driving in Convoy, Research for the Future. Volkswagen AG, Research and Development in cooperation with Public Relations, released information.
36. R. Oshima *et al.* (Japan Government Mechanical Laboratory), Control System for Automobile Driving, *Proceedings of the Tokyo IFAC Symposium,* 1965, pp. 347–357.
37. T. Fujioka, K. Yoshimoto, and S. Takaba, A case study on an automated driving highway system in Japan, Transportation Research Board 72nd Annual Meeting, Washington, DC, 1993.
38. S. Tsugawa, N. Watanabe, and H. Fuji, Super smart vehicle system-its concept and preliminary works, *Proc. Vehicle Navigation and Information Systems Symposium,* Dearborn, MI, Oct. 1991, pp. 269-277 (SAE Publication P-253).
39. M. Nakamura, H. Kawashima, T. Yoshikai, and K. Aoki, Road–vehicle cooperation driving system, *Proc. 1994 Vehicle Navigation and Information Systems,* Yokohama, Japan, pp. 425–430.
40. A. Hosaka, Steering Control of an Autonomous Vehicle Using a Fuzzy Logic Controller, Institute of Mechanical Engineers, Paper No. C391/005 (1989).
41. S. Tsugawa, Japanese IVHS Looks to 2014 and Beyond, *IVHS Rev.* pp. 51–65 (Summer 1993).
42. N. Sato, Advanced Road Transportation Systems in Japan, *IVHS Rev.* pp. 77-87 (Fall 1993).
43. A. Okuno, A. Kutami, and K. Fujita, Towards autonomous cruising on highways. Automated Highway/Intelligent Vehicle Systems: Technology and Socioeconomic Aspects, SAE Publications SP-833, pp. 7–16 (1990).
44. T. Morita, H. Takahashi, J. Satonobu, K. Kojima, and Y. Morimoto, An approach to the intelligent vehicle. *Proc. Intelligent Vehicles 93 Symposium,* Tokyo, pp. 426-431.

45. K. HAYAFUNE AND H. YOSHIDA, Control method of autonomous vehicle considering compatibility of riding comfort and vehicle controllability. Automated Highway/Intelligent Vehicle Systems: Technology and Socioeconomic Aspects, SAE Publication SP-833, pp. 35–40 (1990).
46. K. OHNISHI, J. KOMURA, AND T. ISHIBASHI, Development of automatic driving system on rough road—Realization of high reliable automatic driving system. *Proc. Intelligent Vehicles 92 Symposium, IEEE,* Detroit.
47. T. SUZUKI AND A. TACHIBANA, A fully automated intelligent vehicle system based on computer vision for automated highway system, *Proc. First World Congress on Applications of Transport Telematics and Intelligent Vehicle-Highway Systems,* Paris, 1994, pp. 1925–1932.
48. L. SAXTON, Automated Control—Cornerstone of Future Highway Systems, *IVHS Rev.* pp. 1–17 (Summer 1993).

CHAPTER 2

Reasons for Operating AHS Vehicles in Platoons

Steven E. Shladover

1. INTRODUCTION

The advent of the federal Automated Highway Systems (AHS) program, based on the guidance contained in ISTEA, has brought increased attention to the concept of highway automation within the past year. The AHS Precursor Systems Analyses (PSA) projects and the initiation of work by the National Automated Highway System Consortium (NAHSC), which will conduct the Systems Definition Phase of the program, have provided opportunities for numerous people to think about AHS issues. Prior to these developments, AHS was generally regarded as something of a "fringe" issue in the world of Intelligent Transportation Systems, and only received serious research attention within the PATH Program.

PATH researchers have been studying the design and operation of AHS in depth since 1988, and have already devoted about 150 labor years of effort to the subject. In the course of this study, we have devoted considerable attention to the issue of how to organize the movements of vehicles to maximize both capacity and safety. This has led to serious evaluations of the concept of grouped or platooned operations of vehicles, as well as full-scale testing and demonstration of such opera-

Steven E. Shladover • California PATH Program, University of California, Berkeley, California 94804.

Automated Highway Systems, edited by Ioannou. Plenum Press, New York, 1997.

tions. In platooned operations, vehicles are clustered together in groups of up to 20 vehicles, with very short spacings between vehicles within the platoons and long spacings between the platoons.

New arrivals to the study of AHS tend to regard platooned operations with considerable skepticism. The purpose of this chapter is to explain the reasoning behind the concept of platooning and the knowledge that we have gained about it in the course of the past 6 years of intensive study. In this way, we hope to overcome much of the misunderstanding of the concept that has been evident in recent commentaries on platooning and hopefully thereby address the concerns of the skeptics. The chapter concludes with identification of the remaining uncertainties about platooned AHS and the issues that still need to be studied.

2. REASONS FOR OPERATING AUTOMATED VEHICLES IN PLATOONS

The platoon mode of operation was conceived as a way of expanding the envelope of capacity and safety that can be achieved by road vehicles. It is obvious that the one way of ensuring that there will be *no* vehicle crashes, injuries, or fatalities is to have no vehicles in motion. It is also readily apparent that as the speed and density of vehicles using the road system increase, the likelihood and likely severity of the crashes will increase. The capabilities of drivers are the principal limitation here. Driver errors are responsible for well in excess of 90% of the crashes that occur today, and the limited ability of drivers to follow other vehicles produces the limitation on lane capacity. The limitations of drivers' ability to perceive changes in vehicle spacing, relative motion, and acceleration and their limited speed and precision of response ensure that lane capacity cannot generally exceed 2200 vehicles per hour under manual control.

Electronic sensors, computers, and actuators can provide faster and more precise responses than human drivers. Moreover, they do not get drunk, fatigued, or emotionally upset and therefore perform very consistently compared with human drivers. These performance advantages should make it possible for electronically controlled vehicles to operate at significantly higher levels of capacity (shorter average spacings) and safety than today's manually driven vehicles. However, drivers do not have accurate perceptions of the safety of their vehicle-following distances, and typically drive at closer spacings than they "should" if they are to be able to avoid collisions when malfunctions occur. An automated system must be designed to be able to avoid virtually *all* crashes that could cause serious injuries or fatalities, which is a higher standard than drivers today apply to their own driving.

In order to be able to increase lane capacity, it is necessary to operate vehicles at closer *average* spacings (for the same speed). In order to do this, we must carefully consider the relationship between safety and the size of the gap or spacing. It is clear that if vehicles remain so far apart that they cannot ever collide on occurrence of any single failure, there will be a substantial safety gain. However, if they remain that far apart, the capacity is likely to be significantly reduced below today's level.

As the gap size is reduced, the opportunities increase for collisions to result from malfunctions.

The key issues in the attempt to improve both capacity and safety by use of automation are therefore centered around determination of what conditions (gap sizes, environmental conditions, and vehicle malfunctions) can lead to crashes, how the severity of those crashes varies with the conditions, and what severity of crash could be considered tolerable. If one determines that *no* crash, however mild, is *ever* acceptable on an automated system, one must then abandon further consideration of AHS. This is because it is impossible to design and implement an absolutely perfect system that never suffers malfunctions. However, we can certainly expect to satisfy the requirement for no collisions in the absence of AHS malfunctions.

Under the full range of possible vehicle operating conditions, there could, of course, be a wide variety of crash mechanisms such as steering failures producing lane departure crashes, sudden accelerations of following vehicles, and so forth. For purposes of the capacity/safety analysis here we confine our attention to the rear-end crash caused by a malfunction that makes one vehicle (the "leading" vehicle) decelerate abruptly. Other crash mechanisms also need to be subjected to careful study to assess their implications for capacity and safety.

In considering operations of vehicles within a single lane, for simplicity, the factors that determine the impact speed of a rear-end crash are:

a. Initial spacing between vehicles
b. Deceleration rate of leading (malfunctioning) vehicle
c. Delay time from start of leading vehicle deceleration to start of following vehicle braking
d. Emergency braking deceleration rate of following vehicle
e. Speeds of both vehicles at time of malfunction

Using data from NHTSA's National Center for Statistics and Analysis, Hitchcock has shown how the severity of injury or probability of fatality to vehicle occupants is related to the impact speed (velocity difference between vehicles at time of impact).[(1)] This means that any system design should seek to minimize that impact speed. Hitchcock has also suggested, based on the NHTSA data, that an impact speed of 3.3 m/s represents an approximate threshold between crashes that produce only mild to moderate injuries and those that may produce more severe injuries.[(1)] Of course, these effects are highly dependent on additional factors such as the masses of the colliding vehicles, the occupant restraints, and the health and ages of the vehicle occupants.

There are three conditions under which a rear-end crash can occur, depending on the values of the parameters (a–e) defined above:

1. Follower hits leader before follower has started to apply brakes.
2. Follower hits leader while both are braking.
3. Follower hits leader after leader has stopped.

Condition (1) typically occurs when the initial gap between the vehicles is very small, while condition (3) typically occurs when the initial gap is fairly large. The no-collision conditions occur when the initial gap is large enough that the follower can stop just short of touching the leader or when there is no gap at all, so that the vehicles decelerate in constant contact with each other. The collision impact speeds can become quite large at intermediate spacings, as illustrated in Shladover.[2–4] The safety argument in favor of platoon operations is based on avoiding the intermediate spacings associated with the higher collision impact speeds. These potentially hazardous spacings are unfortunately comparable to the spacings that are often advocated for use with autonomous vehicle-follower systems.

The relationship between the crash impact speed and initial vehicle spacing is highly dependent on the parameters (a–e) defined above, and variations in the values of those parameters can have dramatic effects on the shape and magnitude of the impact speed-versus-spacing curve. The serious discussions about platooned operations need to be focused on defining reasonable ranges for these parameter values and defining the degree of aversion to crashes of different levels of severity.

Under virtually all conditions and parameter values, initial gaps between vehicles of the order of 1 m or less lead to crash impact speeds that are quite modest (even under a "worst case" condition of a 1g deceleration of the leader, with no response by the follower, the impact speed would be 4.47 m/s, or 10 mph). As the gap increases, the impact speed increases up to some maximum [the value of which is extremely sensitive to the parameters (a–e)] before decreasing again until it reaches zero at the point where the follower is able to stop before touching the leader. Depending on the parameter values chosen, this no-crash spacing could vary extremely widely (anywhere from 10 to 100 m). The extreme sensitivity of the impact speed-versus-spacing relationship to the parameters (a–e), several of which are very difficult to select, makes the determination of nominal vehicle-following gaps challenging and controversial.

Until these vehicle-following gaps have been selected, it is impossible to define the capacity that an AHS can achieve. The inherent difficulty of selecting the gap size is the primary reason that AHS capacity estimates by different investigators have varied so widely and have been so difficult to compare with each other. Rather than rederiving capacity estimates, we refer to Shladover,[4] where capacity estimates were derived for several sets of parameter values, representing both "optimistic" and "pessimistic" combinations of assumptions. These illustrate the wide variability in capacity estimates, the limited potential for capacity increases without using platoons (unless some very optimistic assumptions are made), and the phenomenon of diminishing returns with respect to platoon length as the platoons become longer. The guiding principles behind these derivations have been to assume as small a gap as possible within platoons (nominally 1 m) and to use conservative assumptions to select the gap between platoons in order to minimize the probability of any crash involving more than one platoon.

For example, in order for an automated system of single vehicles (without platoons) to reach a capacity of 2000 vehicles per hour per lane at a 30 m/s cruise speed, it is necessary to assume that the failed leading vehicle decelerates at no

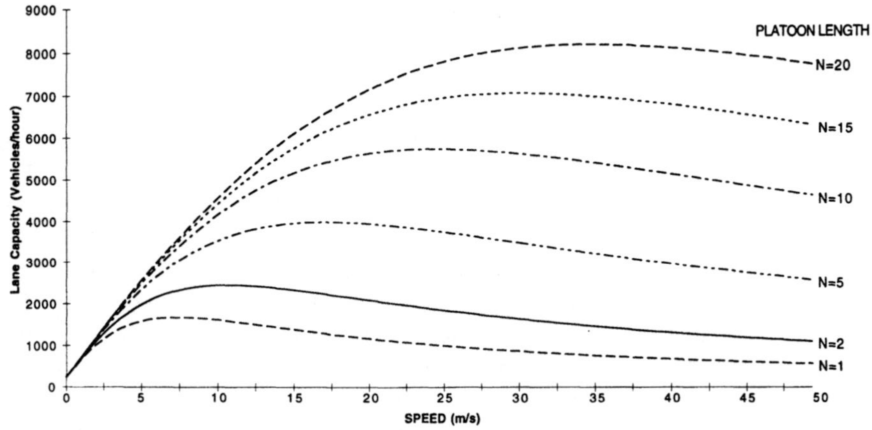

FIGURE 1. Lane capacity of a platoon system under intermediate assumptions (0.7*g* failure, 0.3*g* emergency braking, 0.2-s reaction delay).

more than 0.5*g* and the follower can decelerate at least at 0.4*g* within no more than 100 ms. Under this set of assumptions to govern *interplatoon* spacings, operations in 5-vehicle platoons would provide almost 6000 vehicles per hour, while 10-vehicle platoons would reach 7500 and 20-vehicle platoons would push 8500. On the other hand, if the failed leader decelerates at 2*g* and the follower can only decelerate at 0.3*g* within 300 ms, the single vehicle capacity is only a few hundred vehicles per hour, and even 5-vehicle platoons can only bring it up to about 1500 at 30 m/s. Twenty-vehicle platoons would be needed to get the capacity up to 4500 vehicles per lane per hour under these conditions. The achievable capacities in the end are likely to be somewhere between these extremes, along the lines of the estimates shown in Figure 1, which are based on a failed vehicle deceleration at 0.7*g*, following vehicle braking at 0.3*g*, and a 200-ms reaction time.

3. HISTORY OF AUTOMATED PLATOON CONCEPTS

The idea of operating automated vehicles in platoons did not first arise as part of the discussion about the AHS program, but dates to the period of intense activity on Automated Guideway Transit (AGT) systems in the 1970s. Shladover[5,6] includes a listing of 22 citations to published papers and reports about automated platoon concepts that were available as of 1978, not including any by this author. Much of this work was aimed at mechanical coupling of automated vehicles in dynamically reconfigurable "trains," rather than platoons that would only be coupled by the action of their control systems. However, control-coupled platoons were considered by researchers from organizations such as General Motors[7] and Engins MATRA.[8,9] The MATRA experience is particularly interesting because they built a test track and a group of test vehicles that they operated under automated platooning in 1972–73 and, in a later reincarnation, in 1986–87. This project, called ARAMIS, involved platoons of up to 25 small transit vehicles operating at speeds of up to 80 km/h and 30-cm spacings, using both ultrasonic and optical ranging sensors. The existing papers in the open literature describing the first generation of this system[8,9] do not give many details about the technological implementation, but nevertheless make for informative reading.

Shladover[5,6] describes the findings of a comprehensive analysis of the use of automated platoons or mechanically coupled trains of small transit vehicles to provide personalized rapid transit service on special guideways. This report and a series of technical papers based on its findings (including Shladover[2,3,5,6,10] and others) define the capacity and safety issues, the nonlinear longitudinal control system design considerations, and the implications of merging of streams of platooned traffic. However, this work did not extend beyond analysis and into experimental verification.

More recently, Volkswagen developed an automated platoon system that they called "convoy driving" in association with the PROMETHEUS program. Because of proprietary information restrictions, there has been virtually no public documentation of the technical characteristics of this system although there is a very impressive videotape of a 1989 demonstration of the system on a test track. This shows operation of multiple-vehicle platoons at highway speeds under fully automatic steering and longitudinal control, with spacings that appear to be 1 m or less. The video includes a demonstration of emergency braking on wet pavement and a segment in which the driver of the test van turns on the automatic control system and then leaves the driver's seat to take up a position in the passenger's seat, showing his confidence in the system's ability to drive the vehicle without his intervention. The Volkswagen effort was terminated after the initial demonstration, apparently because of sensitivity on the part of the German public agency officials to the opposition of the Green Party politicians to a system that was perceived as too "pro-automobile."

In 1988, researchers in the California PATH Program picked up from where Shladover[5,6] left off and began more in-depth investigations of the control design and system operational implications of automated platooning for AHS. The principal findings of this research will be covered in Section 5.

4. MISCONCEPTIONS ABOUT AUTOMATED PLATOONING

Much of what has been said and written about automated platooning in recent years has been based on several misconceptions about the concept. When these misconceptions are resolved, it becomes apparent that the problems with platooned operations are not nearly as serious as they might appear to be at first glance.

4.1. Platooning Is Very Complicated and Expensive

The large majority of the hardware and software needed to make platooned AHS work are also needed for fully automated but nonplatooned AHS. It is important to concentrate on those items that would need to be *different* in order to make platoons work. The only features that would be peculiar to a platooned AHS are:

- Vehicle–vehicle communication system capable of transferring reasonably high-bandwidth control information (of the order of 100 kilobits per second, not necessarily megabytes per second)
- Ranging sensors with accuracy of several centimeters within the range of a few meters
- Software logic for joining and splitting platoons

4.2. *The Sensor Requirements for Platooning Are Very Demanding*

Ranging sensors for the very-short-range application of measuring the spacings between vehicles within a platoon appear to be easier to develop than the ranging sensors that would be needed for larger-gap AHS operations, and much easier than those that would be needed for Adaptive Cruise Control or Collision Avoidance applications. The short-range sensors have a very large "target" to see in front of them, and do not need to employ much sophistication to identify that target from among many others within the field of view. Since platoons would only be formed with other suitably equipped vehicles, the vehicles could be equipped with cooperative targets or transponders to make detection and ranging even easier.

Within the past year, PATH researchers have tested ultrasonic, FMCW radar, and infrared (IR) triangulation sensors for vehicle-to-vehicle ranging at a variety of spacings. The ultrasonic system gives extremely good ranging information at gaps of 7 m or less except in dusty conditions, while the IR triangulation system is effective up to a range of about 20 m under static conditions, and the radar works very well between 1.5 and 10 m under dynamic conditions. There are no fundamental reasons why these technologies should not be effective at vehicle-to-vehicle gaps of the order of 1 m (which could correspond to somewhat larger sensed gaps, depending on where the sensors are mounted on the vehicles). The measurements from one or more of these types of sensors can be combined with measurements of vehicle location relative to permanent markers installed in the pavement (which would serve primarily as lateral guidance references) using data fusion software to produce extremely robust estimates of vehicle spacings.

4.3. *Platooned Operations Would Expose Travelers to Frequent Crashes, Which Would Not Be Acceptable*

The platoon operating concept is firmly based on the principle that there shall be no crashes in the absence of AHS malfunctions. There is no intention to have platooned vehicles acting like amusement park "bumper cars." The nominal spacing between vehicles within a platoon must be large enough that it can accommodate normal variability in the response of the vehicles within the platoon based on differences in external force loadings, vehicle condition, and control system inaccuracies. One of the primary goals in the design of the control systems is to minimize the variability in spacing so that the nominal gap can indeed be made small while still avoiding inadvertent contact between vehicles.

Crashes of platooned vehicles should be almost as rare as crashes in nonplatooned automated vehicles, but when they occur they should be much less severe. Crashes could only occur as a result of a malfunction, not as a part of normal operations. While malfunctions are intended to be rare, particularly with the added protection provided by automated check-in functions, they cannot be eliminated entirely. When they do occur, the response of the overall system must be designed to be safe. Among other things, the nominal intraplatoon spacing must be large enough that the occurrence of a "typical" malfunction cannot ensure the occurrence of a crash. In other words, the gap between vehicles within a platoon must be large enough that each follower has the opportunity to reach its full braking rate between the time the leader suffers its malfunction and the time the follower would collide with it. This means that crashes within a platoon should only occur when there are multiple malfunctions or particularly severe malfunctions (ones in which the leader decelerates more rapidly than the follower's maximum braking capability).

Table 1 provides some sample numbers to illustrate the significance of this for intraplatoon gaps. We assume the mild failure (coast down after propulsion system failure) to produce a deceleration rate of 0.1*g*, while the severe failure (multiple tire blowout) would cause the leader to decelerate at 0.5*g*. Using today's automatic braking technology, it would probably take about 250 ms for an automatic braking system to reach full braking effectiveness. We expect that this could be reduced to 100 ms in the next few years based on ongoing laboratory developments. The "manual control" columns in Table 1 assume a highly skilled and alert driver rather than a typical driver. For a typical driver, increasing the sense and initiate time from 500 ms to 1500 ms for the severe failure would increase the minimum spacing from the 2.5 m value shown in Table 1 to 10 m. This kind of difference is consistent with the differences between the spacings seen in normal freeway driving and the spacings seen on racetracks.

As Table 1 shows, the automated systems need spacings of no more than 0.5 m to ensure that the follower is able to initiate braking before touching the failed leader. The spacings for the cooperative system are substantially shorter than for the autonomous system because the failure of the leader can be communicated directly to the follower, rather than having to wait for a change in the spacing to become noticeable to the follower's sensor system. The platoon concept is intended to be inherently cooperative in order to ensure stability of dynamic response, so the worst case in Table 1 that is applicable to platoons is a spacing of 0.24 m, which assumes current braking capabilities.

4.4. *Small-Gap Operations within Platoons Will Be Alarming to Drivers and Passengers*

This contention, although it is taken as an article of faith by some platoon opponents, cannot be proven or disproven until it is possible to conduct realistic human factors experiments on real vehicles. Some preliminary indications should become evident using driving simulator experiments. These can also be used to help

TABLE 1
Braking Lag Times (milliseconds) and Minimum Spacings (meters) to Avoid Crashes When Failures Occur

Failure severity	Manual control (typical driver)	Manual control (highly skilled)		Automatic–autonomous (10-cm sensor resolution)				Automatic–cooperative			
		Severe	Mild	Severe		Mild		Severe		Mild	
Braking lag times (ms)											
Sense problem/initiate action											
Severe failure (0.5g)	1500	500		200	200			40	40		
Mild failure (0.1g)		—	1500			450	450			40	40
Communicate			—	—	—	—	—	20	20	20	20
Apply brake pressure											
Normal system	500	500	500	250	—	250	—	250	—	250	—
Advanced system		—	—	—	100	—	100	—	100	—	100
Total times (ms)											
Normal braking	2000	1000	2000	450	—	700	—	310	—	310	—
Advanced braking		—	—	—	300	—	550	—	160	—	160
Minimum spacings (m) to avoid crash											
Normal braking	10.00	2.50	2.00	0.506	—	0.245	—	0.240	—	0.048	—
Advanced braking		—	—	—	0.225	—	0.151	—	0.065	—	0.013

FIGURE 2. A platoon of four vehicles operating under automatic longitudinal control at gaps of 4 m.

design vehicle maneuvers so that the gap-versus-time profiles can be specified in ways that are comfortable and reassuring. They can also provide useful guidance regarding the types of system status indicators that people prefer to have in order to provide assurance that the control system is functioning properly throughout its maneuvers.

PATH researchers have offered demonstration rides for visitors in an automated platoon of four vehicles operating at gaps of 4 m at speeds between about 12 and 30 m/s. (See Figure 2.) Although this size gap would be considered as serious tailgating in normal driving, it was not disquieting to the passengers of the test cars because the gap was maintained very precisely, and some of them even commented on how unexceptional the experience was. When the control system provides such tight control of gap, it gives the sensation that the vehicles are rigidly linked and the vehicle in front appears to be "pulling" the following vehicle along behind it.

5. RECENT RESEARCH ACCOMPLISHMENTS ON PLATOON OPERATIONS

The PATH research on platoon operations has already considerably expanded knowledge about how platoons could be made to work and how the platoon mode of operation would compare with individual vehicle operations (sometimes referred to as "free agents"). Some of the key findings are in the following areas.

5.1. Control Accuracy and Ride Quality

There is a fundamental trade-off between the accuracy of control response and the smoothness of ride that can be achieved. Control analyses and simulations under idealized conditions have indicated that it should be possible to achieve both high accuracy and high comfort. However, until recently it has not been possible to verify this in testing of full-scale vehicles. Experiments conducted in the summer of 1994 using a platoon of four vehicles have finally shown that we can achieve a very smooth ride (disturbances small enough to be imperceptible to riders), while maintaining spacing to within 30 cm of the desired value under steady driving conditions and within 50 cm under a wider range of conditions, including during acceleration and deceleration maneuvers conducted on both positive and negative slopes of up to 3%.[(11)] These experiments have also demonstrated that disturbances can be attenuated sufficiently that the ride in the fourth vehicle is not perceptibly different from the ride in the second vehicle.

5.2. Actuator Performance Requirements

The key actuators for longitudinal control within a platoon are the throttle and brake actuators. It was demonstrated in 1992 that good engine control could be achieved with a throttle actuator having a response rate of at least 500 degrees per second.[(12)] Braking requirements are somewhat more difficult to achieve because of inherent lags built into existing power braking assist systems. A brake actuation system that acts through the brake booster (by acting on the brake pedal, for example) does not appear to be adequate for platoon use because it has excessive lag time. New designs that act directly at the master cylinder have been shown to have fast enough response to meet ride quality needs for use in closed-loop platoon control systems during tests conducted in 1995.

5.3. Sensor and Communication System Performance Requirements

A combination of analyses, design studies, laboratory tests, and experiments on full-scale vehicles have enabled us to make some preliminary evaluations of the performance that should be needed from ranging sensor and vehicle–vehicle communication systems in order to support platooned operations. Significantly, these appear to be achievable using technical approaches that should not be inherently costly or exotic.

The ranging sensor requirements that are peculiar to platooning are the need to operate down to a very short range (potentially, though not necessarily, as short as 1 m) and to provide range measurements to reasonably high precision at these short distances. In order to provide smooth control response, the range (and range rate) measurements need to be relatively stable and have low noise content. The control system's signal processing software needs to be able to filter out anomalous individual data points, but systematic measurement errors are more difficult to eliminate. The range measurements should be accurate to within 5 to 10 cm of the true range at the closest spacings in order to permit the control system to work well. This appears to be readily achievable using several of the sensors with which we have experimented.

The vehicle–vehicle communication system must be able to supply information about the speed and acceleration of the platoon leader and each leading vehicle to each following vehicle, as well as serving as the conduit for emergency status information. The cycle time for each exchange of information should be in the range of 20 ms to promote good platoon dynamic response. Given the amount of information that must be provided and the need for error detection and correction, as well as provisions for multiple transmissions to reduce the effective communication error rate, the intraplatoon communication system appears to need a capacity of the order of 100 kbps.(13) This appears to be achievable by means such as optical wireless (IR) and microwave mobile radio. Good performance has already been achieved in PATH platoon experiments using a commercial off-the-shelf wireless local area network radio communication system.

5.4. Platoon Collision Dynamics

The common calculations used to study the potential for collisions among vehicles within a platoon are based on simple kinematic analyses. These are not adequate for addressing the complications that arise with multiple-vehicle collisions, diverse vehicle masses, alternative emergency response strategies, and the impact-absorbing effects of bumper systems. These issues have been addressed in a recent PATH study led by Professor Benson Tongue of UC Berkeley.(14)

The dynamics of multiple-vehicle collisions can be quite complex, and often the most severe crash does not occur between the first two vehicles to contact each other. Mass differences among the involved vehicles can produce further "strange" effects. In all cases, though, the availability of information about the platoon leader (or the first vehicle involved in the crash) has been shown to improve the ability of the control system to avoid any crashes (or reduce the number of vehicles involved). Providing information to each vehicle about the movements of the vehicles *behind* it also makes it easier to avoid or mitigate crashes. It is difficult to draw simple conclusions from this study because of the complicated interactions among all of the variables that affect collision dynamics. Further research is continuing in this area in order to lead to the development of robust strategies for handling intraplatoon collisions.

5.5. Probabilities of Injury and Fatality under Different Failure Conditions

Vehicle crashes are rare events, produced by combinations of random phenomena that cannot be represented purely deterministically. For any given type of crash, the severity of an individual occurrence (probability of causing serious injuries or fatalities) is influenced by the following random variables:

a. Deceleration rate of failed vehicle, based on the type of failure that occurred and the condition of the vehicle's tires and the road surface
b. Braking rate of following vehicle, based on its loaded mass, the condition of its brakes and tires, and the condition of the road surface
c. Delay time between failure and following vehicle's response
d. Speed and acceleration of each vehicle at the time of the failure.
e. Masses and structural properties of both vehicles
f. Existence and use of occupant restraints in each vehicle
g. Ages and health of occupants of each vehicle

Some of these factors are obviously so random that they cannot be accommodated realistically in an analysis (f, g), while others can be addressed directly in the AHS design (c, d). A recent PATH study by Hitchcock[1] has explicitly considered probability distributions of factors (a, b, e) in a series of Monte Carlo simulations to estimate the collision impact speeds that would occur under both wet and dry road conditions. By use of data from the Crashworthiness Data System (CDS), the correlation between collision impact speed and injury level was estimated. Combining this with the Monte Carlo simulation results led to a first set of estimates of the probabilities of serious injuries and fatalities under different failure conditions involving platoons as well as vehicles operating under autonomous adaptive cruise control (AACC) and point–follower control systems.

This study showed that the rates of serious injuries and fatalities were lowest for the closely spaced platoon system under almost all conditions, largely because the crashes that did occur within the platoons were at impact speeds lower than those needed to cause serious injuries or fatalities. By contrast, the AACC systems appeared to be vulnerable to much more serious collisions, particularly if operated at high densities. This means that they must have significantly higher reliability against failures that cause abrupt decelerations than the platooned AHS if their overall injury and fatality rates are to be comparable. The results of this study imply that the mean time between abrupt-deceleration failures for a platooned AHS vehicle would need to be of the order of 10^4 h in order for the injury rate for that type of accident to be about 10% of today's freeway driving injury rate.

5.6. *Aerodynamic Drag Reductions*

Recent PATH wind-tunnel experiments conducted by Professor Frederick Browand and colleagues at USC have indicated the potential for reducing aerodynamic drag by operating AHS vehicles in close-formation platoons.[17] This research has shown the relationship between drag force and vehicle spacing under various conditions, including crosswinds, for scale models of GM Lumina APV vehicles. The results indicate the potential for dramatic drag reductions, which improve as the vehicle spacings become shorter. The closest spacings reported thus far are one-half vehicle length (approximately 2.5 m for these vehicles), and even at these spacings the drag reduction approaches 50% for a long platoon, compared with the drag on an individual vehicle. Interestingly enough, all vehicles in the platoon show reduced drag, although the effect is stronger for the middle vehicles than it is for the first and last vehicles. The implications of drag reductions of up to 50% for fuel economy and pollutant emissions are extremely significant and need further exploration.

5.7. Energy Consumption and Pollutant Emission Effects

The aforementioned reductions in drag have been used as inputs to two studies that are considering the environmental implications of ITS. In one of these PATH projects, Dr. Matthew Barth of UC Riverside is using mathematical models to predict the changes in energy consumption and vehicle pollutant emissions that would result from a variety of ITS operations. His initial study of the platooned AHS has indicated that the drag reduction associated with short-headway platooning should reduce by about half the pollutant emissions per vehicle as compared with an AHS using vehicles operating independently (as "free agents").[16] The reductions should be more substantial when compared with conventional manually driven traffic, with its stop-and-go cycles. The second study, by Browand's group, has estimated fuel savings of the order of 25% for large platoons operating at short spacings.[15] These environmental benefits are extremely significant when compared with any of the proposed pollution control measures except for the most draconian demand reduction and alternative propulsion regulations.

6. TECHNOLOGICAL CAPABILITIES NEEDED TO MAKE PLATOONING WORK

The ongoing PATH research on platooned AHS has served to "demystify" the technology that will be needed to make platoons work. There are no dramatic breakthroughs needed in fundamental physics or engineering science, but there will need to be much careful system design work, safety-verified control software, refinement of technologies, and work on cost reductions. In this regard, platooned AHS is no different from any other kind of AHS.

The areas in which platooning imposes more severe performance requirements than other forms of AHS are:

1. Accurate short-range sensing
2. Fast, reliable vehicle–vehicle communication
3. Safety-verified cooperative maneuvering protocols
4. An informative user interface to provide the driver with reassurance about the status of system performance and maneuvers
5. Very fast and precise throttle and brake control actuators

The specifics of several of these have already been addressed as part of the PATH research program and were discussed in Section 5.

7. REMAINING PROBLEM AREAS

The foregoing is not intended to convey the idea that all problems related to platooned operation of AHS have been solved and we are ready to deploy a system tomorrow. There are several open issues that have not been fully addressed yet and that could represent problems peculiar to platooning (rather than to AHS in general). They are as follows.

7.1. Human Factors and User Acceptance

The acceptability of platooned AHS to a representative sampling of drivers cannot be proven until a highly refined prototype is available for human factors experimentation. Some preliminary indications should be given by the results of the ongoing FHWA/Honeywell project on Human Factors Design of AHS, but these must remain preliminary because the experiments will be performed in a driving simulator rather than a real vehicle.

7.2. Achievable Intraplatoon Spacing

Considerable research will still be needed to determine how small a spacing can be achieved between consecutive vehicles within an automated platoon. This must depend on the precision of the sensors, actuators, and control system, the robustness of the systems with respect to external disturbances, and the acceptability to users. The technological issues will require experimentation on test tracks under diverse conditions, while the user acceptability will have to be determined as part of the human factors experiments.

7.3. Intraplatoon Collision Effects

The possibility of having collisions among the members of a platoon following a malfunction raises a number of difficult issues that still need to be resolved. These include the need to develop credible estimates of the probabilities of occurrence of impacts of different severities and the need to define an acceptable threshold for each of these probabilities. A combination of difficult technical and policy issues must be addressed in order to solve these problems.

7.4. Unusual Aerodynamic Effects

Although the general steady-state effects of platooned operations on vehicle aerodynamics have already been evaluated, there are some more subtle effects that are not yet fully understood. These include the transient aerodynamic forces that will occur in maneuvers when vehicles enter and leave platoons, and the potential that very close vehicle following may lead to a loss of cooling air flow to vehicle radiators or the ingestion of exhaust fumes from the leading vehicle into the passenger compartment of the following vehicle. These need to be studied through a combination of planned scale model and full-scale testing.

8. CONCLUSION

Although various uncertainties remain about platooned operation of AHS, much has been learned in recent years. The results of this research certainly indicate that there is some substantial promise to this concept and that it is worth continuing with research to address the remaining uncertainties. Platooned operation appears to offer some significant advantages relative to other implementations of AHS, and therefore it needs to remain under serious consideration for AHS research, development, and demonstration activities.

ACKNOWLEDGMENTS. This work was performed as part of the California PATH Program of the University of California, in cooperation with the State of California Business, Transportation, and Housing Agency, Department of Transportation; and the U. S. Department of Transportation, Federal Highway Administration.

The contents of this chapter reflect the views of the author who is responsible for the facts and the accuracy of the data presented herein. The contents do not necessarily reflect the official views or policies of the State of California. This report does not constitute a standard, specification, or regulation.

REFERENCES

1. A. HITCHCOCK, Intelligent Vehicle Highway System Safety: Multiple Collisions in Automated Highway Systems, Institute of Transportation Studies, University of California, Berkeley, PATH Research Report UCB-ITS-PRR-95-10 (1995).
2. S. E. SHLADOVER, Dynamic Entrainment of Automated Guideway Transit Vehicles, *High Speed Ground Transp. J.* **12**(3), 87–113 (1978).
3. S. E. SHLADOVER, Longitudinal Control of Automated Guideway Transit Vehicles within Platoons, *J. Dyn. Syst. Meas. Control* **100**(4), 302–310 (1978) (also ASME Paper No. 78-WA/DSC 13).
4. S. E. SHLADOVER, Potential Freeway Capacity Effects of Automatic Vehicle Control Systems, *Applications of Advanced Technologies in Transportation Engineering*, ASCE Conference, Minneapolis, MN, 1991, pp. 213–217.
5. S. E. SHLADOVER, Operation of Automated Guideway Transit Vehicles in Dynamically Reconfigured Trains and Platoons, Reports UMTA-MA-06-0085-1 through -3, PB 300513, PB 80-142862, and PB 80-198856 (1979).
6. S. E. SHLADOVER, Vehicle-Follower Control for Dynamic Entrainment of Automated Guideway Transit Vehicles, *J. Dyn. Syst. Meas. Control* **101**(4), 314–320 (1979) (also presented at Joint Automatic Control Conference, Denver, CO, 1979).
7. N. H. TRINER AND L. ROBERTSON, Guideway Design Considerations for the General Motors Dual Mode Transit System, Paper presented at First International Conference on Dual Mode Transportation, Transportation Research Board, Washington, DC, 1974.

8. J. P. COUPLAN, Le Système de Transport Urbain ARAMIS, *Proceedings of First International Conference on Transportation Research*, Bruges, Belgium, 1973 (Transportation Research Forum), pp. 351–354.
9. J. P. MAURY, The ARAMIS PRT System, SAE Paper 740143 (1974).
10. S. E. SHLADOVER, Operation of Merge Junctions in a Dynamically Entrained Automated Guideway Transit System, *Transp. Res.* **14A**(2), 85–112 (1980).
11. S.-B. CHOI AND J. K. HEDRICK, Vehicle Longitudinal Control Using an Adaptive Observer for Automated Highway Systems, *Proceedings, American Control Conference*, Seattle, WA, 1995, pp. 3106–3110.
12. K. S. CHANG, J. K. HEDRICK, W.-B. ZHANG, P. VARAIYA, M. TOMIZUKA, AND S. E. SHLADOVER, Automated Highway System Experiments in the PATH Program, *IVHS J.* **1**(1) (1993).
13. S. R. SACHS AND P. VARAIYA, A Communication System for the Control of Automated Vehicles, Institute of Transportation Studies, University of California, Berkeley, PATH Technical Memorandum 93-5 (1993).
14. Y.-T. YANG AND B. H. TONGUE, Intra-Platoon Collision Behavior During Emergency Operations, *Veh. Syst. Dyn.* **23**(4), 279–292 (1994).
15. M. S. ZABAT, N. S. STABILE, AND F. K. BROWAND, Estimates of Fuel Savings from Platooning, *Proceedings of ITS America 1995 Annual Meeting*, Washington, DC, pp. 1203–1208 (1995).
16. M. J. BARTH, AHS Reduction of Vehicle Emissions: Preliminary Analysis, *Proceedings of ITS America 1995 Annual Meeting*, Washington, DC, pp. 1137–1145.
17. M. ZABAT, S. FRASCAROLI, AND F. K. BROWAND, Drag Measurements on 2, 3 and 4 Car Platoons, SAE Paper No. 940421, International Congress & Exposition, Detroit, MI, 1994.

CHAPTER 3

Integration of Automated Highway Systems into Existing California Freeways*

Youngbin Yim, Mark A. Miller, Paul Hellman, and Mohammad Sharafsaleh

1. INTRODUCTION

Automated Highway Systems (AHS) have the potential for providing improvements over conventional freeways in terms of vehicle throughput, traffic safety, driver comfort, and environmental impacts. Although a number of AHS deployment concepts have been presented in the literature, few studies have attempted to evaluate the issues and risks associated with alternative deployment concepts.[1-3] This research is devoted to an investigation of the issues and risks associated with the integration of AHS facilities into existing California freeways, in both urban and rural environments.

Shared and dedicated space configurations are the two basic configurations examined in the analysis. Shared space or right-of-way refers to the situation where

*The contents of this chapter are partially based on papers previously published in the Proceedings of the 1994 IVHS America Conference and the Proceedings of the 1995 ITS America Conference.

Youngbin Yim, Mark A. Miller, Paul Hellman, and Mohammad Sharafsaleh • California PATH Program, University of California at Berkeley, California 94804.

Automated Highway Systems, edited by Ioannou. Plenum Press, New York, 1997.

at least one of the existing roadway lanes is converted for automated usage. Dedicated space refers to the situation where automated highways are built in a space designated solely for automated vehicles, independent of and isolated from other traffic. Combinations of these two basic configurations were also investigated. Such a shared–dedicated configuration would consist of a separated structure built in the same right-of-way as preexisting facilities, by way of a tunnel (below grade), an overhead facility (directly above grade, or off to the side), or an at-grade parallel facility.

The most heavily congested urban freeway corridors in California (San Francisco Bay area, metropolitan Los Angeles, Sacramento, and San Diego) were examined to compare the freeway characteristics and to assess the feasibility of integrating AHS into the existing freeway configurations. The rural corridors were also examined to assess the feasibility of freeway automation on the existing freeways.

The common problems with the California urban freeways for automation are space constraints regarding the availability of sufficient median space for conversion to automated use and the frequency of overhead bridges with too low a clearance to accommodate an elevated structure for automated vehicle usage. This study is aimed at identifying potential issues and risks likely to emerge with an AHS when it is integrated into existing freeway infrastructures in California.

The primary objective is to identify and analyze site-specific issues and risks associated with AHS deployment concepts that need to be resolved prior to its successful deployment. Although case studies were conducted for site-specific urban and rural area corridors, this study focuses on the issues of automated highway concepts and deals with the physical design compatibility of AHS concepts in an urban and rural environment. Several AHS concepts were evaluated to determine the extent to which these concepts can be integrated into existing freeway infrastructure configurations considering availability of land, safety of operation, and environmental concerns.

This chapter begins with freeway characteristics in California followed by AHS deployment concepts. Potential issues for implementation of the AHS concepts were then identified using evaluation criteria.

2. CALIFORNIA FREEWAY CHARACTERISTICS

Physical features of the most heavily traveled urban freeway corridors in California were investigated in order to gain an understanding of the opportunities and constraints associated with AHS deployment in urban areas. The physical parameters examined in this baseline study were overpasses (vehicle, pedestrian, and rail), soundwalls, retaining walls, frontage roads, land use features (e.g., structures, parking lots, bodies of water, rail lines), tunnels, presence of high-occupancy-vehicle lanes, number of lanes, and median and shoulder widths. Physical features of selected rural freeway corridors in California were also investigated; in addition to the above physical features, variations in terrain among rural corridors were examined. This study of freeway physical features was essential to the feasibility analysis of alternative AHS deployment concepts.

3. AHS DEPLOYMENT CONCEPTS

The principal concern of this research is to identify the issues and risks associated with the integration of AHS into existing California freeways and to determine what such facilities might look like. The two basic deployment concepts evaluated in this study are the shared space and dedicated space concepts. The shared space concept is both automated and manual facilities occupy a common right-of-way and utilize common entry and exit facilities. Shared space AHS facilities can be built by converting existing freeway lanes to automated lanes or through the construction of additional lanes. Under the shared space concept, all vehicles would enter and exit the freeway under manual control via existing, and possibly newly constructed, on- and off-ramps. A transition lane, in which vehicles shift between manual and automated control, would serve as the transition zone between the manual and automated portions of the freeway. The dedicated space concept is building an AHS facility either within the same right-of-way as a conventional freeway or within its own dedicated right-of-way and is served by exclusive entry and exit facilities. Under the dedicated space concept, automated and manual facilities would be physically separated by elevation (vertical separation), based on conventional technology, or by a continuous lane barrier where both facilities are at the same grade level.

Under the shared space concept, the automated facility would be constructed at the same grade level as the adjacent manual facility. Under the dedicated space concept, the automated facility could be constructed at, above, or below grade with respect to the adjacent manual facility or, when located within an exclusive right-of-way, with respect to the surrounding terrain.

3.1. At-Grade AHS Concept

Two at-grade AHS concepts are evaluated in this study: the shared space at-grade concept and the dedicated space at-grade concept. Under either of these concepts, several studies indicate that lane barriers are necessary to separate the automated and manual portions of the freeway for reasons of safety and operational efficiency.[4] Under the shared space concept, a minimum of two manual lanes would be required to adequately accommodate weaving movements between the manual lanes and the transition lane and merging and diverging movements near on- and off-ramps. A minimum of two manual lanes would be needed under both the shared space and dedicated space concepts to enable passing maneuvers among faster and slower manually operated vehicles.

Under the shared space at-grade concept, all vehicles would enter the freeway under manual control through conventional on-ramps. After entering the freeway,

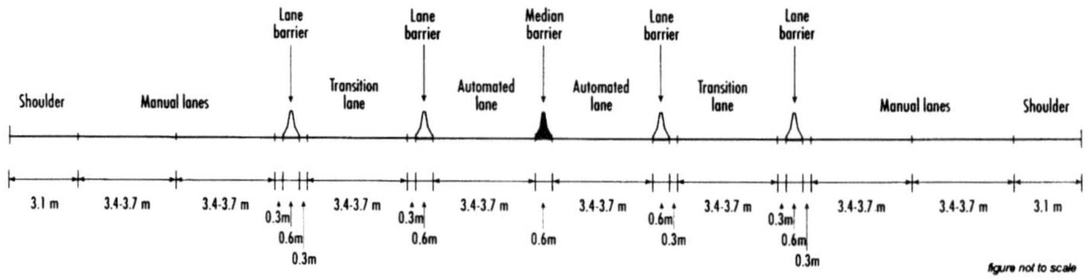

FIGURE 1. Typical cross section of shared at-grade concept.

properly equipped vehicles intending to travel on the automated portion of the freeway would first enter the transition lane under manual control. Following a successful check-in procedure to ensure that the vehicle is properly equipped and is in satisfactory operating condition for automated travel, the system would shift the vehicle to automated mode. The automated vehicle would then enter the adjacent automated lane when an adequate merging gap becomes available. For vehicles to shift back to manual control, essentially the reverse of this procedure would be followed. Lane barriers between the transition lane and the adjacent manual and automated lanes would physically separate automated and manual traffic to the extent possible. Strategically located openings in the lane barriers would allow vehicles to maneuver between the automated and manual portions of the freeway. Figure 1 shows a typical cross section of how an existing eight-lane freeway might look after being converted to a shared space at-grade facility. Existing on- and off-ramps would be utilized by vehicles using both the manual and automated portions of the freeway.

Lane barrier openings between the transition and automated lanes can serve as either entry or exit points for the automated facility. It is suggested that entry points be located approximately 3.5 km apart and that exit points be located approximately 2 km following entry points to provide adequate space for vehicles to execute entry and exit maneuvers.[(5)] Since it would only be possible to exit the automated facility at designated exit points, drivers wishing to exit the freeway must exit the automated facility far enough upstream of the desired freeway off-ramp to allow sufficient distance in which to weave through manual traffic before reaching the off-ramp. Entry/exit zones, containing an entry point and the immediate downstream exit point, should be approximately 2 km in length. Entry/exit zones should be spaced approximately 1.5 km apart to allow for the movement of vehicles between the transition and manual lanes. Within this 1.5-km zone, no lane barrier would be present between the transition lane and the adjacent manual lane and no barrier openings would be provided between the transition lane and the automated lane (Figure 2). Barrier openings should be offset in such a way so that at all points along the automated facility at least one barrier separates the automated and manual lanes. Under the shared space at-grade concept, a minimum of four lanes in each direction may be required, assuming at least one automated lane, one transition lane, and a minimum of two manual lanes.

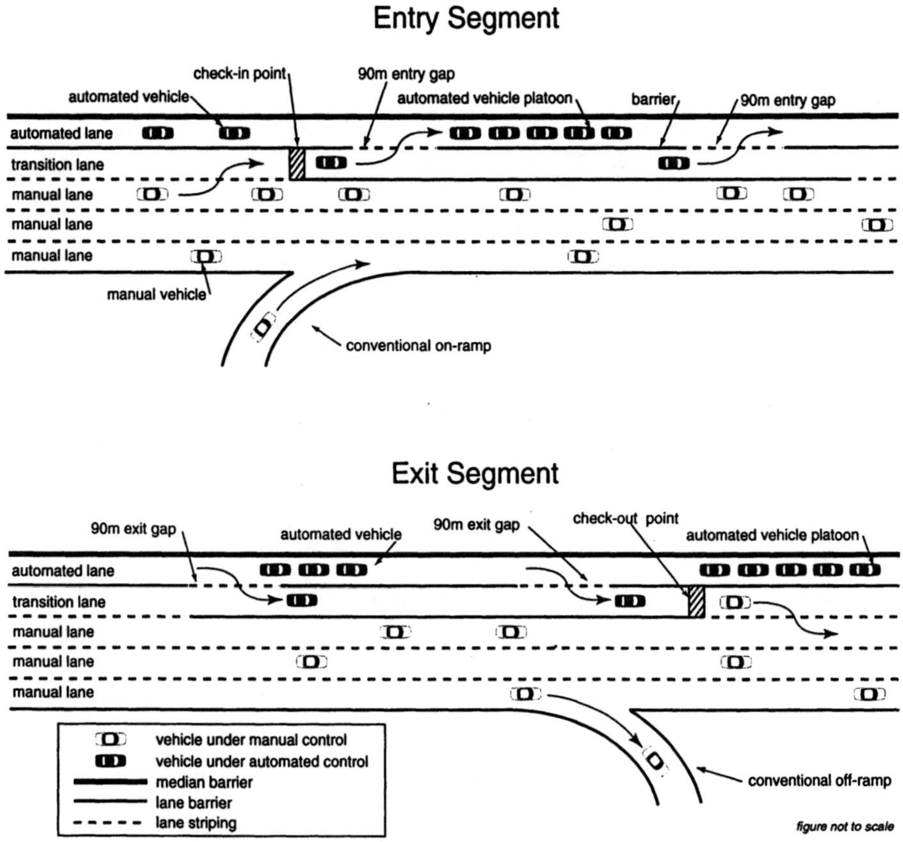

FIGURE 2. Entry/exit configuration of shared at-grade concept.

Under the dedicated space at-grade concept, the automated and manual portions of the freeway would be physically separated from one another by a continuous lane barrier. Figure 3 shows how an existing eight-lane freeway might look if converted to a dedicated space at-grade facility. A minimum of three lanes in each direction of travel may be sufficient since a transition lane is not needed; however, exclusive entry and exit facilities for the automated lane would be nesessary. Entry and exit facilities for the automated lane could consist of ramps connected to either existing or new overpasses. One of the possible entry and exit configurations for the automated facility would be to locate on- and off-ramps within the median as shown in Figure 4. Because of the spatial requirements of these on- and off-ramps, lateral expansion of the freeway may be necessary where on- and off-ramps are located.

Because of the width of lane barriers, additional space for some degree of lateral expansion of existing freeways may be necessary to accommodate at-grade AHS facilities. Under the shared space at-grade concept, lane barriers and barrier shoulders may require the widening of freeway roadbeds by 2.1 m in each direction.

FIGURE 3. Dedicated at-grade concept.

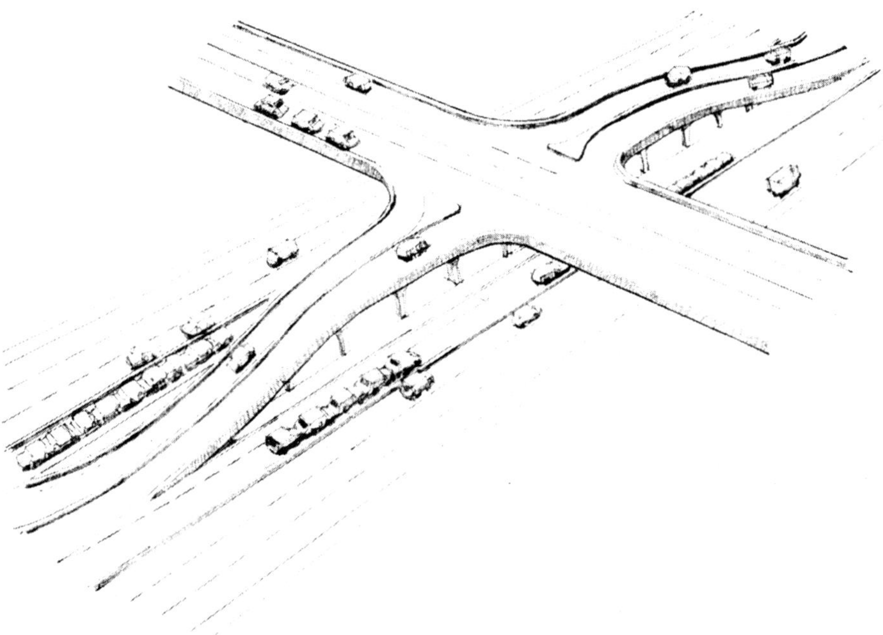

FIGURE 4. Entry/exit configuration of dedicated at-grade concept.

The need for lateral expansion of the freeway beyond the outside shoulder is dependent on the availability of median space and existing lane widths. The standard minimum freeway median width in California is typically considered to be 1.2 m, comprised of a concrete median barrier (0.6 m wide at the base) and 0.3 m inside shoulders on each side of the barrier. On many existing freeways, inside shoulders are not wide enough for emergency parking; however, a minimum inside shoulder width of 0.3 m (1 foot) is typically maintained. Since vehicles in the median lanes of an AHS would be under automated control, it may not be necessary to provide an inside shoulder. Many freeway medians are wider than 1.2 m and in many cases are significantly wider, particularly in rural areas. On conversion of a freeway to the shared space at-grade concept, it is preferable that available median space be used to the extent possible in order to minimize the extent of lateral expansion beyond the outside shoulder.

The standard lane width for freeways designed to accommodate heavy-duty vehicles is 3.7 m (12 feet). However, lanes of 3.4 m (11 feet) in width are common on many California freeways. Lane widths of 3.4 m are particularly common along freeway segments where lanes have been added and where existing rights-of-way are limited. Along those freeway segments where lane widths are 3.7 m, it appears that lane barriers for automated facilities can be erected without the need for lateral expansion of the freeway. For example, if a ten-lane freeway with 3.7 m lane widths and a median width of 1.8 m is converted to a ten-lane shared space at-grade facility, by reducing the lane widths to 3.4 m and reducing the median width to 0.6 m, no lateral expansion beyond the existing outside shoulders would be necessary. In most cases, it would not be desirable to reduce the outside shoulder width of a freeway in order to accommodate AHS facilities because adequate shoulder space is necessary for efficient traffic operations. On freeways with 3.7 m lane widths and median widths of at least 4.9 m, available median space is sufficient to accommodate the spatial requirements of the shared space at-grade concept without the reduction of lane widths to 3.4 m.

The need for either a continuous or intermittent emergency lane along the automated facility is an issue requiring further consideration. The provision of an emergency lane would serve two primary purposes: to facilitate emergency vehicle access to the automated facility and to temporarily store disabled vehicles to avoid the blockage of an automated lane. The inclusion of emergency lanes would increase the spatial requirements of an AHS facility significantly, resulting in increased costs and construction difficulties, especially in urban areas.

3.2. Above-Grade AHS Concept

Above-grade, or elevated, AHS facilities could be built above existing freeways within existing rights-of-way in most cases. Above-grade AHS facilities may be appropriate in densely developed downtown areas where lateral expansion of freeways may be economically infeasible.

Figure 5 shows how an existing eight-lane freeway might look with an added above-grade AHS facility. Structural supports for above-grade facilities could be

FIGURE 5. Dedicated above-grade concept.

located within the existing freeway median if space is available. Where existing median widths are not sufficient to accommodate such structures, the median could be widened either by reducing lane widths or by expanding the freeway beyond the outside shoulder. For a two-lane elevated facility, it may be necessary to provide an emergency lane, or shoulder, adjacent to each automated lane to facilitate emergency vehicle access to the automated facility and to temporarily store disabled vehicles to avoid the blockage of an automated lane (Figure 6). Alternatively, a single emergency lane serving both lanes would reduce the width of the elevated structure. Movable barriers or openings in the barriers between the automated lanes and the emergency lane could be used to enable emergency vehicle access to any portion of the automated facility. Since vehicles would not be able to enter the automated facility from the manual lanes, dedicated entry and exit facilities would be required. Such facilities could consist of ramp flyovers constructed over the manual lanes connecting the elevated facility to either existing or new overpasses (Figure 7).

3.3. Below-Grade AHS Concept

Below-grade AHS facilities could be built as tunnels beneath existing freeways. Below-grade facilities may be feasible only for short distances and where existing free-

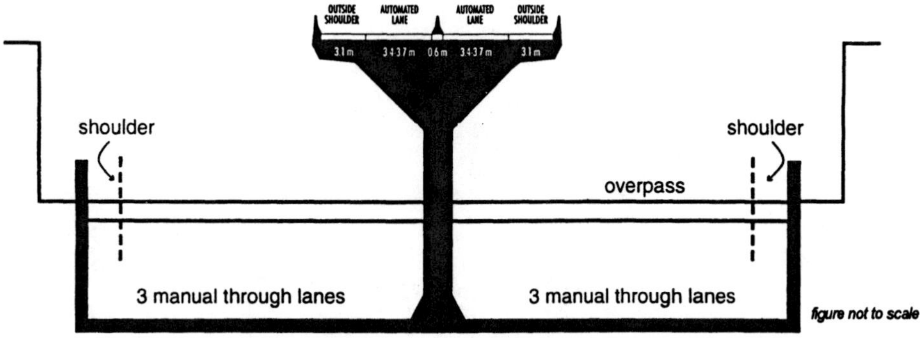

FIGURE 6. Cross section of dedicated above-grade concept.

ways are not depressed, or constructed below the grade level of the adjacent topography. The below-grade AHS concept is most appropriate where a freeway merge or split occurs in the leftmost lane of a shared space at-grade facility, permitting the uninterrupted flow of automated traffic and preserving all movement options for both automated and manual traffic. Figure 8 shows how a tunnel can be used to enable continuous automated travel when there is a freeway split in the leftmost lane.

4. EVALUATION OF AHS CONCEPTS

The at-grade, above-grade, and below-grade AHS concepts were evaluated based on criteria established with respect to the physical features of existing

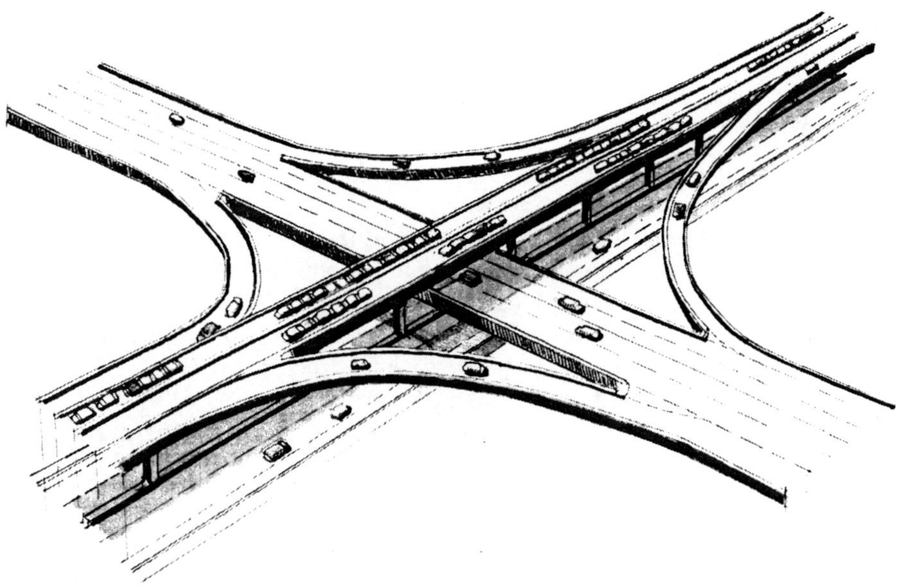

FIGURE 7. Entry/exit configuration of dedicated above-grade concept.

FIGURE 8. Dedicated below-grade concept.

California freeways. The purpose of this evaluation was to determine the viability of these concepts from both economic and environmental perspectives. Based on these considerations, a hierarchy of AHS concepts, from the easiest to most difficult to integrate into existing freeways, was developed.

4.1. Evaluation Criteria

The following evaluation criteria were established to evaluate the alternative AHS deployment concepts:

1. *Rights-of-way*—Minimize the acquisition of real estate and maximize the use of the existing rights-of-way
2. *Environmental impacts*—Minimize the visual, air quality, and noise impacts of AHS facilities and maximize the compatibility of the AHS facilities with adjacent land use characteristics
3. *Cost*—Minimize the capital and operating costs of AHS construction and modifications to the existing roadway infrastructure
4. *Natural hazards*—Minimize seismic hazard risks
5. *Construction impacts*—Minimize the interruption of traffic during construction

4.2. Deployment Criteria

Based on the above evaluation criteria, AHS facility deployment criteria were identified:

1. The implementation of an AHS facility through the conversion of existing freeway lanes should be given priority over all other concepts. This concept is the least costly to implement in terms of right-of-way acquisition and modifications to the existing infrastructure, introduces minimal seismic safety hazards, and visual and noise impacts would be minimized.
2. At-grade lateral expansion of existing roadways should utilize available median space to the extent possible. This is preferable since this land is already part of existing rights-of-way, minimizing the need for additional real estate. Right-of-way acquisition in urbanized areas can be extremely costly and/or politically unpopular, particularly when the condemnation of residences is necessary.
3. The use of elevated structures should be avoided to the extent possible. Elevated structures are extremely costly to construct, require the construction of costly entry/exit facilities, introduce seismic safety hazards, and may result in significant visual and noise impacts. However, elevated facilities may be the most preferable alternative for freeway segments that do not contain adequate rights-of-way to accommodate at-grade facilities and where restrictions to lateral expansion of the roadway make it infeasible to do so. In such cases, the cost of at-grade lateral expansion may be significantly higher and less acceptable to the public than the use of an elevated facility.
4. The use of tunnels should be avoided to the extent possible and should only be considered for relatively short distances. Like elevated structures, tunnels are extremely costly to construct and introduce new seismic safety hazards. In addition, tunnel construction may require the relocation of underground utilities, a potentially costly and cumbersome undertaking. Sufficient ventilation systems would be required in order to avoid the accumulation of unsafe levels of vehicle emissions within tunnels. While tunnel construction is extremely expensive at present, some experts believe that the cost of tunnel construction may decrease dramatically as tunneling technology improves.

4.3. Hierarchy of AHS Concepts

Each of the AHS concepts was qualitatively evaluated in terms of the above deployment criteria. Based on this evaluation, the following hierarchy, from easiest and most preferable to implement to most difficult and least preferable to implement, was established:

1. *At-grade (lane conversion)*: Existing lanes are converted for automated and manual use. Minimal lateral expansion of the roadway is necessary.
2. *At-grade (median)*: Available space within the median is used to accommodate lateral expansion of the roadway in order to provide a sufficient number of lanes for an at-grade AHS facility.

3. *At-grade (median and shoulder)*: A combination of available space within the median and beyond the outside shoulders is used to accommodate lateral expansion of the roadway in order to provide a sufficient number of lanes for an at-grade AHS facility.
4. *At-grade (shoulders)*: Available space beyond the outside shoulders is used to accommodate lateral expansion of the roadway in order to provide a sufficient number of lanes for an at-grade AHS facility.
5. *Above-grade (median)*: Available space within the median is used to accommodate the structural supports for an elevated AHS facility.
6. *Above-grade (shoulders)*: If available median space is insufficient to accommodate the structural supports of an elevated AHS facility, available space beyond the outside shoulders is used to expand the roadway, thereby freeing up the required space within a widened median area.
7. *Below-grade (tunnel)*: An AHS facility constructed below the existing grade level of the roadway.

The at-grade lane conversion concept appears to be the easiest concept for integrating AHS facilities into existing freeways, based on economic and environmental considerations. With the shared space at-grade concept, existing on- and off-ramps could be utilized and in many cases only minimal lateral expansion of the roadway would be required. Despite these advantages, this concept has the following disadvantages:

1. Unauthorized manually driven vehicles could enter the automated portion of the roadway more easily than under the other concepts.
2. Lane barriers could present a safety hazard to vehicles changing lanes through barrier openings.
3. A decrease in the level of service of the manual lanes could result from vehicles weaving through these lanes to access the automated lane.
4. Underutilization of the automated lane(s) could result when manual lanes are congested, since it would be difficult for vehicles to weave through the manual lanes to reach the automated lane(s).

Some of these problems may be alleviated through the implementation of the dedicated space at-grade concept, in which the automated and manual lanes would share a common right-of-way but are physically separated by a continuous lane barrier. Under this concept, exclusive entry/exit facilities would be necessary, eliminating the need for a transition lane. There are various trade-offs that must be considered to determine which of these at-grade concepts would be preferable to implement. At this point, it is not possible to discern which of these concepts would be more preferable to deploy. While it may be possible to determine which concept is superior when comparing the two on the basis of a single criterion, such as safety, given the wide array of performance criteria that must be considered it is not possible to make a determination of overall superiority at this time. Some of the issues that require further study include comparisons of cost, traffic operations, and safety. Furthermore, because of the site-specific nature of many of the issues in question, it is difficult to make generalizations regarding the superiority of one concept over the other. Depending on the specific attributes of a given freeway corridor, the selection of the "best" concept to implement may vary.

Exclusive AHS entry/exit facilities, consisting of flyover ramps or ramps from existing overpasses, would provide for access and egress to and from a dedicated space at-grade facility. Exclusive AHS entry/exit facilities would provide improved safety and operation through the physical separation of manual and automated traffic. Because of the cost of exclusive entry/exit facilities, automated facility access and egress opportunities would likely be less frequent than existing on- and off-ramps and, therefore, less frequent compared with the shared space at-grade concept. In addition, the construction of exclusive AHS entry/exit facilities may require the acquisition of additional right-of-way and may result in adverse traffic impacts to the local street network in the vicinity of such facilities. Since the widespread deployment of AHS is not likely to occur in the immediate future, the development of lighter-weight construction materials and methods to reduce the spatial requirements of elevated structures may have a significant impact on the feasibility of constructing above-grade AHS entry and exit facilities.

5. DETAILED DISCUSSION OF AHS IMPLEMENTATION ISSUES

The specific issues associated with each of the aforementioned AHS concepts would vary significantly depending on site-specific conditions. The trade-offs between the shared space and dedicated space concepts, such as entry/exit, cost, safety, traffic operations, and implementation phasing issues, must be studied further in order to better understand the advantages and disadvantages associated with these alternative concepts. Specific issues that emerged in the process of evaluating the alternative AHS concepts were the operational features of the shared space concept, design requirements for above- and below-grade facilities, and environmental concerns.

5.1. Operational Features of the Shared Space Concept

The operational features of the shared space concept include: (1) operational efficiency, (2) lane barriers, (3) safety hazards, and (4) the psychological impacts of lane barriers.

5.1.1. Operational Efficiency. Under the shared space at-grade concept, vehicles would use existing on- and off-ramps, thereby requiring vehicles to weave through the manual lanes prior to entering the automated facility and to weave through the manual lanes after exiting the automated facility in order to exit the freeway. The capacity of an automated lane is expected to be significantly greater than the capacity of a manual lane. Therefore, when the automated lane is well utilized, weaving activities associated with entering and exiting the automated system may result in a decrease in the level of service in the manual lanes. When the manual lanes are congested, this may result in underutilization of the automated lane(s) because of the difficulty involved in weaving through the manual lanes to access the automated facility. Lane barriers between manual and automated systems could further amplify these problems, since vehicles would be allowed to maneuver between the transition lane and adjacent manual lane only at points where barrier openings are provided; this restriction to free lane-to-lane movements could result in a further decrease in the level of service in the manual lanes.

5.1.2. Lane Barriers. If lane barriers are to be used, an appropriate barrier type must be identified prior to implementation. A candidate barrier design investigated for this study was the Caltrans' Concrete Barrier, Type 50, which is representative of the standard concrete median barrier used on freeways with narrow medians in California. This barrier is 0.61 m wide at the base contouring to a width of 0.15 m at the top with a total height of 0.81 m. This barrier requires little maintenance and is able to prevent major accidents. With this barrier, it has been found that most vehicles that collided with the barrier were redirected by the "safety shape" with no damage and were able to drive away. Furthermore, this is the cleanest barrier in terms of design, with no projections to collect debris.[6] Because of the aforementioned attributes of this barrier type, it is believed that a barrier of similar design would be appropriate for the shared space concept. As a result of the space requirements of this barrier type, in addition to any buffer space provided between barriers and lane boundaries, some degree of lateral expansion of roadways would likely be necessary to accommodate these barriers.

5.1.3. Safety Hazards. In order to allow vehicles to shift among the manual, transition, and automated lanes under the shared space at-grade concept, strategically positioned openings in the lane barriers would be necessary. Because of these barrier openings, it would be impossible to prevent unauthorized manually driven vehicles from entering those portions of the roadway designated exclusively for vehicles under automated control. Drivers under the influence of drugs or alcohol may inadvertently enter the automated facility. Thrill-seekers and troublemakers may purposely enter the automated facility under manual control. Debris from accidents in the manual lanes could potentially reach automated portions of the roadway, depending on the location of openings in adjacent lane barriers and the angle of approach of the debris. Under this infrastructure type, there are no apparent design options available to ensure that manually operated vehicles would be unable to enter automated portions of the roadway. Considering the expected significant increases in lane capacities and vehicle speeds attainable under automated control, the effects of unauthorized vehicles entering the automated facility could be severe.

A major safety hazard associated with lane barriers is the design of barrier end treatments. Isolated freestanding ends of lane barriers following openings could present a safety hazard to approaching traffic and therefore must be shielded from impacts. This may be done by placing an appropriate crash cushion at the approach end of the barrier. Warning pylons or other devices placed in front of crash cushions can be used to alert drivers of the approaching hazard. Reflector units or electric lights along the top of barriers can be used to provide enhanced delineation at night and during adverse weather conditions. While such warning devices are not needed for vehicles under automated control, they would be needed by those vehicles that fail automated lane check-in and must proceed under manual control before returning to the manual lanes. Despite the use of crash cushions, isolated freestanding ends of barriers would remain a hazard. Barrier openings must be of sufficient length to provide ample opportunities for lane change maneuvers under all traffic conditions. Furthermore, openings in the two adjacent barriers separating the automated, transition, and leftmost manual lanes should not be adjacent to one another to ensure that the automated facility is protected from accident debris by at least one barrier to the greatest extent possible.

5.1.4. Psychological Impacts of Lane Barriers. Along stretches of the automated and transition lanes, barriers would be located along both sides of these lanes. In the automated lane, since vehicles will be under automatic control at all times the barriers on both sides of the lane would not necessarily pose a safety hazard. However, these barriers may give motorists a feeling of confinement. In the transition lane, vehicles that fail check-in would have to manually drive through a short section of roadway with barriers on both sides prior to returning to the manual lane. If transition lanes are 3.7 m in width, the standard freeway lane width, it may or may not be necessary to provide buffer space between the barriers and the lane boundary since 3.7-m lanes provide ample space for even the largest passenger vehicles. However, since drivers are unaccustomed to driving through such confined spaces at freeway speeds, they may be uncomfortable if required to do so. If heavy-duty trucks or buses are permitted to use the automated system, the need for buffer space in areas where vehicles are under manual control would become more apparent because of the wider girth of these vehicles. A narrow buffer, e.g., a third of a meter, could be used to mitigate such psychological impacts of driving under manual control between physical barriers.

5.2. Design Requirements for Above- and Below-Grade AHS Facilities

The design requirements for above- and below-grade AHS facilities include: (1) the structural engineering of elevated structures, (2) location of exclusive AHS entry/exit facilities, and (3) lane splits and merges.

5.2.1. Structural Engineering of Elevated Structures. Where at-grade AHS facilities may be deemed infeasible, elevated structures may be a viable alternative. In choosing an appropriate structure type for an above-grade AHS facility, advantages and disadvantages of various design options need to be assessed. Cost, visual impacts, and seismic safety are the major concerns associated with elevated structures. Considering present construction technologies, one of the possible options for the construction of an above-grade facility is a precast high-strength concrete structure, such as the Harbor Freeway Transitway (Interstate 110) in downtown Los Angeles.

Further advances in materials such as high-strength concrete will likely enable the design of elevated structures with smaller structural members than are used today. Advances in construction methods should enable the assembly of such structures in less time and with less manpower. Although the design of elevated AHS structures in California would incorporate the most up-to-date California State earthquake codes, any structure of this type would represent a potential seismic hazard. Technological advances in high-strength lightweight concrete, steel, and plastic materials may enable designers to reduce costs and lessen the visual and seismic impacts of elevated structures by the time the widespread deployment of AHS becomes possible.

5.2.2. Location of Exclusive AHS Entry/Exit Facilities. Exclusive entry and exit facilities for elevated facilities could consist of flyover ramps connected to vehicle overpasses. A recent study suggests that, because of operational, cost, and space requirements of such facilities, access and egress opportunities exclusively designated for an elevated AHS should be provided less frequently than is typical on conventional freeways. The design of appropriate structures for the construction of extensive ramps is an important issue. Elevated entry/exit ramps would also represent a seismic safety hazard. Technological advances in lightweight high-strength concrete, steel, and plastic materials may help future designers to develop safer, more aesthetically pleasing structures at a lower cost than is possible today.

5.2.3. Lane Splits and Merges. Splits and merges at freeway interchanges become more challenging with the presence of automated lanes. With the conversion of the leftmost lane for automated use under the shared space at-grade concept, on- and off-ramps and freeway-to-freeway splits and merges in the leftmost lane could no longer be used by manual traffic. Significant roadway modifications may be required to allow continuous automated travel while providing all necessary options for both automated and manual vehicles in these situations. Tunnels or flyovers can be introduced to provide continuous automated travel when there is a split or merge in the leftmost lane. Despite the availability of options to remedy many of these problems, reconfiguring freeway interchanges can be costly, especially if additional right-of-way must be acquired. In addition, tunnel construction may require the relocation of underground utilities and would require the provision of sufficient ventilation to prevent the accumulation of vehicle emissions. Although tunnel construction is costly at present, some experts believe that the cost to tunnel construction may decrease dramatically as tunneling technology improves.

5.3. Environmental Concerns Associated with AHS Facilities

The environmental concerns associated with AHS facilities are generally associated with: (1) compatibility with adjacent development, (2) land use considerations, and (3) local street network traffic impacts.

5.3.1. Compatibility with Adjacent Development. When an elevated AHS is built, a new structural element is introduced within the freeway corridor. Even if lighter construction materials were used, the visual impact of an elevated structure could be significant, although the degree of the impact would vary depending on the location. When an elevated structure is integrated into a corridor containing many high-rise buildings, its visual impact would be less significant than in an area containing predominantly low-rise buildings. Therefore, elevated structures in suburban and rural corridors would be far more obtrusive than in corridors containing a higher intensity of land uses. Noise impacts associated with above-grade AHS facilities might be more severe than those generated by at- or below-grade facilities.

5.3.2. Land Use Considerations. At-grade AHS facilities can be built either by converting portions of existing roadways for automated use or through the lateral expansion of existing roadways. Where existing roadways are not wide enough to adequately accommodate both automated and manual traffic, lateral expansion of the roadway may be required. For freeway segments with less than four lanes in each direction, implementation of the shared space at-grade concept would require the lateral expansion of the roadway. In other situations such as for elevated structures where the existing right-of-way does not allow the erection of supporting structural components or the construction of exclusive entry and exit facilities, the acquisition of additional right-of-way would be necessary. Lateral expansion of existing freeways may be difficult, potentially impacting residential neighborhoods and other established land uses and requiring the displacement of physical impediments such as retaining walls, soundwalls, and overpass bridge supports. Depending on the availability of usable space within the existing right-of-way and the severity of restrictions to lateral expansion beyond the right-of-way, lateral expansion can be costly and politically unpopular.

5.3.3. Local Street Network Traffic Impacts. In addition to the difficulty in designing an extensive flyover ramp structure and potentially high real estate acquisition costs to accommodate such facilities, adverse traffic impacts to the local street network could be a major issue. One of the potential adverse impacts of exclusive AHS entry and exit facilities is the potential for significant queuing of vehicles on local streets in the vicinity of entry/exit facilities. The basis for stating such concerns is that such queuing problems also exist under current traffic conditions at certain on- and off-ramps. With fewer exclusive AHS entry and exit facilities relative to existing on- and off-ramps, this situation could be exacerbated. Suitable sites such as large parking lots or undeveloped parcels where vehicles can queue up without interrupting local traffic should be investigated for exclusive entry and exit facilities. The availability of land would ultimately dictate site selection for entry/exit facilities.

5.4. AHS Issues Specific to Urban and Rural Environments

Although the discussion to this point has been specific to urban and rural environments only to a limited degree, there are significant fundamental differences that exist between urban and rural environments as they relate to AHS implementation. These fundamental differences pertain to freeway geometrics, restrictions to lateral roadway expansion, and terrain.

In California, rural freeways typically have fewer lanes than urban freeways; four- and six-lane rural freeways are common. Depending on the existing number of lanes and the AHS concept being deployed, the degree of lateral roadway expansion required could vary considerably. Although rural freeways typically have fewer lanes than do urban freeways in California, the availability of median space and space beyond the outside shoulders in rural areas is typically much greater compared with urban areas. Furthermore, restrictions to lateral expansion are not as prevalent in rural areas; soundwalls, retaining walls, land uses, and other physical impediments to lateral expansion are much less common. On the other hand, mountainous rural terrain can be a significant obstacle to lateral expansion not encountered in urban areas. In mountainous areas, steep cuts and slopes are sometimes very close to the inside and outside shoulders. Where lateral expansion is required along a steep cut or slope portion of the roadway, costs and construction difficulties associated with lateral expansion become greater. Where separated roadbeds are present in rural areas, the elevation differential between the two roadbeds can be significant, making it difficult to utilize the median area for lateral expansion. Where the roadbeds consist of bridge structures (e.g., at underpasses and above drainage channels), widening of such facilities would be difficult. Where lateral expansion beyond the outside shoulders is necessary, existing overpasses may not be wide enough to accommodate such expansion and therefore may need to be reconstructed. In mountainous rural terrain, if three manual lanes are provided rather than the minimum of two, the level of service in the manual lanes would improve drastically because of the great variation in speeds of light- and heavy-duty vehicles on steep grades.

In many cases, available median space in rural areas may be sufficient to accommodate on- and off-ramps for dedicated space at-grade AHS facilities, which would connect to overpass entry/exit facilities (Figure 9). Where such entry/exit facilities are located, lateral expansion equivalent to two additional lane widths would be required to accommodate on- and off-ramps. Depending on the existing number of lanes, the combined number of manual and automated lanes (a minimum of three total lanes under the dedicated space at-grade concept—two manual lanes and one automated lane), and the width between the supports of the overpass structure, it may or may not be necessary to reconstruct an existing overpass when converting it to an entry/exit facility.

Since restrictions to lateral roadway expansion in rural environments are typically much less severe than in urban environments, it is not likely that either the above- or below-grade concepts would be considered for implementation in rural areas. In most cases, the shared space or dedicated space at-grade concepts would be more preferable to implement than the other alternative concepts presented.

FIGURE 9. Entry/exit configuration of rural shared at-grade concept.

6. SUMMARY AND CONCLUSIONS

In this study, alternative AHS deployment concepts were examined and issues associated with the integration of AHS into existing California freeways were identified. In both urban and rural environments, it appears that at-grade AHS facilities would be most compatible with the existing environment. Compared with at-grade AHS facilities, above- and below-grade AHS facilities would typically be more costly to construct and would introduce greater environmental hazards. Despite the apparent advantages of the at-grade concept, there are situations in which the deployment of either the above- or below-grade concepts may be more sensible. In densely developed urban areas, the acquisition of additional right-of-way is often costly; therefore, either an above- or below-grade AHS facility should be considered. The overall capital cost of above-or below-grade AHS facilities in high-density urban areas can be less costly when compared with land acquisition costs and this approach can also avoid disruption of existing land use activities. Freeway-to-freeway splits and merges at complex interchanges may require additional infrastructure improvements in order to facilitate the uninterrupted flow of automated traffic streams. Construction of multilevel interchanges may be necessary using either the above- or below-grade deployment concepts.

Under the shared space at-grade concept, vehicles would enter the automated lane(s) from the manual lanes through a transition lane and, therefore, existing on- and off-ramps would be utilized. Under the dedicated space at-grade concept, no movement would occur between the manual and automated lanes and entry/exit to and from the AHS facility would be provided via dedicated on- and off- ramps. In terms of safety and operational efficiency, the complete separation of manual and automated traffic appears to be preferable. However, there are trade-offs between these two concepts that should be examined further. Comparisons between the two concepts that should be investigated further include cost, safety, land use displacement, freeway traffic operations, local street network impacts, and accessibility to the automated facility by emergency vehicles.

In addition to the fairly microscopic issues of AHS infrastructure deployment addressed in this investigation which require further study, many macroscopic issues must be addressed prior to AHS deployment. The incremental development, or phasing, of AHS facilities in relation to market behavior needs to be explored. Whether or not AHS facilities should accommodate both light-duty and heavy-duty vehicles should be determined. Additional multidisciplinary studies addressing a wide range of issues are needed in order to adequately assess the overall desirability and feasibility of wide-scale AHS deployment.

ACKNOWLEDGMENTS. This research was supported by the California Department of Transportation (Caltrans) under Interagency Agreement 65H998 and the Federal Highway Administration (FHWA) of the United States Department of Transportation under Contract DTFH61-93-C-00199. The authors would like to thank Terry Quinlan, Jas Bhullar, and especially Matt Hanson of the Caltrans Division of New Technology, Materials, and Research with whom this project was conducted for their valuable technical support. The authors would also like to thank J. Richard Bishop of FHWA, William Stevens and John Harding of the MITRE Corporation, Steven Shladover of the California PATH Program, and Randolph Hall of the University of Southern California for their valuable inputs to this research.

REFERENCES

1. R. HALL, *Time Benefits of New Transportation Technologies: The Case of Highway Automation,* University of California at Berkeley, Institute of Transportation Studies, PATH Research Report, UCB-ITS-PWP-91-4, 1991.
2. Southern California Association of Governments, *Highway Electrification and Automation Technologies—Regional Impacts Analysis Project Phase II Report*, Los Angeles (1992).
3. P. VARAIYA, AND S. SHLADOVER, Sketch of an IVHS Systems Architecture, University of California at Berkeley, Institute of Transportation Studies, PATH Research Report, UCB-ITS-PRR-91-3 (1991).
4. A. HITCHCOCK, *Layout, Design, and Operation of a Safe Automated Highway System*, UCB ITS-PRR-94-17 (March 1994).
5. P. VARAIYA, *Activity Area J: AHS Entry/Exit Implementation, FHWA AHS Precursor Systems Analyses* (1995).
6. California Department of Transportation, *Traffic Manual*, Chapter 7 (1987).

CHAPTER 4

System Configurations

Evolutionary Deployment Considerations

Randolph W. Hall

1. BACKGROUND

The goal of this chapter is to develop a framework by which system configurations can be defined for the purpose of Automated Highway System (AHS) cost/benefit analysis. This framework was used in the Federal Highway Administration (FHWA) Precursor Systems Analyses (PSA) program. This framework was applied to a range of scenarios, to capture benefits and costs on a regional and national basis. The framework reflects risks and uncertainties with respect to such factors as market penetration, construction and vehicle costs, and public acceptance of new facilities.

The specific purpose of this chapter is to document the evolutionary deployment strategies used in the research. These strategies focus on technologies that are essential to AHS and, as a consequence, should be included in AHS cost and benefit estimates. While technologies for advanced traveler information and advanced traffic management are related to highway automation, they are not essential.

Randolph W. Hall • Department of Industrial and Systems Engineering, University of Southern California, Los Angeles, California 90089-0193.

Automated Highway Systems, edited by Ioannou. Plenum Press, New York, 1997.

Therefore, they will not be covered in the evolutionary deployment strategy presented here.

The overall philosophy is to develop a plausible scenario for AHS development that is responsive to market forces and does not stretch technological limits in early stages of deployment. In developing this scenario, many uncertainties exist. Hence, this document represents only one of many possible plans.

The remainder of this chapter is divided into six sections. First, publications on evolutionary deployment are reviewed. Next, criteria are presented for development of evolutionary strategies, and the dimensions of evolution are defined. Then the evolutionary strategy to be used in the cost/benefit task is introduced, first at a functional level, and then in terms of specific technologies. Finally, the next steps of the research are provided.

2. RELATED PUBLICATIONS

Varaiya[(1)] identifies five aspects of Intelligent Transportation Systems (ITS) development: function, architecture, design, evolution, and evaluation. Evolution is defined as the timing of system development and deployment, and the extent to which the architecture should accommodate new functions not included in earlier designs, while evaluation is defined as the effectiveness, costs, and benefits of different ITS proposals. Varaiya's paper focuses on the first three aspects of design.

Though not addressed explicitly, many authors have discussed general premises regarding evolutionary development of ITS. For example, auto manufacturers[(2,3)] have focused on the extent to which individuals are likely to purchase automation equipment prior to the construction of the AHS infrastructure (or at least prior to conversion of lanes from manual to automatic). Heinrich states that "the ultimate success [of Intelligent Vehicle–Highway Systems (IVHS)] will be highly dependent upon the acceptance and continued use of in-vehicle IVHS equipment by the vehicle driver." The author also links "smart vehicles" and "smart highways" and argues that smart highways have to exist before smart vehicles become a reality. He also asserts that car buyers are generally conservative regarding what they buy and are looking for practical solutions for their needs, and argues that "the capability of the IVHS infrastructure to provide timely and credible traffic advisories will play a key role in forming and more importantly maintaining the buyer's interest in IVHS."

Johnston *et al.*[(4)] proposed five stages for deployment of automated freeways:

1. Voluntary on-board navigation and route-guidance devices
2. On-board longitudinal control
3. Lateral control and dedicated lanes
4. Full automation of some lanes
5. Full automation of all lanes

Their paper, however, didn't specify how these stages are to be deployed nor did it address the stimulus by which the development would move from one stage to the next.

Hall[5] provides a progression in the development of the highway infrastructure. Different versions of the AHS, including AHS without automated lane change, or with automated lane change at reduced speed, are presented. These are evaluated with respect to their ability to reduce congestion at highway bottlenecks. The paper does not address evolution in vehicular technology. Hall and Tsao[6] developed an influence diagram representing uncertainties affecting the deployment of AHS. These uncertainties include: technological feasibility, opposition of special interest groups, and inability of auto makers to provide the needed automation. The study synthesizes expert opinions "to identify the critical issues, technical or not, that need to be resolved to ensure timely and efficient deployment of AHS." The influence diagram reflects the staged aspect of AHS deployment, as progress on deployment cannot occur until institutional issues are resolved.

Ward[7], as part of the AHS PSA effort, developed an evolutionary strategy in some detail, emphasizing motivations for consumers to purchase at each stage of evolution. He hypothesizes that Intelligent Cruise Control will be the first step toward the automated highway. He also suggests that vehicles, if appropriately designed, might be enabled to form "spontaneous platoons," providing capacity increases.

Ioannou *et al.*[8] define a series of evolutionary configurations as part of their AHS PSA effort. The authors emphasize the evolution of lateral and longitudinal control, taking technological feasibility and capacity and safety benefits into consideration. The result is a series of five evolutionary stages, which gradually automates the vehicle and shifts control to the roadway.

Al-Ayat and Hall[9] developed a framework for planning evolutionary deployment of ITS technologies. It defines an evolutionary deployment sequence, identifies baseline assumptions, and develops an evaluation framework, consisting of strategy development, strategy evaluation, technology and barrier identification, and strategy refinements. This framework provides a foundation for the evolutionary strategy presented later.

As a whole, the literature on evolutionary deployment aims to develop plausible scenarios under which today's conventional highway might evolve toward full automation, without incurring unreasonable costs along the way. The strategy developed in this report continues in this vein, with emphasis on building up a fleet of equipped vehicles prior to major investments in AHS infrastructure.

3. CRITERIA FOR EVOLUTIONARY DEPLOYMENT

This section outlines ground rules (i.e., constraints) for evolutionary deployment, as defined in Al-Ayat and Hall.[9] These ground rules are partially based on the FHWA PSA Broad Area Announcement (BAA). The FHWA ground rules are supplemented to explicitly account for the evolution of AHS functions. The complete set of ground rules are the basis for the evolutionary deployment strategy used in the cost/benefit analysis.

The BAA constraints, together with others that are necessary for developing evolutionary deployment strategies, are presented here (constraints 1 through 4 are adapted from the FHWA list):

1. All vehicle types (automobiles, buses, trucks), although not necessarily intermixed, must be supported in the mature system. Initial deployment emphasis will be on automobiles and vehicles with similar dynamics and operating characteristics.
2. Not all vehicles or roadways will be instrumented.
3. An AHS will perform better than today's roadways in all key areas including safety, throughput, user comfort, and environmental impacts.
4. ITS technologies will operate in a wide range of weather conditions typical of those experienced in the continental United States. These include snow, low-visibility fog, and heavy rain conditions.
5. Vehicle equipment provides substantial user benefits, even where AHS is not implemented. User benefits could include enhanced performance, or driver comfort. These benefits are needed to motivate drivers to acquire the equipment and to provide sufficient incentive for manufacturers to invest in tooling and infrastructure needed to produce the needed technologies.
6. Automation does not require significant demolition or relocation of houses/businesses or result in unacceptably negative impacts on neighborhoods surrounding a freeway.

In summary, automation should strive for the capability to operate at much higher capacity, without increased delays, with much higher safety and energy savings, and with a decrease in pollution. Moreover, automation should be introduced in an equitable manner and, ideally, should not cause any loss or penalties to any individual or group.

An evolutionary deployment strategy should also provide users with increased functionality over time, according to the following guidelines:

7. Full vehicle automation requires minimal retrofit to vehicles designed to operate in low-traffic environments in order to operate within a high-capacity AHS environment. However, the functions available at each step are useful by themselves, and do not require retrofit to provide significant benefits.
8. Each development step has a high likelihood of acceptance by the user and by the public.
9. Success in each step increases the chance of public acceptance of the following development step.
10. Development steps take into consideration the long lead-time requirements for research and development prior to deployment.

11. Increasing functionality can be achieved without the need for discarding major portions of the system.

As a group, criteria 7–11 provide one strategy in which AHS is built incrementally, with each step carefully planned, from both a technical and an institutional perspective, to lead toward the endpoint.

4. DIMENSIONS OF EVOLUTION

Evolutionary deployment can be defined along two principal dimensions:

1. Product development cycles
2. Market penetration cycles

The *product development cycle* leads to the technology's introduction into the marketplace, and includes the following steps (some of which may occur simultaneously):

- Research
- Proof of concept development
- Operational testing
- Design for manufacture
- Education
- Technical standards development
- Legislation

The *market penetration cycle* begins when the product is first introduced, and can be divided into the following steps:

- Introduction
- Growth
- Maturity
- Decline

For any component technology of AHS, it is possible, and desirable, to plan for an evolutionary sequence through the steps of product development and market penetration. Such a plan could be used to develop cost estimates for AHS research and development, and to assess the effectiveness of governmental and private expenditures over time. However, to keep cost analyses within bounds, our emphasis will be on the evolutionary steps associated with market penetration.

With respect to market penetration cycle, each component technology may progress through a distinct sequence of introduction, growth, maturity, and possibly decline. At any point in time, one technology may be in the introduction phase (perhaps autonomous lane keeping), another in the growth phase (perhaps automated inspection), and another in the mature phase (perhaps Advanced Integrated Cruise Control). Hence, the evolutionary deployment strategy is not just concerned with the initial introduction of the technology, but also with how the technology spreads across classes of users and across regions.

Put another way, the ultimate success of AHS, or any intermediate technology, hinges on introducing technologies into the right markets so that future growth can be sustained. This in turn depends on achieving sufficiently large up-front benefits to justify initial investments while not stretching the technologies beyond safety limitations. For example, autonomous vehicle control might first be implemented on highly traveled, but uncongested, intercity freeways, frequented by business travelers. Comfort would then be a major benefit, which may induce motorists to invest in necessary in-vehicle technologies. Once market penetration becomes sufficient, the technology could then spread to lower-traffic roads and lower-end users. Once safety is proved, the technology might spread to congested roadways, with the goal of increasing capacity. The underlying concept is that operation under less stringent low-traffic conditions can lead to refinement of the technology, for eventual operation under stringent conditions. Most importantly, a critical mass of users is likely needed before, first, special facilities are constructed and, second, existing roadways are converted to automation (Table 1).

AHS comprises a vast set of component technologies. Because these component technologies must ultimately work together as a system, our evolutionary deployment strategy will be defined by how the system evolves with respect to the functions it performs, using a framework developed in Hall.[10] According to this framework, the system contains three principal entities: vehicle, driver, and roadway/infrastructure. Each entity contains five basic elements: (1) sensing, (2) intelligence, (3) memory, (4) actuators, and (5) communication. These elements

TABLE 1
Market Penetration Evolution

	Introduction	Growth	Maturity
Incentives for user to purchase	Comfort Convenience Energy savings Enjoyment Performance Safety (subsidy?)	Travel time savings Others (subsidy?)	Travel time savings Others (subsidy?)
Infrastructure	Minimal (autonomous vehicles)	Limited special facilities	Roadway conversion & special facilities
Local market penetration	Very small (< 5% of vehicles on the road)	Small (5–30%)	Moderate to high (20% +)

and entities work together to perform the tasks of driving on a highway, which are principally the following:

- *Cruising:* Maintaining desired separation and velocity; steer to follow center of lane; accelerate and decelerate to accommodate other vehicles or to avoid obstacles.
- *Lane maneuvering:* Steer into desired gap in adjacent lane without conflicting with adjacent vehicles or objectives. Select most efficient lane of travel.
- *Exiting/entering:* Set trajectory that allows splitting from current roadway and merging into another roadway while avoiding vehicle or objects, or perform transition to/from automated driving. Check in or check out, with associated inspections, as necessary.
- *Path choice:* Select most efficient roadway between a selected origin and destination.

This last task is largely independent of highway automation, falling more in the category of Advanced Traveler Information Systems. Therefore, it will not be covered in the evolutionary deployment strategy.

Using the framework, the highway system of today can be depicted as in Table 2. Most of the intelligence resides in the driver, who is responsible for all of the driving tasks. This same framework will be used to define the steps of evolution in the following sections.

5. EVOLUTIONARY STRATEGY FOR COST/BENEFIT STUDIES

To satisfy the criteria for evolution outlined in Section 3, it is desirable to build up a critical mass of AHS vehicles prior to extensive deployment of AHS infrastructure. We propose accomplishing this goal through the development and sale of an ARV. An ARV is modularly designed, allowing easy installation, after point of sale, of supplemental sensors and communication devices. Its controller is reprogrammable, either via software or plug-in replacement of integrated circuits, to respond to additional sensor and communication inputs. At point of sale, the ARV does possess electronic steering, braking, and throttle control; a flexible driver interface; safety diagnostics; and adaptive cruise control (ACC), and automated steering capabilities (the latter of which can only function on designated intercity roadways because lane referencing systems must be installed, and to minimize safety risk).

The ARV is capable of operating on a fully automated highway, if retrofitted at modest expense. At the same time, an ARV provides immediate driver benefits, namely, comfort and performance. These benefits must be sufficient to motivate a substantial percentage of drivers to purchase an ARV, whether or not extensive AHS infrastructure is ever deployed. These benefits must also motivate purchases

TABLE 2
Conventional Vehicle on Conventional Roadway

	Capabilities		
	Vehicle		
	Machine	Human	Roadway
Sensors		Vehicles/adjacent lanes Vehicles/same lane Obstacles Hazards Weather conditions Vehicle condition	Traffic conditions
Intelligence		Lane keeping Lane maneuvering Lane choice Speed regulation Exiting/entering	
Memory		Rules of the road Driving skills	
Actuators	Hydraulic brake Conventional throttle Hydraulic steering		
Communication	*To drivers* Vehicle warning lights	*To drivers* Turn signals	*To drivers* Traffic conditions (CMS, HAR)

across the country, and not just in the locations where fully automated guideways are constructed, in order to achieve scale economies in ARV manufacture.

From the standpoint of roadway infrastructure, we propose that AHS be first deployed on roadways connecting urban centers. The "Auto-Steering AHS" would provide a lane referencing system, which would enable ARVs to operate in a fully automated mode with large intervehicle spacing (assuming that a minimum spacing is regulated in manufacture). These initial AHS would not provide capacity gains, but would, at relatively modest cost, motivate ARV purchases. They would not, for instance, have inspection facilities, and vehicle–vehicle communication would not have roadside controllers to regulate speeds or lane changes. They would, however, provide an opportunity for testing and refining AHS technologies, prior to their deployment under more demanding conditions.

The next stage, AHS1, would enable vehicles to operate under close headways within platoons on isolated AHS lanes. Vehicles would enter the lane through manual lane changes into interplatoon gaps. Capacity would increase, but not nearly so much as possible with automated lane changes. As with ARV, AHS1 would provide a testing ground for advancement to the next stage of evolution.

The final stage, AHS2, enables automated lane changes as well as side-by-side operation of automated and manual lanes (perhaps separated by noncontinuous barriers). This is the most technologically challenging stage, which might only be achieved after years of experience operating AHS1. It would also provide the biggest capacity gain, as well as cost savings in construction, because AHS lanes would not have to be completely isolated. Optimistically, if AHS technology advances rapidly, then the AHS1 stage might be skipped, moving directly to AHS2.

5.1. Evolutionary Scenario

Building from the "catalysts" of the ARV and the Auto-Steering AHS, the evolutionary strategy is filled out below, from the vehicle and roadway perspectives. At any point in time, we recognize that the fleet of vehicles operating in the United States will contain a mix of capabilities, categorized into the following four stages:

Vehicle Types

1. Basic: No automation capabilities (other than basic cruise control), and no possibility of retrofit for automation.
2. ARV: Possesses automated adaptive cruise control and drive-by-wire, and can be retrofitted for full automation at reasonable cost.
3. AHS1: Able to operate in fully automated mode under high volume, but not able to automatically change lanes.
4. AHS2: Able to operate in fully automated mode, and to change lanes under full automation.

In addition, at any point in time, the U. S. highway system will contain a mixture of roadway types, categorized into the following five stages:

Roadway Types

1. Basic: Conventional highway, with ITS functions limited to ramp metering, loop detectors, and changeable message signs (at most).
2. Roadway: Roadway–vehicle communication is provided, primarily to Vehicle Comm support traveler advisor and route planning functions, but upgradeable to support AHS.
3. Auto-steering: In addition to roadway–vehicle communication, a lane referencing system is installed to support automated steering. Fully automated driving is allowed under low-volume conditions for all vehicle types (i.e., lane referencing is only installed on designated roadways).
4. AHS1: AHS lanes are isolated from conventional, and allow fully automated driving for automobiles and light trucks and vans under high-volume conditions. Automated inspection is provided. Automated lane changing is not supported.

TABLE 3
AHS Ready Vehicle on Auto-Steering Roadway

	Capabilities		
	Vehicle		
	Machine	Human	Roadway
Sensors	Brake Engine Velocity, acceleration, jerk Yaw angle Vehicles condition Lane reference Major obstacles Distance (to leader)	Vehicles/adjacent lanes Roadway hazards Weather conditions	Roadway conditions Traffic conditions
Intelligence	Lane keeping	Lane/maneuvering Lane choice Speed regulation Exiting/entering Path choice	
Memory		Rules of the road	Lane reference system
Actuators	Electronic brake Electronic throttle Electronic steering		
Communication	*To drivers* Obstacle/speed warnings Vehicle condition	*To drivers* Turn signals *To roadway* Hazards	*To vehicles* Steering preview *To drivers* Roadway conditions Traffic conditions Hazards

5. AHS2: Automated lane changing and lane merging is provided. Furthermore, automated lanes do not have to be completely isolated from conventional.

5.2. Assignment of Functions

For the Stage 1 vehicle, and Stage 1 and 2 infrastructure, the vehicle will operate more or less as it does today, under manual control. Stage 2, 3, and 4 vehicles, and Stage 3, 4, and 5 infrastructure, demand a shift in control from the driver to the vehicle, and to the roadway.

The following outlines functions for fully compatible systems, that is, where the vehicle and the infrastructure are at compatible stages of development. It should be recognized that the functionality is limited by the minimum of the vehicle and road-

TABLE 4
AHS1 Vehicle on AHS1 Roadway (Manual Lane Change)

	Capabilities		
	Vehicle	Driver	Roadway
Sensors	Brake Engine Velocity, acceleration, jerk Yaw angle Lane reference Major obstacles Distance (to leader) Vehicle condition	Unusual hazards	Roadway conditions Weather conditions Traffic conditions
Intelligence	Lane keeping, with close headway vehicle following	Lane maneuvering Lane choice Exiting/entering	Speed regulation Entrance metering System shutdown
Memory		Rules of the road	Lane reference system
Actuators	Electronic brake Electronic throttle Electronic steering		
Communication	*To roadside* Vehicle condition *To drivers* Enter/exit commands Vehicle condition *To vehicles* Acceleration, velocity, jerk, location	*To roadway* Enter/exit requests Hazards *To drivers* Turn signals	*To vehicles & drivers* Permissions to enter Enter/exit commands Roadway conditions Weather conditions

way capabilities (e.g., an ARV cannot operate on AHS1 or AHS2 without upgrade, and an AHS1 or AHS2 vehicle performs the same as an ARV on an auto-steering roadway).

Table 3 outlines the ARV operating on an auto-steering roadway. The vehicle has actuators for electronic steering, throttle control, and braking. It also has associated sensors and intelligence to allow automated driving under controlled, low traffic, conditions (e.g., on rural highways). On the other hand, the vehicle does not have a complete array of sensors, or vehicle–vehicle communication, to enable automated travel under high-flow conditions. This would require retrofit, through installation of sensor and communication modules within a defined architecture.

Table 4 outlines the AHS1 vehicle operating on the AHS1 roadway. This stage adds automated check-in/check-out, with vehicle diagnostics. Vehicle–vehicle communication is also provided within lanes, between leaders and followers, to exchange information on acceleration, velocity, jerk, and location. It also provides

TABLE 5
AHS2 Vehicle on AHS2 Roadway

	Capabilities		
	Vehicle	Driver	Roadway
Sensors	Brake Engine Velocity, Acceleration, jerk Yaw angle Lane reference Major obstacles Distance (to leader) Vehicle condition	Unusual hazards	Roadway conditions Weather conditions Traffic conditions
Intelligence	Lane keeping, with close headway vehicle following Lane maneuvering Exiting/entering (partial)	Exiting/entering (partial)	Speed regulation Entrance metering System shutdown Lane choice Lane change initiation
Memory		Rules of the road	Lane reference system
Actuators	Electronic brake Electronic throttle Electronic steering		
Communication	*To roadside* Vehicle condition *To drivers* Enter/exit commands Vehicle condition *To vehicles* Acceleration, velocity, jerk, location Lane change data	*To roadway* Enter/exit requests Hazards	*To vehicles & drivers* Permissions to enter Target speeds Enter/exit commands Roadway conditions Weather conditions

additional sensors installed on the roadway to check for weather conditions. Finally, it provides for speed regulation, through commands issued from the roadway.

Table 5 outlines the AHS2 vehicle operating on the AHS2 roadway. This stage adds communication to and from vehicles in adjacent lanes, in combination with improved side sensors, to enable automated lane changes. In addition, the infrastructure now has the capability to issue lane-change commands, based on global traffic conditions.

5.3. Operating Concepts

For the purposes of cost/benefit analysis, it is not necessary to completely define the AHS operating concept, with respect to such issues as platooning versus no platooning, lane assignment strategies, and system architecture. These factors only indirectly affect costs and benefits, which are more closely tied to highway capacity, operating speed, and hardware/software. Instead of specifying precise values for capacity and speed, ranges were incorporated in sensitivity analyses for costs and benefits.[11] Hardware and software cost, on the other hand, are relatively insensitive to the operating concept, and can therefore be calculated independently.

The analysis recognizes that an AHS1 has inherently less capacity than an AHS2 because of the disruptive effects of manual lane changes. These manual lane changes may significantly limit the capacity gains of automation. Hence, a principal motivation for upgrade from AHS1 to AHS2 will be the resulting capacity gain.

5.4. Justification for Evolutionary Steps

The evolutionary steps are intended to achieve the objectives outlined in Section 3. Justification is provided below.

1. Support for all vehicle types
   - Strategy will not replace conventional roadways, which would be available for all vehicle types.
   - Low-volume intercity roads would allow all vehicles to be automated.
   - High-volume roadways would only allow automobiles and light trucks/vans. This is justified by safety considerations, improved performance, and the high cost of providing separate automated facilities for all vehicle types. Buses could be accommodated on ordinary roadways through HOV preferences.
2. Not all roadways will be instrumented
   - Strategy explicitly allows for a range of roadway types.
3. Performs better with respect to all key areas
   - Requires completion of cost/benefit analysis and other system analyses: to be determined.
4. Able to operate in wide range of weather conditions.
   - Requires technical analyses: to be determined.
   - However, strategy allows flexibility for different guideway types in different regions.
5. Provides substantial user benefits where AHS is not implemented
   - ARV provides potential benefits of fuel economy, improved maintenance, better performance, and access to auto-steering roadways.
6. No negative impacts on neighborhoods
   - Will be used as a constraint in construction scenarios to be created later, provided that cost is not prohibitive.
7. Minimal retrofit cost
   - The ARV contains electronic steering, braking, and throttle control, longitudinal sensor, and controller. Upgrade to Stage 1 only requires installation of communication modules, software upgrade, and perhaps supplemental sensors.

8. High likelihood of acceptance of each step
   - By beginning with low-volume roadways, the fear factor is diminished, and safety risks are reduced.
   - By not providing capacity gains initially, environmental concerns are less likely to block initial implementation.
   - By keeping governmental costs low initially, taxpayer complaints may be reduced.
   - Direct user benefits, in terms of comfort, performance, and possibly energy savings and pollution reduction, are provided in each step.
9. Public acceptance for subsequent steps
   - High-volume roadways, and later lane changes, are only implemented after users are familiar with the technology. By then, support could be generated.
   - By beginning small, the system can be tested without embarrassing failures.
   - Strategy does not require conversion of conventional roadways to automated use until there is a sufficient number of equipped vehicles.
10. Consideration of long lead times
    - Simpler technologies are implemented first, so that benefits can be generated while research proceeds.
    - Initially implementations can be used to test and refine more advanced technology.
11. No need to discard major portions of the system
    - The strategy is modular and incremental, without requiring the vehicle purchaser, or government, to discard investments.

5.5. Market Penetration

As already mentioned, evolutionary deployment must account for the fact that market penetration takes time. When a technology is introduced, only a small percentage of new vehicles, and an even smaller percentage of vehicles on the road, will contain it as an option. Taking air bags as an example, in 1992, five years after their introduction, less than 60% of new vehicles sold contained a driver-side air bag, but fewer than 20% of the vehicles on the road had an air bag. These data represent a product that had unusually fast market growth.

6. IMPLEMENTATION SCENARIOS

Specific implementation scenarios were created for the purpose of cost analysis. These scenarios translated the AHS functions specified in the prior section into specific technologies. Because AHS technologies still require considerable research, the implementation scenarios are somewhat speculative, and represent a best guess based on the state of the art today. Rockwell International was responsible for specifying the electronics for each step in the evolutionary sequence, both for the vehicle and for the roadway. Bechtel was responsible for specifying the construction, including means for adding lanes (if necessary), barriers, conduits, and installation of electronic devices on the roadway. The following subsections summarize the scenarios investigated. Details can be found in Refs. 12 and 13.

6.1. Electronics

Electronics were specified for three configurations: two Minimal Infrastructure Modification (MIM) configurations and one Infrastructure Intensive (II) configuration. In the MIM configurations, as much of the electronics as practical are placed on the vehicle, and in the II configuration, as much of the electronics as practical are placed on the roadway.

In all configurations, ARV are allowed to mix with conventional traffic on auto-steering roadways, under low-traffic conditions. However, in AHS1 and AHS2, the MIM configurations are differentiated according to operating environment. In one case, automated and manual vehicles operate in mixed traffic, and in the other, automated vehicles operate on dedicated lanes. Note that mixed traffic in AHS1 and AHS2 is inconsistent with the evolutionary scenario presented earlier. They are included here with the purpose of bounding cost, without presuming feasibility and safety. The II configuration, on the other hand, is intended to be deployed on a completely dedicated roadway in AHS1, and on joint-use roadways in AHS2. This means that in AHS2 certain lanes are designated for fully automated vehicles, but these vehicles must pass through manual traffic to reach these dedicated lanes. Joint-use roadways are technically more challenging. Hence, for II, they are only used for the last evolutionary stage (AHS2).

Details of the technological implementation can be found in Rockwell.[13] Table 6 summarizes assumed technologies.

6.2. Roadway Construction

In order to transition to a fully mature AHS facility, either existing roadways need to be modified or entirely new facilities need to be constructed. This section describe various alternatives that allow conventional and automated vehicles to travel on the same right-of-way. Completely separated AHS facilities can be extrapolated from the scenarios presented here.

The costing of these scenarios was based on a reference roadway: the Hollywood Freeway located in Los Angeles (additional details in Bechtel[12]). We determined that a single "one-modification-fits-all-locations" scheme would not be possible, and that various alternatives would need to be investigated. For instance, where right-of-way is severely restricted, it may be impossible to widen the existing highway, and an entirely new facility might have to be constructed. As another example, it may be feasible to add dedicated AHS on-ramps in some cases; in other cases, AHS vehicles may be forced to use conventional on-ramps, because of construction limitations. To a degree, then, construction considerations might also drive the selection of an electronics configuration.

TABLE 6
Electronics Implementation

Nomenclature/description	Vehicle equipment							Purpose/function
	Mixed traffic			Dedicated lanes				
	Minimum infrastructure modification					Infrastructure intensive		
	ARV low volume	AHS1, lane keep	AHS2, lane change	AHS1, lane keep	AHS2, lane change	AHS1, lane keep	AHS2, lane change	
Vehicle								
Sensors								
Multi beam millimeter radar	1	1	1	1	1	1	1	Range to vehicle ahead
Magnetic field sensor	1	1		1		1		Lane keeping
Vision-based sensors		2	4	1	3	1	3	Ranging, visible turn signals, tracking lane markers
Rain/snow sensor		1	1	1	1			Detect wet road conditions
Beacon emitters			1		1		1	Distance measuring reference, ID code transmission
Mag filled sensor w/code			1		1		1	Lane keeping and location code reading
Intelligence								
Processor/diagnostics (1)	1						1	Sensor data processing and diagnostics
Processor/diagnostics (3.4)							1	Sensor data processing and diagnostics
Processor/diagnostics (4.0)						1		Sensor data processing and diagnostics
Processor/diagnostics (4.4)						1		Sensor data processing and diagnostics
Processor/diagnostics (5.9)					1			Sensor data processing and diagnostics
Processor/diagnostics (6.6)		1						Sensor data processing and diagnostics
Processor/diagnostics (8.5)			1					Sensor data processing and diagnostics

Communication								
Road to vehicle (TMS), Receive		1	1	1	1			Speed command inputs, delta to AM/FM radio
Vehicle/vehicle, fore & aft		1	1	1	1	1	1	Platooning communication only, forward car to following car
R/T, auto check-in and control						1	1	Vehicle to/from control communications, voice and digital
Actuators								
Brake actuator	1	1	1	1	1	1	1	Automatic vehicle control
Throttle actuator (engine control)	1	1	1	1	1	1	1	Automatic vehicle control
(Less standard direct drive)		1	1	1	1	1	1	Removal of non drive-by wire components
Interfaces								
Standard actuator interface unit	1	1	1	1	1	1	1	Standard interface, actuator/processor
Driver interface unit	1							On/off
Drive-by-wire driver int. unit		1	1	1	1	1	1	On/off, routing input, encoders on controls

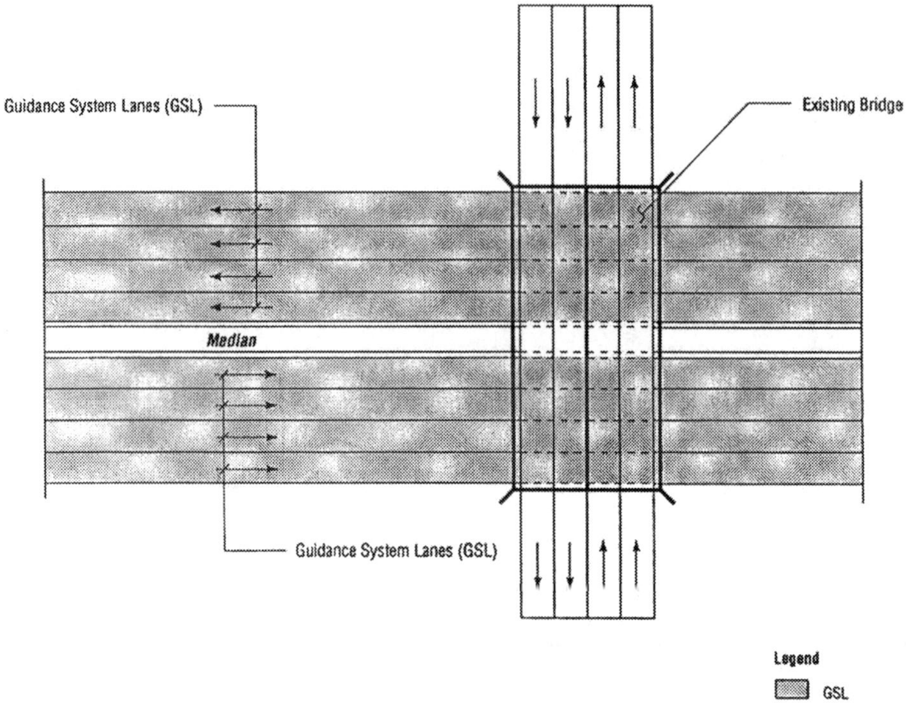

FIGURE 1. Option 1: All lanes automated.

The following roadway options describe possible adaptations to incorporate AHS traffic.

1. *All four lanes automated with AHS technology* (Figure 1). These may appear and function as normal lanes to nonautomated vehicles, or may be entirely dedicated to automated vehicles. Since all lanes are automated, there is no need to physically separate manual from automated traffic.
2. *Add one lane to create a total of five lanes per direction: one lane automated, one buffer lane, and three lanes to remain conventional traffic* (Figure 2). Use of this option is severely restricted in some cases, especially well-developed areas, because of the high cost of additional right-of-way. However, it is expected to constitute a particularly useful option in growing areas where traffic volume can be expected to steadily increase with or without the addition of AHS. The buffer lane is largely used for the necessary merging between AHS and conventional traffic and could be the site of some check-in motion testers (assuming they are flush with the roadway) of AHS requisite equipment, should that equipment become available in the future.
3. *One lane automated, one buffer lane, and three conventional lanes, as described above, but dedicated on/off-ramps for automated lane entrances and*

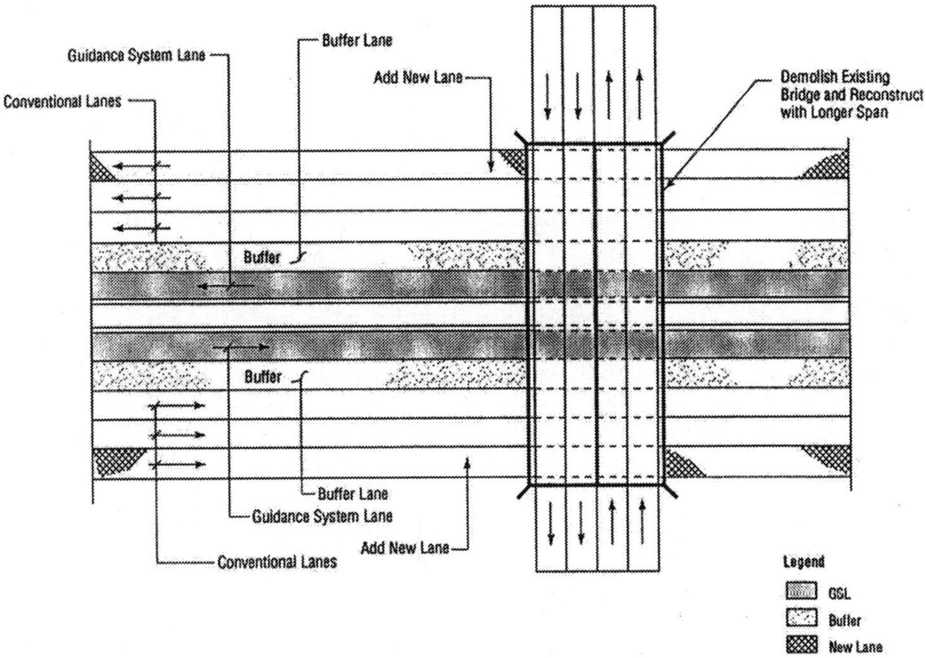

FIGURE 2. Option 2: One lane automated, three lanes conventional, buffer lane added between automated and conventional lanes.

exits into buffer lane from overhead structures such as bridges (Figure 3). This option has many of the advantages and disadvantages of the preceding option, but reduces weaving on the roadway to access the automated lane. The buffer lane remains essentially unproductive except for its capacity to feed the automated lane. In this option, continuous barricades might be possible between the automated section and the conventional section.

4. *One lane automated, no buffer lane, three conventional lanes, with automated lane delineated from conventional lanes by a rumble stripe or special striping* (Figure 4). This option follows a precedent in Los Angeles, where striping and signs provide the only demarcation of certain HOV lanes. Exclusion is by traffic enforcement, where substantial fines are levied for violation of the specially designated lane.
5. *Automated lane created on a dedicated structure with its own exit/entry ramps* (Figure 5). This structure is anticipated to follow the existing highway right-of-way for the most part, but exit/entry ramps may be more widely spaced than typically found for conventional roadways to minimize costs and traffic flow interruptions.

New roadway construction along alternate right-of-ways (e.g., utility corridors) can, in general, be extrapolated from the last option (dedicated AHS structure) presented above.

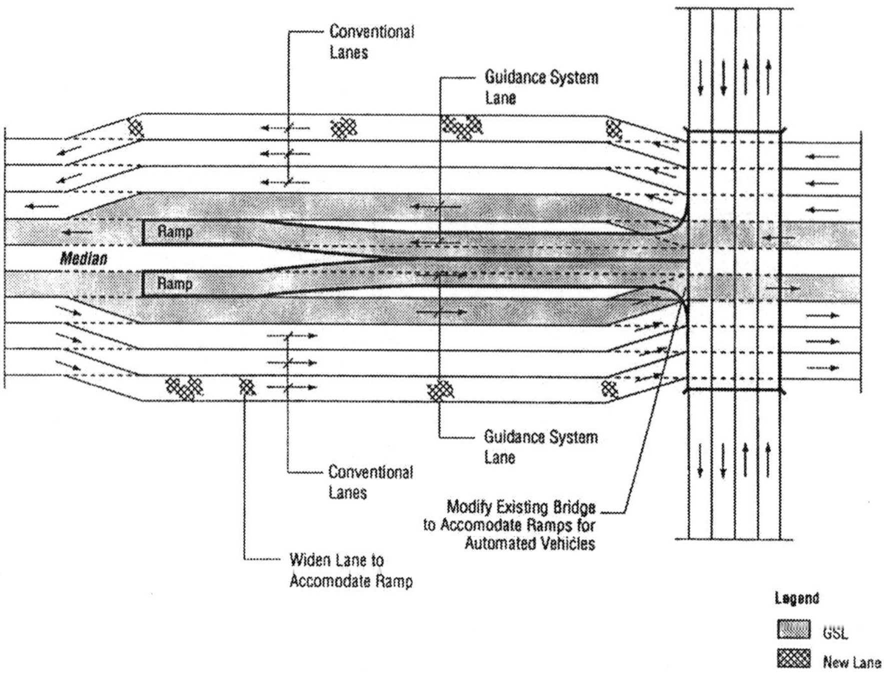

FIGURE 3. Option 3: Similar to Option 2 but without buffer; on/off-ramps added with bridge structure.

Table 7 indicates which roadway scenarios are applicable to which electronics scenarios. Question marks appear where the roadway scenario is possibly applicable, given the following qualifiers.

TABLE 7
Applicability of Roadway Scenarios to Electronics Scenarios

	Electronics scenarios						
	Mixed traffic			Dedicated lanes			
	Minimum infrastructure modification					Infra. intense	
Roadway Scenarios	ARV	AHS1	AHS2	AHS1	AHS2	AHS1	AHS2
1. All lanes allow automated		X[a]	X	X			
2. 1 auto + 1 Buffer				?[b]	X	?	X
3. 1 auto + 1 buffer + dedicated ramps				X	?	X	?
4. 1 auto, no buffer				?	?	?	?
5. Dedicated facility				X	?	X	?

[a]X: Roadway scenario is definitely applicable to electronics scenario.
[b]?: Roadway scenario is possibly applicable to electronics scenario.

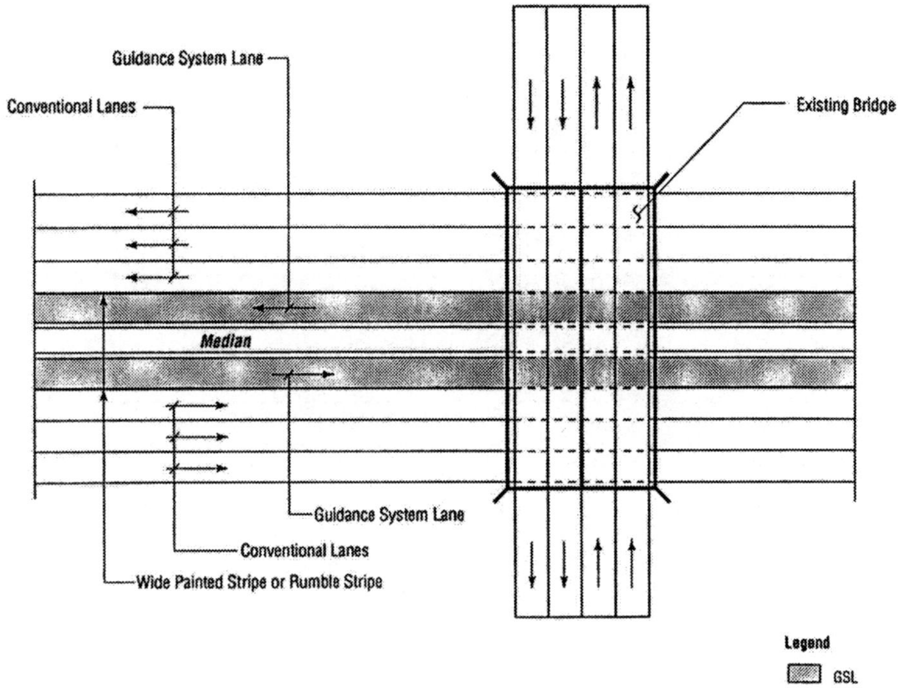

FIGURE 4. Option 4: One lane automated in each direction, three lanes to remain conventional, separation by wide striping or rumble stripes.

Roadway Scenarios 2 and 4 do not physically separate conventional vehicles from manual vehicles, but do provide dedicated lanes. This is possibly feasible for AHS1, depending on advancements in surveillance and enforcement technologies, to keep conventional vehicles out of automated lanes.

Roadway Scenario 3, which provides dedicated ramps for AHS, is applicable to AHS2 (which allows conventional and automated on the same facility), provided that access is also provided from conventional lanes. Due to added cost, this may be unrealistic.

Roadway Scenario 4, which does not provide a buffer, either requires a manual lane change directly into the automated lane (which is counter to the intention of AHS2, especially if there is a single automated lane), or an automated lane change, in which one of the conventional lanes accommodates automated vehicles before or after lane change (which is counter to the intention of AHS1). Either case is problematic.

Roadway Scenario 5, which provides a dedicated facility, is counter to the intention of AHS2, which allows automated and manual vehicles to travel on the same roadway. However, it's possible that dedicated facilities constructed for AHS1 may be upgraded to AHS2, once automated lane change is provided. Entirely new facilities, on the other hand, are not intended to be constructed for AHS2. The

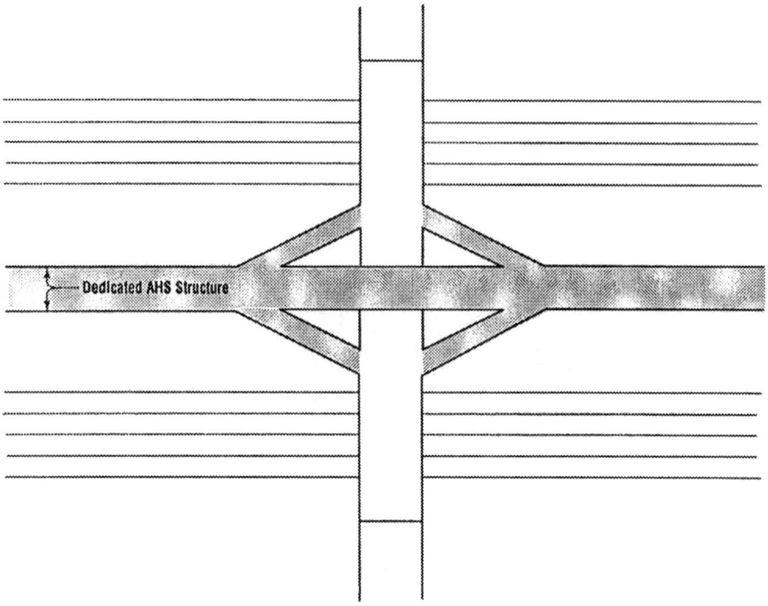

FIGURE 5. Option 5: AHS access at overcrossing—plan.

rationale is economic: if AHS technology is sufficiently advanced, then it would not be worthwhile to incur the added cost of completely segregated facilities.

7. DISCUSSION

This chapter has presented an evolutionary deployment strategy for AHS based on the objectives of responding to market forces and not stretching technological limits in early stages of deployment. The strategies have been translated into specific technologies, for which costs have been estimated.[10] While the ultimate system may prove to be quite different from the assumptions made here, this chapter provides a plausible and specific strategy for evolution, which could be the basis for planning AHS research and deployment efforts.

ACKNOWLEDGMENTS. The electronics scenario contained in this report is based on the contributions of Rockwell International. The roadway scenario is based on the contributions of Bechtel. This chapter benefitted greatly from the ideas of Dr. Rokaya Al-Ayat, who was the principal author of "A Conceptual Approach for Developing and Analyzing Alternate Evolutionary Deployment Strategies for Intelligent Vehicle/Highway Systems." The research was supported by the Federal Highway Administration, Caltrans, and PATH.

REFERENCES

1. P. VARAIYA, Smart Cars on Smart Roads: Problems of Control, *IEEE Trans. Autom. Control* **38**(2), 195–207 (1993).
2. B. F. HEINRICH, IVHS—An Automotive Perspective. Chrysler Corporation, SAE Technical Paper 912780, presented at the Vehicle Navigation & Information Systems Conference, Oct. 1991, Dearborn, MI.
3. R. A. PLACE, IVHS—Auto Industry Perspective. Ford Motor Company, SAE Technical Paper 912781, Presented at the Vehicle Navigation & Information Systems Conference, Oct. 1991, Dearborn, MI.
4. R. A., JOHNSTON, M. A. DELUCHI, D. SPERLING, AND P. P. CRAIG, Automating Urban Freeways: Policy Research Agenda. Presented at the ASCE International Conference on Applications of Advanced Technology in Transportation Engineering, San Diego, CA, 1989.
5. R. W. HALL, Time Benefits of New Transportation Technologies: The Case of Highway Automation, *Transp. Plann. Technol.* **19**, 1–17 (1991).
6. R. W. HALL and J. H.-S. TSAO, AHS Deployment: A Preliminary Assessment of Uncertainties, PATH Program, Institute of Transportation Studies, University of California, Berkeley, PWP 94-2 (1994).
7. J. WARD, A Hypothesized Evolution of an Automated Highway System. For Federal Highway Administration Precursor Systems Analyses Program (1993).
8. P. IOANNOU, M., LAI, J. DICKERSON, AND A. KANARIS, Evolutionary Representative System Configurations and Roadway, Vehicle, Driver Functions. Center for Advanced Transportation Technologies, University of Southern California, Report 94-06-01 (1994).
9. R. AL-AYAT AND R. W. HALL, A Conceptual Approach for Developing and Analyzing Alternate Evolutionary Deployment Strategies for Intelligent Vehicle/Highway Systems, PATH Working Paper 94-5 (1994).
10. R. W. HALL, The Architecture of Transportation Systems, *Transp. Res. C* **3**, 129–142 (1995).
11. R. W. HALL, Cost/Benefit Analysis of Automated Highway Systems. Prepared for Federal Highway Administration, Contract DTFH61-93-C-001-99 (1994).
12. Bechtel. Roadway Costs. Prepared for Federal Highway Administration, Contract DTFH61-93-C-001-99 (1994).
13. Rockwell, Electronics Cost Analysis. Prepared for Federal Highway Administration, Contract DTFH61-93-C-001-99 (1994).

CHAPTER 5

Step by Step to an Automated Highway System—and Beyond

Jerry D. Ward

1. WHY?

Why on earth would anyone want to "automate" our highway system? Even if we had "autopilots" on every car, would we trust them enough to leave the driving to them? And what good would they do anyway?

There are simple answers to these questions, even if at this stage we have to extrapolate a bit. Let us start with the last one: what good would it do? While there are lots of things people can do far better than machines and computers, there are two areas in which people come out distinctly second best. First, electronic sensors and computers don't get tired and fall asleep—at least very, very rarely—and they don't get distracted by the kids in the back seat or the need to tune the radio. While it's doubtful that an automated system will be any safer than a human driver when he or she is alert and prudent, it will be a lot safer than the bored, tired, distracted, or careless driver.

The second area in which the automated system handily beats the human is in speed of reaction. The average alert driver takes something like 1 to 1.5 s to hit the brakes after detecting the need to do so. An automated system is easily five to ten times faster. This means that cars can safely drive closer together, which in turn

Jerry D. Ward • Ward Consulting, El Cajon, California 92019.

Automated Highway Systems, edited by Ioannou. Plenum Press, New York, 1997.

means that our freeways can carry more cars—something like a factor of two more cars is thought to be a realistic expectation. Automation offers a way to effectively turn our four-lane freeways into eight-lane freeways without pouring more concrete.

But would we trust this marvelous new system? Probably not much at first. That is one reason why we would want to start small with driver-aid kinds of things, and add new capability a step at a time so we creep, not jump, to the fully automated system. The other reason to advocate an evolutionary approach is that it will lead to a better system: it gives a chance to learn as we go, to correct mistakes, to improve approaches.

Back to our first question, "Why?" We've already partially answered it: the potential for an increase in safety, and the ability—without new concrete—to markedly increase the effective capacity of our freeways and, later, our surface streets. A third reason is sheer convenience, like relief on long trips or in heavy traffic. The prospect of being able to drive safely while reading a book is a little hard to swallow today, but over time we will get used to automation and it will happen. In fact, for most people, this truly revolutionary convenience will be the dominant motivation for buying an AHS-equipped car. And vehicle automation can make driving easier and safer for our elders. We're afraid to talk about the very long run just yet, from our perspective today the possibilities defy credibility.

We will begin at the beginning, thinking about how we might start designing these remarkable systems.

2. NOTIONAL BASIS FOR SYSTEM DESIGN: EMULATE THE HUMAN

Consider how a person drives in a freeway lane. She gets most of her information on the state of traffic around her vehicle through her eyes, and she keeps track of the location of her vehicle on the road the same way. She uses all of the visual cues available: the position and movement of the traffic itself, lane markers, rain on the windshield, and so forth. She looks most closely at the relatively few vehicles that could endanger her vehicle, like the one in front of her that might stop suddenly, or the one in the next lane if its turn signal is on or if it starts to pull in front of her.

The driver is continuously interpreting all of this information. He decides what is a safe following distance from the vehicle in front. He has some "rules" for deciding, like it is safe to follow closer at low speeds than at high speeds; in wet weather the braking is not as good, so the gap—the following distance—should be larger. And so on. (It is apparent that not all drivers subscribe to the same rules.)

Drivers also have rules for taking action. If he is gaining slowly on the car in front, he backs off slightly on the accelerator—or thinks about passing. If he is closing a bit faster, he applies gentle braking. If he has glanced down to retune his radio, and looks back up to see the car in front of him suddenly closer and closing rapidly, he slams on his brakes, hopes, and pumps adrenalin. And maybe cuts into the next lane.

If we set out to automate this "gap control" function, then emulating this pattern of sensing and interpretation, decision, and action seems a sensible approach to mechanization.

The human eye, along with the brain that interprets its signals, is far more capable and versatile than any artificial sensor plus interpreter—any "computer-brain"—that could be built today. But a function like gap holding does not require all of that capability; nearly all that is needed is an unambiguous measure of the distance to the car ahead and to the few vehicles in the immediate vicinity that might move into the intervening gap, thus becoming the new "target."

The point is that if we deal with just a few functions at a time, it is only necessary to emulate very specific and limited functions of the eye–brain, not duplicate its total capability.

This is not intended to imply that the job is easy, nor that we won't want even more information for more capable systems in the future. Even now, just to get unambiguous range, the system must be smart enough to avoid measuring the distance to the wrong car: for example, to confuse the car in front with the one in the next lane because the road is curved. It should also be able to recognize when another vehicle (or motorcycle) starts moving into the lane. Or to react appropriately when the car ahead moves out of the lane.

It is still to be determined if the best way to handle this general sensing function is with vision-based sensors roughly analogous to a video camera, or radar, or laser ranger, or variant of these. Our guess is that production systems will not be based on just one type of sensor, but on some combination of sensors to improve versatility and robustness. But the issue is not whether adequate sensing is feasible, it is rather how best to do it: we have lots of options, and costs will come down rapidly with both improving technology and with high-quantity production.

The more complex part of the computerbrain's job is interpreting the sensor outputs to provide a basis for choosing the best vehicle response to the situation. The best approach to carrying out this task is also yet to be defined. In fact, the most difficult technical challenge facing these new systems is less the sensors themselves than this job of interpreting what the sensor "sees."

The decision rules that govern vehicle responses are easier to visualize. It is well within current capability to devise a set of rules, probably couched in "fuzzy" terms, that adequately decide when and how vigorously to put on brakes or change the throttle setting.

So, first, we don't have to match the versatility of the eye or the data processing capability of the brain in its entirety, we only have to emulate a very tiny portion of it to carry out a few specific and limited functions. And, second, for these limited functions it may be possible to actually improve on our human capabilities. We can measure distances and rates of closure with great precision. We can have "eyes in the back of our head"—and wherever else we want them. In time we should have practical, affordable sensors that can see through fog.

This perspective on system mechanization is consistent with the notion of incremental evolution alluded to earlier. The start is with simple functions, like automatic braking, in fairly orderly traffic, like on freeways and interstates. As experience deepens our understanding of the problems, as sensor-interpretation capabilities grow, and as our ability to define behavioral rules for more complex control situations improve, the scope of automation could be expanded a step at a time within the same basic mechanization framework. As the need for information broadens beyond that which can be acquired by vehicle-mounted sensors, then communication links to infrastructure intelligence can be introduced; we have conjectured that this will occur when auto-operation goes beyond just freeways onto surface streets and intersections—still a bit in the future.

While the "eyes and brains" of the system we have been discussing are the key challenge, it is obvious that it will also be necessary to replace human muscle in the operation of the vehicle's throttle, brakes, and steering. It will probably turn out that we will also use this system for human driving, so-called drive-by-wire. Like everything, this will require careful engineering, but it is relatively straightforward and well understood. Drive-by-wire is already beginning to come into being on a few cars.

3. HOW SAFE IS "SAFE"?

Of course these systems must be safe. But does that mean safe from a piano falling from an aircraft? (Statistically, an exceedingly rare event.) Or from the car in the next lane that suddenly pulls into your right rear door? (While more frequent than the piano, still a rare event.) Or from being rear-ended by a car with an inattentive driver? (Unfortunately, not nearly as rare as we would like.)

The answer we finally arrive at to questions like these may very well determine if we will ever have AHS. If we say yes to the piano criterion, we have precluded any AHS; absolute safety is impossible to achieve for either human driver or automated system. But it seems perfectly feasible to achieve levels of safety with an automated system that are substantially better than those that obtain today with human drivers, even with automated vehicles mixed with manually driven vehicles. But nothing is free, and, in general, the more kinds of accidents we attempt to protect against, the higher the costs of the system. Setting the safety standards unrealistically high could easily drive up costs and complexity sufficiently to kill the market for the systems. The perfect is the enemy of the good.

It is our judgment that the standards should be set in terms of some level of improvement over that achievable today by an average human driver. Actually spelling this out will require careful judgment supported by thorough analysis of all of the reliability, operational, cost, and market penetration trade-offs implied. It is not a trivial issue.

4. DEDICATED LANES . . . OR JUST ANOTHER CAR ON THE FREEWAY

There are many people who envision the AHS as having its own freeway lanes exclusively dedicated to these new automated vehicles; vehicles under manual control would be excluded from these facilities. These lanes would either be taken from the existing lanes and dedicated to the AHS, much as many HOV lanes have been created, or made available through new construction, possibly even new minifreeways completely separate from existing freeways. In some concepts, two dedicated lanes would be set aside, one to be used as a transition lane to the automated lane. There are some who advocate physical barriers around the AHS lane to protect it from the possibility of non-AHS vehicles entering.

The alternative to dedicated lanes is to let AHS-equipped vehicles share the freeway or interstate with nonequipped vehicles—just another vehicle in mixed traffic.

4.1. Dedicated Lanes: Advantages

There are advantages to dedicated lanes. For one, occupants of such lanes are assured that they are not being followed by a human driver who might rear-end them, because all vehicles in the lanes are AHS-equipped. Second, if physical barriers are in place, or if the AHS lanes are on their own separated freeway, there is very little chance of a scofflaw or rogue vehicle entering from the adjacent lane.

Another advantage is that "platooning" is easier. The platooning concept will be discussed later, but it is a technique whereby cars form electronic "trains" on the highway, thus markedly increasing the number of vehicles that can share the lanes and therefore the effective capacity of the lanes. Since only equipped vehicles can form into platoons, mixed traffic reduces the opportunity for having two equipped vehicles next to each other. This disadvantage is greatest when a moderate percentage of total traffic is AHS-equipped: when the percentage is low there aren't enough equipped vehicles to fill the lane and platooning is not needed; when the percentage is high the odds are also high that the adjacent vehicle is equipped and therefore capable of platooning, so the penalty is small.

4.2. Dedicated Lanes: The Fatal Flaw

There is one major disadvantage to dedicated lanes, a disadvantage that in this author's judgment is fatal: the problem of the chicken (vehicles equipped with AHS) and the egg (dedicated lanes on which to operate them). We will outline the essential facets of the difficulty.

It is currently accepted thinking that the development of the AHS-equipped vehicles for sale will depend on the initiative of the private sector. So the decision to invest the funds for development, production, and marketing is a private sector decision. While it is conceivable that the AHS dedicated lanes and access ramps could also be a private sector initiative, it is much more likely to be provided by the public sector, just as most of the concrete in the nation is initiated and built today. We examine the problems raised.

4.3. The Investment Decision to Develop AHS-Equipped Vehicles

Offering the AHS option for sale on vehicles will require the investment of very substantial sums both by the automobile manufacturers themselves and by the electronic equipment manufacturers who supply the sensor-electronic systems to them. A necessary condition to risk these funds is that the anticipated market will be very large. There are clearly other considerations, but the issue of market size is fundamentally important to the investment decision.

And fundamentally important to market size is the availability of dedicated lanes on which these new AHS-equipped vehicles can be operated; clearly there is no market where there is no place to use the option. Thus, the size of the market is in a very real sense hostage to the allotment or construction of the lanes to be dedicated to AHS use. It would take almost foolish bravery on the part of these manufacturers to go ahead with even a remote doubt in their minds that many, many miles of dedicated AHS lanes will be in place by the time their new AHS vehicles are ready for sale.

Can they get this assurance? These lanes depends in all likelihood on public sector commitments and investments made specifically to support the AHS. Given the long lead times, the vagaries of politics, and the sheer number of cities that have to make these commitments to create the requisite market, it will not be easy. Let us look at the decision from the perspective of the public agencies that will have to furnish the dedicated lanes.

4.4. The Investment Decision for Dedicated AHS Lanes

While people will consider buying AHS vehicles primarily for the convenience of automated driving, the primary motivation of public agencies to invest in AHS is to gain increased freeway capacity. But if they have to dedicate, say, one lane out of every four, they will not break even on capacity until one out of four—25%—of the vehicles using the freeway are AHS-equipped. If, say, the total fleet turns over every 8 years, and half of the new cars purchased every year are AHS-equipped—a very optimistic assumption for the early years—it will take nearly 4 years to reach 25%. So they have a net loss on capacity for 4 years. And if they don't dedicate the lane, no one will buy the AHS vehicles. The manual drivers who must continue to endure the unrelieved and worsened congestion in full view of this underutilized road space will probably be somewhat less than understanding.

If they have to build new lanes they still face this same problem, along with the fact that they have had to make a major investment when it is still a gamble that the public will buy the vehicles in sufficient quantity to fill the new lanes.

And even if a municipality is enthusiastic about supplying new AHS facilities, they will be constrained by the consideration that the market for AHS vehicles they alone create is too small to induce the private sector to develop the AHS vehicles. Many others must also commit to precipitate a go-ahead from the private sector. One is led into thinking that if a market is to be created, it will have to depend largely on new infrastructure built as a federal initiative, because taking an existing lane is politically too dangerous for a city before AHS vehicles are available, and it is too much to expect that individual city-by-city promises to build new infrastructure can create the confidence needed by the private sector to go ahead.

All this appears to add up to the conclusion that the initial development step for any AHS concept that depends on dedicated lanes will be very, very steep.

5. THE MIXED TRAFFIC SYSTEM

In the author's judgment the problems of initial deployment facing the dedicated lane AHS are a powerful incentive to aim development at the mixed traffic system, a system that does not depend on either dedicated lanes or complex entry/exit procedures.

Here we describe and broadly analyze an evolutionary scenario in which the equipped vehicles operate in mixed traffic with unequipped, manually operated vehicles. We believe the scenario described is both technically feasible and operationally realistic. A gradual evolution of incremental change in the scope and capability of the automated features provides a progressive buildup of operational experience to guide each new step.

We believe the scenario described offers the prospect of an essentially seamless and nondisruptive path to an AHS—and beyond.

In devising this scenario we have attempted to meet two primary criteria:

- The evolution of the system should be technically sensible, with each step building on previous steps. We do not want Step #8 to obsolete Steps #2 and #3.
- At each step in deployment, there should be reasonable correlation in time and degree between costs and benefits. Investment, either by individuals or by political bodies, is only motivated by perceived benefit.

Our first step has been to hypothesize what is believed to be a sensible technical evolution for an AHS, using our best judgment and knowledge from already available analyses. In our hypothesized scenario, the initial AHS is designed for operation where our current congestion problems are most obvious: on urban freeways. In addition to the freeway analysis, the scope of the potential usefulness of the particular function for operation in other venues is briefly addressed: arterials and other surface streets, interstates, other rural highways, and intersections of various types.

5.1. The Evolutionary Scenario

There are experimental cars on our highways today that are already taking the first steps toward an AHS: these are cars with their cruise control units augmented by a forward-looking sensor that measures the distance to the vehicle ahead and lets up on the throttle, or perhaps even downshifts, when one might be getting too close. This adds a kind of minimum-gap-control function to the normal speed-hold function of today's cruise control to produce what is being called intelligent cruise control (ICC). It is pretty minimally "intelligent," just enough to prevent cruise control

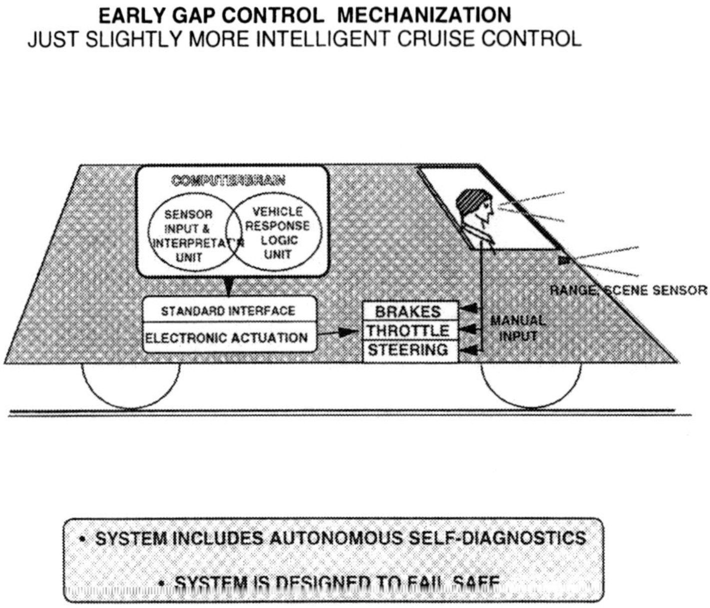

FIGURE 1. Slightly intelligent cruise control.

in the hands of an inattentive driver from "cruising" into the rear of the car in front. And because it can't react very vigorously, the gap it maintains must be fairly large—large enough that other cars can keep cutting in front. But it should offer a driver some relaxation in low-density traffic, as well as provide our first operational taste of situation-responsive automation.

Figure 1 depicts the general mechanization of this (slightly) intelligent cruise control. The heart of the system is the what we have been calling the computerbrain, which includes the forward-looking range sensor and the interpretation and vehicle response computations. At this stage, it's a pretty limited "brain," but it will evolve. And at this point we are controlling only the throttle through electronic actuation. We could have shown the driver also controlling the throttle electronically rather than through his conventional controls; we've saved that step to true drive-by-wire for a later stage of evolution.

5.2. *Truly Intelligent Cruise Control*

We can raise the IQ of this ICC by letting this same sensor and computerbrain—actually a somewhat smarter one—also provide automatic emergency braking. Now our smarter ICC would be capable of two functions, automatic emergency braking (Autobrake) and automatic use of both throttle and brake to maintain a safe gap behind the car in front (Autogap). We show this truly intelligent cruise control with full automatic control of the total drivetrain in Figure 2.

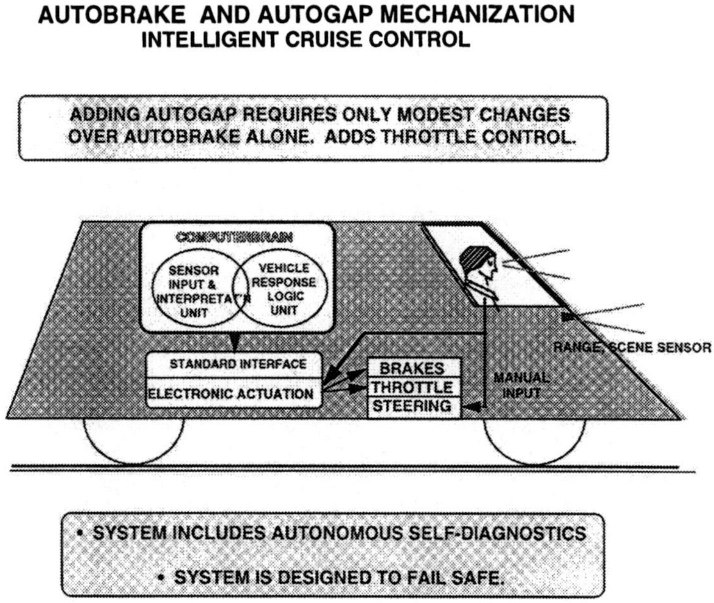

FIGURE 2. Intelligent cruise control.

We will henceforth use ICC to refer to the smarter version, including both the Autobrake and Autogap functions. From the driver's perspective it will operate much like Cruise Control today, except that when her car overtakes another vehicle, the system will automatically revert from holding speed to holding a fixed-but-driver-adjustable distance behind that vehicle. Thus, at some moderate transition distance the speed controls become position controls, varying the gap between the vehicles in response to the driver's desires. With full braking capability and very fast automatic reactions, this gap can safely be smaller than that most drivers would hold if driving manually, and probably smaller than they would choose even with ICC holding the gap. Like cruise control today, a tap on the brakes or the throttle gives drivetrain control back to the driver.

The Autobrake feature operates even with cruise control off, invisible to the driver. In fact, if the driver is alert enough to never have a lapse in concentration and lucky enough to never have someone cut directly in front of him and abruptly slow, the system will never be actuated, because the driver's braking will always occur before the point that emergency response is needed. In addition to sensing when emergency braking is needed, the system will also be designed to recognize when it is not needed, such as in sharp turns when objects may suddenly appear dead ahead, but are not threats to safety.

We anticipate that these systems will incorporate a self-testing and self-diagnostic capability, and will be designed to be completely fail-safe and fail-soft. Such designs are relatively commonplace in aerospace engineering, but they do require sophisticated design techniques and extensive verification testing. We suspect that the major portion of the developmental engineering and much of the testing will be focused on providing and verifying these features.

Given this self-check capability, these systems create no special requirements for entering the freeway; the driver behaves as he does today. If there had been a malfunction in the system, the self-test would have already indicated it, and the driver would know that auto-operation was precluded until the system was repaired. The system will have been designed to fail softly if a malfunction occurred during operation.

We have already mentioned the motivations for ICC. The obvious one is safety. Artificial sensors can be very reliable, and their attention doesn't wander. Lapses in attention are the largest source of human accidents; automated systems could virtually eliminate these. This one trait of almost eternal vigilance is worth a large investment in itself, and may well be the motivation for many purchasers of this option.

The second advantage is the potential for increased throughput. Automated controls are potentially very fast; it is reasonable to expect that automated systems could react to emergencies on the order of five to ten times faster than a human driver. This capability offers the potential to increase the effective throughput of the freeway, because it enables vehicles to drive closer together safely. We suspect that this feature is unlikely to motivate buying; the connection between increased throughput and personal benefit is too obscure. And the improvement won't be dramatic until we've taken a few more steps in our evolution.

A third advantage, one that should have great appeal to drivers who have to spend long hours on the freeway, is relief from the constant brake–throttle–brake jockeying needed to hold position in heavy traffic. ICC will also offer a bit more relaxation on long trips (but not nearly what the next step will bring). We conjecture that this added convenience may be the most powerful motivation for purchase of the system.

There will be some acceptance problems to overcome. For example, because of the potential for shorter headways, some drivers will perceive themselves to be the victims of uncomfortably close tailgating. This new driving behavior will need to be made acceptable to fellow drivers. Simple techniques may be enough. Perhaps a small green light visible in a rearview mirror could tell drivers that the vehicle behind them has automatic braking. Information programs could help drivers understand what is going on. Time and gradualism should help.

We visualize this introduction of ICC as the key step toward the AHS. It will require no modification of the infrastructure, and equipped vehicles will operate in mixed traffic with unequipped vehicles. The primary motivations for purchase are improved safety and more relaxed driving in freeway traffic.

We think this new ICC would be first offered on new, upper-end cars. Our guess is that it would be very popular, and that a lot of operational experience with automated systems would be gained quickly. This experience should permit a myriad of successive improvements, producing cheaper and better systems as the market broadens.

It is purely a conjecture, but we suspect that the reason the first versions of ICC are of the low-IQ variety—that is, without Autobrake, using throttle control only—is, first, that the more capable system does represent a somewhat more difficult technical problem, but more importantly, the specter of product liability is a much greater concern.

5.3. Automatic Lane Holding and the Mark I Automated Highway System

We envision that the next major new function is automatic lane holding (Autolane). Autolane will require on-board sensors that can accurately determine vehicle position relative to the roadway lane. This implies the need for sensing the road itself, perhaps by tracking the white lines, or perhaps guiding on embedded markers such as the magnetic "nails" being used in the PATH program experiments. There may be other techniques. We conjecture that an operational system will combine several such methods to enhance robustness and fail-softness.

We also conjecture that lane modifications such as painting brighter lane demarcations or installing magnetic nails will be sufficiently inexpensive and practical that all lanes in a freeway can be modified. This would enable drivers with equipped vehicles to operate anywhere on the freeway, choosing the lane that matches her speed preference and trip distance.

It is our opinion that Autolane will be used only in conjunction with Autogap and Autobrake, not alone. With this combination, the driver would manually enter the freeway as he does today, drive to the lane desired, and activate these modes, perhaps with a single control. Now the vehicle is under completely automatic control: Autogap is holding a safe distance from the vehicle in front, Autobrake is providing the safety backup to react quickly to a violent slowing of the vehicle in front or unsafe intrusion into the lane, and Autolane is steering to keep one's vehicle in the center of the lane.

The vehicle would remain in this mode until the driver indicates her intent to take control by turning the system control back to "Off," and actually disengaging the system by grasping the steering wheel and introducing some small manual input, overriding the automatic steering.

This step in evolution is shown in Figure 3. The basic architecture hasn't changed, but the computerbrain is more capable, and we now have electronic control actuation in brakes, throttle, and steering. By this time, the driver is also using the electronic actuation for his manual driving—true drive-by-wire; we suspect that most cars will no longer have the conventional controls for brake and throttle.

These new capabilities need not be restricted to just freeway use. Since the infrastructure modifications needed to support the Autolane feature should be relatively inexpensive and nondisruptive, its use could readily be expanded to interstates and many lesser rural highways. The payoff would be a marked improvement both in safety and in driver convenience. It should be a particular boon on long intercity trips. Just imagine reading a book or watching TV on your new dashboard-TV while your automatic pilot pays more focused and reliable attention to the road and traffic than you ever could.

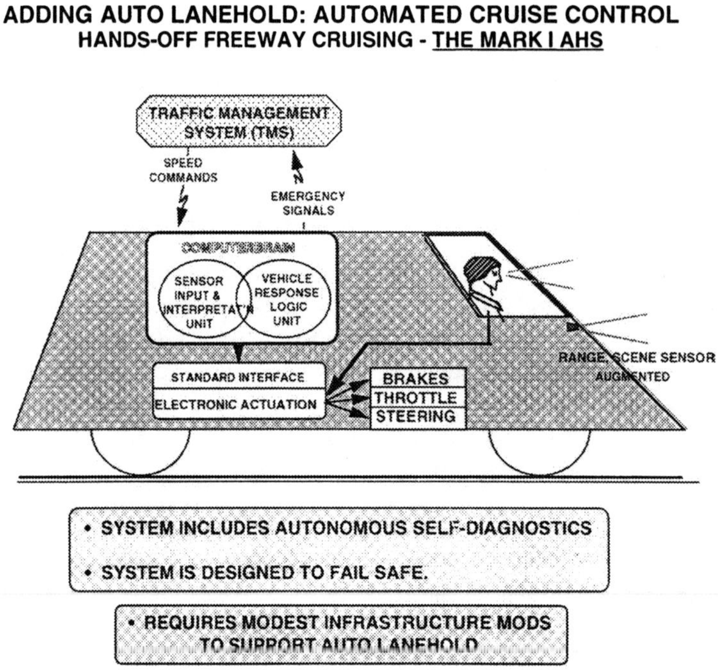

FIGURE 3. Hands-off freeway cruising—the first automated highway system vehicle.

We suggest that a submode of Autolane is automatic lane departure warning. This could be used during manual control to warn of accidental drifting out of the lane, a significant safety feature for the sleepy driver, a particular problem for long-distance truckers.

We have now attained the "Mark I" AHS, which we might appropriately call Automatic Cruise Control (ACC). It will require modification of lanes to support the automatic lane holding. Equipped vehicles will be capable of operating in mixed traffic, sharing lanes with unequipped vehicles. The primary motivation for purchase is the fully automatic, hands-off cruising on urban freeways and intercity trips, and improved safety.

5.4. An Aside: System Standardization and Competition

The alert reader may have noticed that the output of the computerbrain—the signals that actuate the controls—feeds into a "standardized interface." The reason for this architecture is to permit competition in the computerbrain marketplace. It appears likely that the electronic actuators of the brakes, throttle, and steering will have to be tailored to the particular vehicle, i.e., they will be vehicle-unique. We see no reason, however, that the computerbrain must also be vehicle-unique, given reasonable standardization in the way sensors are integrated into vehicles, and some design flexibility in matching computerbrain performance expectations with vehicle

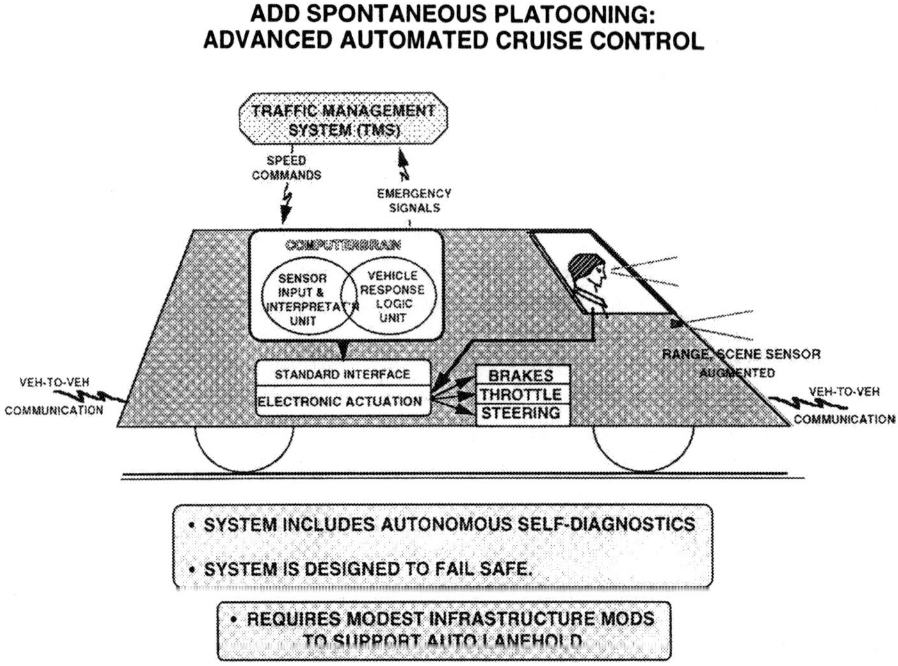

FIGURE 4. We add platooning capability.

performance capabilities. Allowing the computerbrain to competitively evolve independently of the vehicles will result in much faster evolution of capability and reduction in prices. Further, once drive-by-wire is available in all three vehicle controls, new capabilities in a particular vehicle can be obtained through upgrading or replacement of the computerbrain, possibly switching to a new manufacturer if the car owner perceives an advantage in doing so.

5.5. Platooning—"Highway Togetherness"

The next step in our hypothesized evolution is the augmentation of ACC with the capability to "platoon," to move to within 2 to 3 feet of the adjacent vehicle so the two vehicles—or three or four—now operate as one, like a very short train. Now the vehicles take up much less freeway space, so the freeway can carry more vehicles. To enable platooning we have to add fore and aft vehicle-to-vehicle communication so the two vehicles can automatically synchronize their controls and smarten up the computerbrain a bit more. See Figure 4.

We have hypothesized an approach in which the platooning takes place spontaneously, with the vehicles themselves "deciding" when to platoon. In this mode, the individual vehicles under full ACC (Autobrake plus Autogap plus Autolane) will be sensing and deducing the state of the traffic. This is done simply by passing back and forth to each other the size of the gaps between vehicles: if the gaps on

average are large, there is plenty of room for new entrants into the lane, and therefore there is no need for platooning to provide more space. If the gaps are small, vehicles deduce the need for more capacity, and if two adjacent vehicles are both under autocontrol, they will spontaneously platoon. (If the adjacent vehicle doesn't answer the communication signal, then it is not under autocontrol.) This platooning creates more space in the lane for additional vehicles: new capacity is automatically and spontaneously provided.

Autoplatooning is envisioned to also work in reverse: deplatooning occurs when there is no longer need for it to increase lane capacity. It also occurs when the driver signals that she wants to assume manual control, probably in order to move out of the lane and exit the freeway.

The sole purpose of platooning is to increase the effective capacity of the freeway. It is dubious that people will like it for themselves: the notion of driving within just a few feet of the next vehicle doesn't naturally appeal to most people. One can think of several possibilities for overcoming this reluctance. First, since platooning substitutes for more concrete, it may be rational to pay people to equip their cars that have ACC with Autoplatoon.

Second, it may be justifiable to require that all ACC systems sold beyond some date include Autoplatoon. Third, even though it is actually safer to be 2 feet apart rather than 20 feet, it may be decided that the risk is so low that we can operate with wider spacings until people grow accustomed to the new proximity—or perhaps forever: the penalty on capacity is modest.

5.6. Fully Automated Freeway Trips

This evolution would not stop here, although our crystal ball begins to get much cloudier. Probably the next step is to provide automatic lane change (Figure 5). This enables the entire freeway trip to be automated: entry into the freeway, movement to the desired lane, then subsequent exit from the freeway. This is fairly ambitious technically, but in a decade should appear much less so. Fully automating the freeway portion of a trip may have considerable appeal to our growing contingent of elderly drivers.

5.7. Toward Fully Automated Travel

Somewhere in this evolution the step will be taken that we believe will be the true watershed in surface transportation: the integration of the automated vehicle with the Traffic Management System, the smart car with the smart road. In our opinion, this step will be as significant to the future of ground transport as the original introduction of automatic controls with ICC. This integration will begin to happen when we first start to introduce the automated vehicle to surface streets. The first steps may be undramatic, but they will open the door to the fully automated origin-to-destination vehicle.

Freeways and highways basically constitute one long "link" on which the phenomenon of primary interest is the longitudinal spacing between vehicles, a vari-

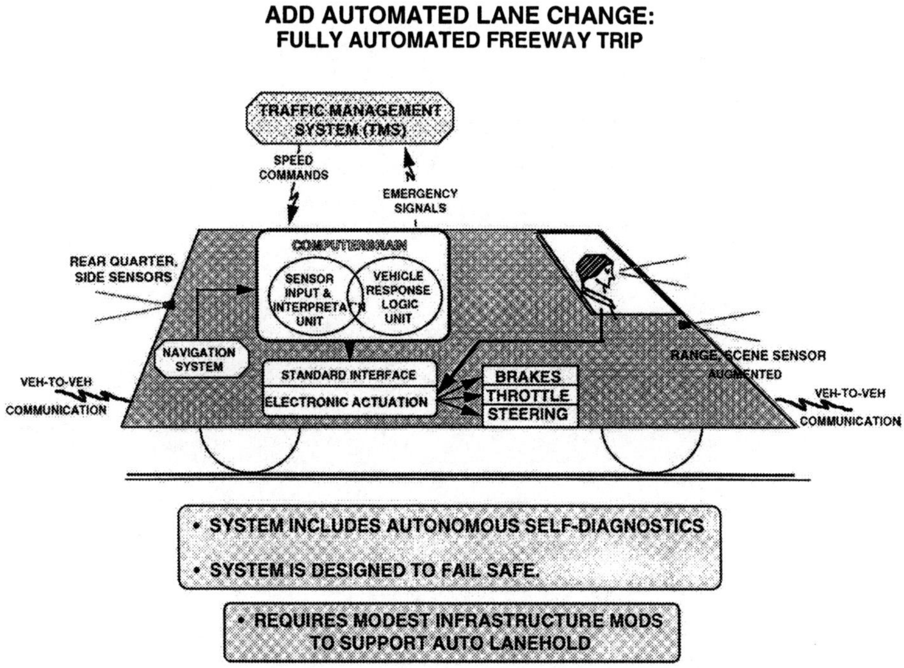

FIGURE 5. Fully automated freeway trips.

able most easily measured and manipulated from on board the individual vehicle. This leads naturally to the essentially autonomous mechanizations we have been describing.

On surface streets, the primary interest shifts to the "nodes," the intersections. Links are short, and particularly in high-density regions the behavior on the links is largely dictated by the conditions at the intersections. The node, not the link, is the dominant influence on surface street traffic behavior. The need now is not just smart cars, but smart intersections.

This shift from link-behavior-dominated phenomena to node-behavior-dominated phenomena leads to a different mechanization for an automated vehicle system. Now two functions are needed to effectively control flow at an intersection. First, we continue to need to ensure safe separation between vehicles in a vehicle stream; we are already getting the necessary information from the vehicles' Automated Vehicle Control System, and it seems sensible to continue to carry out this safe-separation function with an on-board system.

The second function is the macrocontrol of the traffic streams through the intersection, in essence the allocation of time slots to each of the desired travel directions to optimize total throughput and perhaps bias some flow directions to prevent jam densities in other parts of an areawide system. The information to perform this function is beyond the reach of any individual vehicle, but is the bread and butter of

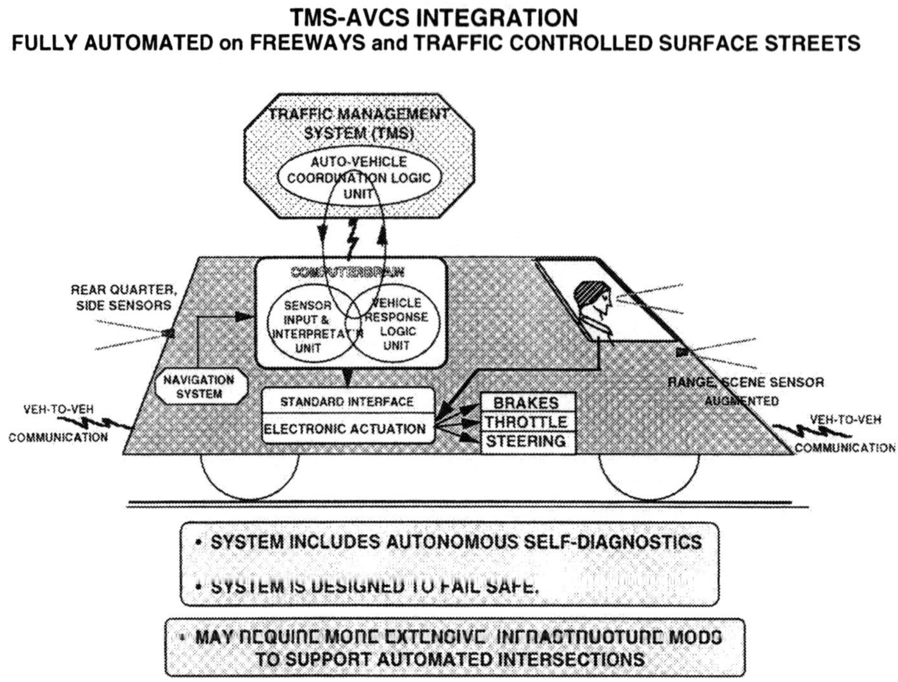

FIGURE 6. Automated travel on surface streets.

the Traffic Management System (at least in advanced systems). It only makes sense to continue with this functional allocation, but rather than send traffic commands to a driver using traffic lights, we will send them to the now much more versatile computerbrain on board the vehicle through real-time digital communication. The driver need no longer be in the loop.

We have made a fundamental change in the AHS architecture from essentially autonomous vehicles behaving in a self-organizing system, to an integration of the smart street with the smart car. The Advanced Traffic Management System (ATMS) provides macrocommands, and the on-board Automated Vehicle Control System executes them and provides safe microcontrol. Figure 6 illustrate the notion of this new integration.

This coming integration cannot be ignored in the development of an Intelligent Transportation System architecture. There was little interplay between the essentially autonomous vehicle and the infrastructure Traffic Management System when operation was confined to the freeway, but when it graduated to surface streets the AVCS system becomes inextricably intertwined with the other elements of ITS. In fact, in the author's judgment, the smart car integrated with the smart street and smart freeway/highway will become the heart of the Intelligent Transportation System.

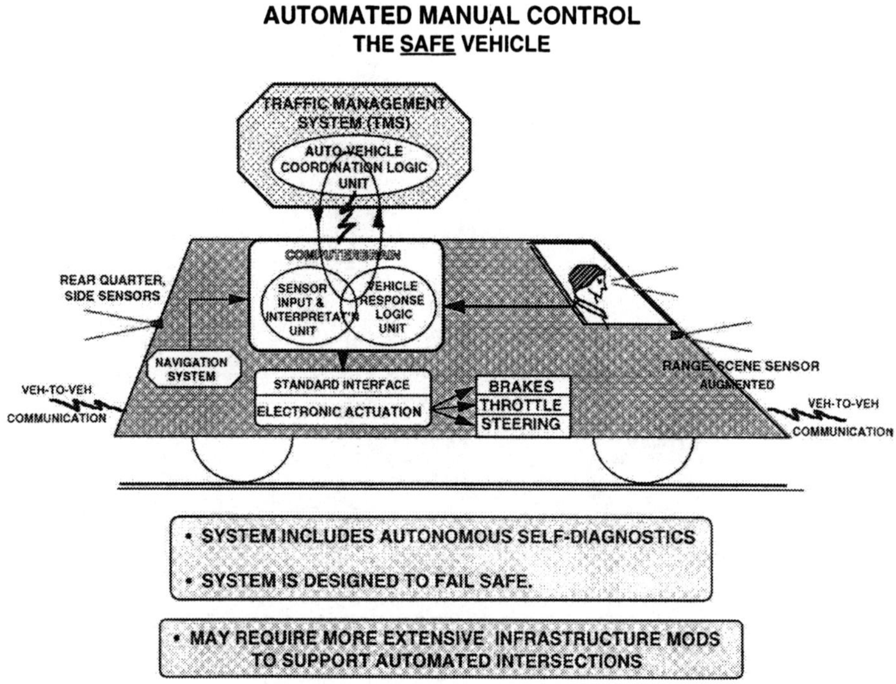

FIGURE 7. Automated manual control.

5.8. Fully Automated Trips—Origin to Destination

We now have the capability for fully automated driving on both freeways and surface streets. But who navigates? There are three options. The first is an on-board navigation system that is heading the vehicle toward the destination selected by the "driver passenger," but reacting to congestion information furnished by the ATMS (although we really should have eliminated congestion by now, or at least made it exceedingly rare).

The second option is guidance by an even-more-capable ATMS: the "driver passenger" informs the system of his desired destination and the ATMS furnishes the necessary macrocommands to the vehicle.

The third is the manual "driver," whose control inputs now go to the Computerbrain, not the wheels. It is quite possible that the nature of the driver's controls in this mode of operation will have changed dramatically, since there is no need for anything more than "turn-left-at-the-next-corner" type of inputs. Figure 7 reflects this very important step in the metamorphosis of the Automated Vehicle System.

Clearly, once we moved to surface streets, we were on the path to the driverless vehicle, with all of its implications.

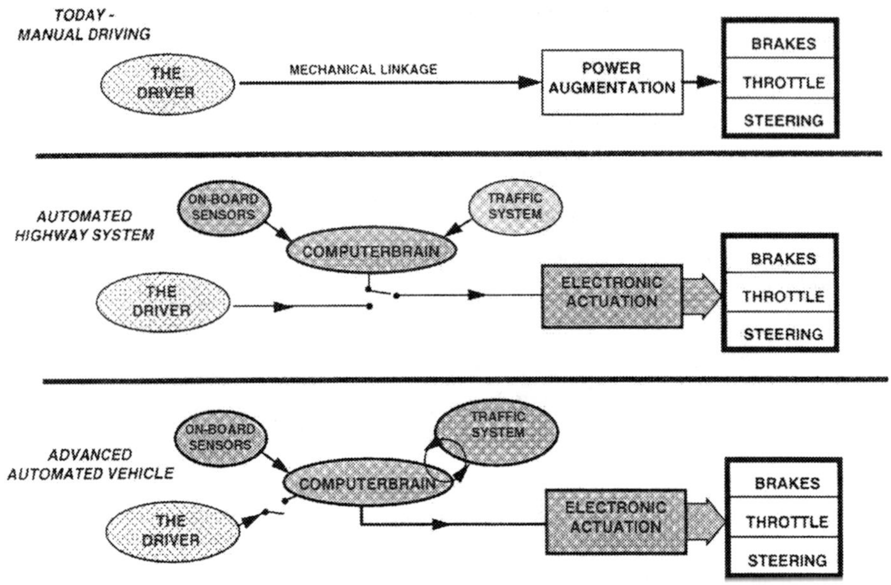

FIGURE 8. The long-range evolution of the automated vehicle.

6. RECAPITULATION

Here we briefly reiterate the broad pattern of this evolution we have described.

Beginning with Autobrake and Autogap, we have modified the vehicle so that the steering and drivetrain controls are actuated by electrical signals, not muscle: we have replaced drive-by-muscle with drive-by-wire. With drive-by-wire, the driver, by the manipulation of her controls, is sending an electrical signal to the actuators, not a mechanical movement.

We have then progressively taken the control of these actuation signals away from the human eye–brain system and substituted, function by function, a "computerbrain" system fed by artificial sensors. We have built into that computerbrain all of the responses to keep the vehicle safe in a very wide variety of driving situations.

We began to put exogenous inputs into the computerbrain when we added the automated lane change function and began to automate the whole freeway trip. But the real watershed was when we moved automation to surface streets and introduced the smart intersection. Now the infrastructure sensors and brains begin to exert major control over the vehicle. This is the step that starts us down the road to full trip automation.

We found that the computerbrain system could ensure safety during manual driving. Now when the driver wants to manually control the vehicle he does so not by sending electrical signals to the steering and drivetrain controls, but by sending his signals to the computerbrain system—he drives *through* the artificial system, not *around* it. The vehicle follows his gross steering signals, but doesn't permit him to do anything unsafe. Now almost anyone can drive. This should be a particular boon for the elderly portion of our population. Now the car can take dad to the transit station, return home and then take Susie to school, and on the way back home stop for its annual service checkup.

This broad pattern of evolution is illustrated in Figure 8. We can be almost sure, however, that it will not happen in the three neat steps we have laid out there; rather, like everything else, it will happen in bits and pieces, in small, almost experimental steps.

There is an almost seamless path available to a revolution in road transport. The technology is either in hand, or within stretching distance. If we just play our cards right.

REFERENCE

This chapter is based on WARD, J., *An Hypothesized Evolution of an Automated Highway System* (Aug. 1994), published by Rockwell International under contract to the FHWA as a part of the Precursor Systems Analyses of Automated Highway Systems.

CHAPTER 6

THE EVOLUTION OF AHS AND CURRENT VEHICLE TRENDS IN LIGHT OF AEROSPACE SYSTEMS EVOLUTION

DAVID A. BLANCETT, GREGORY H. DAVIS, SHERYL A. PAYNE, AND CHARLES E. TAYLOR

1. INTRODUCTION

Digital electronics, head-up displays, electronic systems to indicate maintenance requirements, two-way digital communications data bus, flat panel displays, vehicle condition monitoring, navigation systems and displays: components of an advanced military or commercial aircraft? or tomorrow's automobile? The same technological advances being heralded to improve safety, efficiency, and capacity for air travel are slated for use in Intelligent Transportation Systems (ITS), such as telecommunications and computer capabilities linked with satellite technology. There are broad connections between the technological advances made in aerospace and the enabling technologies for ITS. The Automated Highway System (AHS) is one ITS program that will benefit and can learn from advances made in aerospace systems.

DAVID A. BLANCETT, GREGORY H. DAVIS, SHERYL A. PAYNE, AND CHARLES E. TAYLOR • Advanced Technology and Development Center, Northrop Grumman Corporation, Pico Rivera, California 90660-0158.

Automated Highway Systems, edited by Ioannou. Plenum Press, New York, 1997.

The primary AHS objectives are to make travel safer and more efficient, improve the mobility of people and goods, improve productivity of surface transportation, and contribute to a better quality of life.[1] Similar objectives have existed within the commercial airline industry for more than 30 years with substantial gains made to those ends. Military aircraft and space systems development programs have provided the technological advances for commercial aircraft to make great strides in achieving those objectives and have resulted in development of a thriving air transportation infrastructure and manufacturing base for the entire United States.

Aerospace programs provide a technological "toolbox" with the key components necessary to develop a fully capable AHS. The challenges lie in integrating these components into a cost-effective system that fully meets the desired goals of the AHS. These challenges are focused in the areas of system engineering and societal impacts. This chapter will focus on the former with only brief forays into the latter, although the latter contains the challenges that will have the greatest impact on the rate at which the system evolves and the nature of the "fully" developed AHS.

The United States needs to improve highway safety and efficiency now more than ever. More than 40,000 people die on our highways annually and it is estimated that 80% of our urban highway interstates will be congested by the year 2000.[2] Numerous ITS initiatives, of which AHS is a part, are being pursued to address these issues.

Conceptually, the AHS has been with us for some time. The 1939 New York World's Fair had an exhibit by General Motors called "Futurama," which included an automated highway, to depict life in 1960. However, neither the technology base nor the socioeconomic impetus has existed until now to lead to an organized effort to develop an AHS.

For purposes of discussion for this chapter, the ultimate AHS envisioned is one that has positive control of all vehicles including those that will operate in platoons. The vehicles themselves are assumed to be "smart" with much of the safety monitoring and the vehicle control laws on board. Fore and aft sensor systems will provide inputs to the steering and controls computer. Lane changing sensors will provide input to the control computer to help ensure collision avoidance. A beacon transponder and RF communication will provide an avenue for vehicle-to-vehicle communication/control for platooning as well as vehicle-to-infrastructure-to-vehicle communications and entry/exit infrastructure checks and controls. The vehicle will be cognizant of its position and provides this information to the infrastructure. It is also assumed that a fully implemented AHS will permit operation of all vehicle types and AHS operations will be possible in all weather conditions.

The technologies required to develop the components and overall automated system described above exist and will not be the primary pacing factor. The pacing factors will be cost, acceptance, level of safety, accessibility, maintainability/reliability, and the political climate. A strong perception by the public of a need for in-

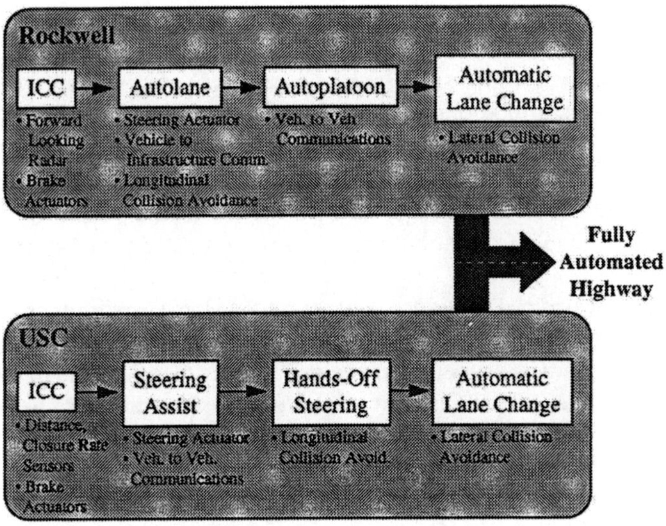

creased safety measures, congestion relief, pollution reduction, more effective resource use, travel time reductions, and capacity increases will be required along with the perceived ability to achieve these needs for a successful AHS.

Implementing an AHS is almost universally envisioned to be an evolutionary process rather than revolutionary. That is, there will be several interim, developmental stages that the AHS will experience from both technological and public acceptance standpoints. This evolution will begin with a more "intelligent" use of the existing infrastructure, since it is not feasible in many areas to build new roads or expand present ones. Several AHS evolution concepts have emerged. Two popular theories come from researchers at Rockwell International Corporation and the University of Southern California (USC) and are shown in Figure 1. There are many similarities between the two evolution theories, especially in their overall phased implementation.

Specifically, both theories begin with the introduction of intelligent cruise control (ICC), which is a technology that builds on today's cruise control function and is expected to be commercially available by the end of the decade. With ICC, the driver does not have to adjust his speed during increasing traffic conditions. The system is "intelligent" enough to recognize closure rates of a vehicle immediately in front and is constantly sensing its distance and velocity. ICC enables the "feet-off" feature of AHS. Although this capability is seen as a required AHS capability, it is also useful as a stand-alone function. Therefore, as in most systems associated with AHS evolution, the impetus behind bringing this capability to market is not the development of the AHS, but rather its stand-alone merit. In this case, the capability would provide improved safety and convenience for the driver and passengers.

There are no requirements for communications between vehicles or between vehicles and the infrastructure in this early phase, although this point is not agreed on between USC's and Rockwell's evolution definition. However, communications (vehicle to vehicle, road to vehicle, road to central control) will eventually be a major element of the AHS, and the architecture and standards to which the communications are developed will be critical to the ultimate development of a fully functioning AHS.

Next, both evolutions suggest an automated steering capability in a single lane only, where there is no lane changing operation while the vehicle is in automatic mode. Electronic actuated steering, on-board sensors (for the road and other vehicles) and sensor interpretation, and vehicle-to-infrastructure communications are all technologies that are applicable to this phase and will need to be integrated within the vehicle as well as with the infrastructure.

Eventually, both evolutions require automatic lane changing, which requires lateral collision avoidance. This technology enables a vehicle to automatically enter and exit the AHS lane and facilitates a multiple-lane AHS concept.

The salability of a system capability, such as ICC or a vehicle safety check system for entering a freeway, is the primary driver for the pace at which the AHS evolves. This is one of many sociological questions that also includes consumer acceptance. However, for the purpose of this chapter, the more critical question is, how are these systems being developed in terms of architectural approach and to what standards? The answer to these questions will determine how easy the final jump to a full AHS will be and what it will cost.

The AHS evolution of the ground transportation infrastructure is under way. Near-term automobile improvements that the automobile manufacturers are putting into production, or will in the near term, include: collision warning, telemetrics (vehicle navigation systems and vehicle location systems), ICC, stability enhancement systems, and driver impairment detectors. Infrastructure enhancements include automated toll collection, weight to roadside readers, and route guidance systems. These are all systems that can be considered upgrades to current vehicle systems, including trucks and buses. These improvements are primarily driven by safety and convenience considerations for the vehicle operators. None of these are AHS specific, but all contribute to the ability to implement an AHS. The new Mitsubishi Diamante is a good example of this evolution. It will be one of the first production vehicles to have Preview Distance Control. This is a laser-based collision-avoidance system that measures the distance to the nearest vehicle ahead, instructing the cruise control to ease off the throttle if the vehicle is too close in order to keep a safe distance. Several millimeter wave (MMW) radar collision avoidance systems are currently in development.

The Diamante also represents the continuing progression toward smart cars. "The gearbox sends signals to the traction control, power steering, variable dampers, and optional four-wheel drive which all react in concert to quicken the hardware's response to edgy performance, or ease it up for laid-back cruising."[3]

It is easier to envision a clear developmental path to an AHS-capable vehicle if this type of system is successful in a production car. A vehicle that can react to unexpected conditions, whether the information is coming from on board or an off-board source (i.e., the AHS control center), is a major step toward the level of vehicle control required for AHS. Adding lane holding control is not a major adaptation from this technology. Therefore, it is conceivable that by the time the AHS is fully implemented, automobiles will be close to having capabilities needed for AHS operation.

Of course, these technological advances come at a price. Like many products, these types of features are first introduced into high-end models to be proven and then flow down to the lower market vehicles as costs are reduced. As an example, by the year 2013, it is expected that 100% of vehicles will be equipped with antilock brakes.[(4)] Interestingly, the Boeing 707 commercial airliner was the first passenger aircraft to utilize antilock brakes more than 35 years ago, which demonstrates the availability of an aerospace technology versus the marketability and affordability of the technology to the consumer.

Other required system elements will be AHS specific, such as control laws for automatic lane control. Introducing these elements will be the defining characteristic for an automated highway since it is at this point that vehicle operation will be both "feet-off" and "hands-off." For these latter capabilities to evolve will require buy-in by the operating community leading to a strong demand for an AHS.

The AHS will be similar to the airline business in that there is major resistance to technology implementation based on the desire to minimize the frequency of capital expenditures, a strong suspicion of the operations and support (O&S) costs associated with new technologies (e.g., composites), and the need to demonstrate that a technology improves safety (or gives the perception of increased safety) in a business where safety standards are extremely high. An example of a national program with some of the same goals as the AHS that failed in this arena is the supersonic transport (SST) program.

Aerospace technology had reached a point in the 1960s to make it feasible to build an SST and a contract was awarded to Boeing in 1966. The primary drivers for the SST were very similar to those of the AHS including reduced travel time, increased throughput, and reduced costs both to the airline and to businesses. There were no technological obstacles to building an SST, although there were many obstacles to building an environmentally acceptable SST, such as sonic booms, noise pollution during departure and approach, and high-altitude pollutants. The limitation on flight routes and speeds necessitated by sonic booms would have reduced time benefits and economic viability. Another obstacle was that the development cost was beyond the ability of private industry, which meant government financing of a private "for profit" venture; politicians believed the program should be funded through private capital instead of utilizing taxpayers' money. Consequently, the program was stopped in the early 1970s before a prototype was completed, the reasons being the "environmental movement and an economic downturn."[(5)] Similarly, environmentally imposed limitations may limit the ability of the AHS to fully achieve its envisioned benefits. For instance, these limitations may include the inability to build on/off-ramps optimized for an AHS and the acceptance of traffic flow off the AHS in town centers.

2. AEROSPACE TECHNOLOGY APPLICATION

There is no complement to a motivating force as strong as war, including a cold war from which the Apollo space program emerged, that will accelerate military systems development. These strong forces result in major technological advances enabling systems evolution and development. The impetus for the military to stay number one in the world produced great advancements in sensor technologies, such as radar, Global Positioning Satellites (GPS), laser optics, headup displays, fly-by-wire, and acoustics. Money was not the controlling factor in the technology development because of the need to be militarily superior to adversaries.

The ITS initiative intends to bring proven information, communications and control technologies to improve efficiency and reduce the negative impacts of surface transportation. The aerospace industry is one of the major contributors to the technology "parts bin" for AHS systems development. From the aerospace industry perspective, the components required to develop an AHS exist and, as demonstrations such as those at PATH have proven, the development of an automated highway is more than theoretically possible. Table 1 illustrates a few examples of aerospace technologies that are applicable to the AHS.

Current generation automobile technology matches current military aircraft technology virtually feature for feature, varying in scale, maintainability, reliability, and cost. Aircraft systems such as multiple dedicated digital processors communicating on serial digital buses, built-in test, fault sensing, indication and recording, automated controls, and augmented steering are either baseline elements or optional features in today's automobiles. As a result, there is significant fallout from aircraft technology and flight test instrumentation design practices which can benefit the planning and design of vehicles for automated highways. However, having key separate components does not necessarily ensure an easy or successful path to developing a "fully integrated" AHS that can achieve the necessary levels of safety and the advertised travel time and pollution reductions at an affordable cost. Developing a fully capable system will be a significant challenge with many pitfalls to be avoided.

Three things need to occur to transition the technologies into the AHS environment: (1) adaptation of the basic technologies to the new environment, (2) less expensive implementation of the technologies, and (3) establishing standards that will permit successful integration of the components into an integrated system. The first item has been steadily occurring.

The second item is also progressing, and there is much aggressive research and development (R&D) under way to achieve this goal, although with more difficulty. The high cost of these components in aerospace-related products tends to be driven by two primary factors. The first is that in most cases the initial R&D cost is completely absorbed by sales in the aerospace market which tend to be low volume. The

TABLE 1
Aerospace Technologies Applicable to AHS

Aerospace technologies	AHS applicability
Built in test (BIT)	• Real-time vehicle status
Control theory, algorithms, and software	• Control the dynamics of a moving vehicle
Sensors to monitor outside world and health of vehicle	• Sensors for traffic, weather, heat, vehicle health • Establish safe performance envelopes for travel and/or platooning
System integration	• Incorporate all vehicle design systems and functions into a single system
Data Links	• Readily adaptable to AHS entry and highway travel
Electronics	• Heads-up displays, gyros, accelerometers adaptable to AHS
Computers (processors)	• Key to system integration • Driver work load reduced for increased safety
Actuators	• Smart actuators needed for full-up drive-by-wire control • Active servo control with built-in feedback
Materials	• Aerospace materials can improve the stiffness in torsion, which is a safety issue
Human system integration	• Situation awareness

second factor is that the level of reliability and safety required from these components is extremely high. The consumer base is much greater in the ground transportation market given widespread acceptance of the desirability of a new system and a price that is matched to the perceived value. However, the levels of safety and reliability cannot be lowered for the AHS, and in fact may even be greater for some components than in the aerospace industry. From a technological perspective, the pace of introducing key AHS components will be driven by the ability to develop a system that has an acceptable acquisition cost while maintaining the same level of reliability and safety that is required in the airline industry. These are levels beyond what is presently required in ground vehicles since breakdowns on an AHS can have much more severe consequences in terms of injury and disruption.

The last item, establishing standards, is probably the most significant problem facing AHS development and the furthest from resolution. The ITS Architecture Development Program is under way to address many of these issues, but by their own admission the "decisions as to the nature and extent of ITS deployment will be made primarily by state, regional and local agencies, transit and commercial fleet operators, consumers and public interest groups,"[2,p.i] rather than a central controlling agency such as U. S. DOT or ITS America. The success of this program in meeting its goal of promoting nationwide standards of ITS is crucial to the success of the overall AHS concept.

2.1. System Standards

Industry standards are vital to the integration of on-board vehicle systems with external supporting systems. There is an immediate need to gain acceptance for national standards for ground vehicle systems that will eventually be part of the AHS system. These will provide the means to manage the evolution. Without managed evolution, the final step to an AHS will be difficult and expensive. Specifically, national standards are needed for:

1. Serial bus communication protocol and software governing data word lengths, frame structure, engineering units algorithms, and synchronization strategy
2. RF communication
3. Pressure, position, and temperature sensors
4. Vehicle designs that affect system safety and vehicle certification

The ability to broadly implement standards has been demonstrated by the aerospace industry. The Aerospace Industries Association (AIA), consisting of several aerospace companies, was formed to create panels that would develop standards for air vehicle design, such as telemetry. This concept has not been properly embraced in the transportation industry. For example, vendors in the Electronic Toll Collection (ETC) industry manufacture transmitters that operate at 2.45 GHz or 902–928 MHz band. The selection between bands seems to vary by country, but more importantly, by various state departments of transportation. The Federal Communications Commission (FCC) has failed to make a ruling on this issue thus far.[6] However, there are some positive indicators that standards implementation is being actively pursued. For example, "the ITS America CVO Architecture & Standards Subcommittee has begun work that will lead to standards that would allow trucks to pass through numerous jurisdictions using one interoperable transponder."[7]

Another major issue is data anomaly detection and prevention. Erroneous data may lead to vehicles not being approved for entry to the AHS, being "kicked off" the highway for a nonexistent problem, or the possibility that a real problem is not detected. High false alarm rates in built-in test (BIT) reduce the system's intended effectiveness as a maintenance tool. All possible environmental extremes must be considered (e.g., temperature, vibration, transient voltages) to reduce BIT false alarm potential. Data tolerances will have to be defined as well as means for data anomaly detection. Data system software checks can be employed for detecting calibration inconsistencies, mux-bus protocol errors, and periodic data sampling rate anomalies.

In the absence of standardized systems, or during the evolution leading to standardization, intelligent data acquisition interfaces will be required to normalize data flowing between the vehicle and the AHS infrastructure. This would impose another processing layer over the desired level, complicating the design process and adding cost. Lack of operating standards presents a challenge for interfacing individual vehicle types with ITS infrastructure nodes in a common format. For example, this problem existed for a preproduction prototype aircraft containing intergeneration avionics subsystems with serial buses designed to a variety of different protocols. The avionics buses each conformed to one of five different serial bus communication standards: ARINC 420, IEEE-488, RS-422, RS-232, and MIL-STD 1553B. The problem had to be solved in the test data acquisition system by implementing bus data collectors that reformatted the captured data into a commonly used standard format, thus normalizing the output data product into a form acceptable to the affected contractor and government test facilities. This normalization approach would offer a method for interfacing Advanced Vehicle Control Systems (AVCS) autos with nonstandardized serial bus systems to the ITS infrastructure during the evolutionary stages of highway automation.

2.2. *On-Board System Monitoring*

Current-generation military aircraft integrate sophisticated flight controls, aircraft management systems, offensive and defensive avionics, and weapons management systems, all supported by a network of high-speed digital processors and serial bus communications systems. On-board diagnostic instrumentation and an intelligent warning/caution advisory system continuously apprise the air crew and test control center of system failures or degradation. BIT diagnostics in major subsystem elements perform self-test functions on system initialization and/or periodically during ground and flight operations. A maintenance recorder logs anomalies and generates out-of-tolerance reports for postflight review by ground crew personnel.

The flight test system monitoring approach has direct application to system monitoring for AHS-capable vehicles. A major premise of flight test instrumentation philosophy is to capitalize on the air vehicle's on-board data resources to supply required test data, and install dedicated test sensors only when data are unavailable from the operational systems. In a recent program, approximately 8000 active test measurands were continuously captured for telemetering and on-board recording. During later parts of flight test for this aircraft, 90% of the required test data were provided by serial data buses. Of special interest is that, except for the electrical interfaces, the data collection for test purposes was totally transparent to the operational hardware/software systems supplying the data.

Additionally, it is desirable that the test data system acquire its information from the system under test in a processed, rather than raw, signal form. This avoids the potential of introducing noise into or electrically loading the monitored system. Data collection is facilitated in both aircraft and automobiles by widespread use of serial digital communication buses ("mux buses") by the on-board systems. In present-generation military and commercial aircraft, mux buses are designed to industry standards (MIL-STD 1553B and ARINC 429, respectively) which define items such as bus control protocol, data word length, and synchronization strategy.

This extensive diagnostic capability needs to test system functionality, not just individual system state, and it should be designed in from the start to be truly effective and unintrusive. BIT check must ensure that not only the individual components are operating properly but also that they are communicating and responding appropriately to each other so that the system as a whole will operate in the manner intended. In this case, using a "federated" monitoring system, with each vehicle monitoring and reporting on the health of many of its key systems, will greatly simplify the overall AHS design and development and will reduce the O&S costs of the highway itself. The preferred approach is to have on-board processing of system health information since the processing is trivial and the infrastructure only needs to consider the findings and not process all of the data. Hence, it is a simpler, cheaper, lower-risk approach. Extensive communications to or from the infrastructure to support verification impose a proportionately heavier work load on the control center, with attendant time-sharing complications to accommodate other vehicles in the system. This also supports consigning test intelligence and responsibility to the vehicle such that the reliability and maturity required can be achieved on the vehicle.

There is a strong trend within the automotive industry to increase on-board system monitoring for purposes of maintenance, safety, and engine monitoring associated with emissions. For example, in 1996 all cars sold in the United States will have on-board diagnostic systems that can detect excessive emissions. This also confirms that the number and sophistication of system test and monitoring capability are growing and will continue to grow.

2.3. On-Board Electronic System Architecture

There have been many approaches for military aircraft avionics architectures in recent years. Table 2 portrays this progression and some of the reasons behind it. The goal behind each change was to build a simpler, lighter, more reliable and maintainable system that at the same time had greater capability to handle the increased processing demands resulting from advanced offensive and defensive avionics suites. Their progression represents a change in philosophy based on experience relative to how computing is handled for large integrated systems.

Federated systems, which require separate processors and displays for each system, involve increased software development and weight problems. This reflects accommodating all of the interfaces between the federated systems (e.g., radio, navigation system, radar) as well as all of the cables necessary to accommodate the interfaces. On a recent aircraft development program, 80% of the effort went into developing redundancy management because of the overwhelming amount of duplication in federated systems. Software, in turn, has been found to be one of the

TABLE 2
Military Avionics Architectures Progression

Federated Archit. (mid-1970s)	1553 Data Bus (early 1980s)	PAVE PILLAR (mid-1980s)	PAVE PACE (early 1990s)
• Primarily separate processors and displays • Connected by electrical power only	• Shares channels, not processors • Fewer cables • Timing requirements, asynchronous • MFD • Weight savings	• Reduced number of processors to maximize common assets • Front-end and core processing, VMS • Trusted system • Reconfigurable assets • Improved LCC and supportability	• Uses more common assets (front-end) • Shares antennas, processors (front-end) • Smaller, faster, fewer modules (core)

most expensive elements to develop, and accurate software development cost has been difficult to predict.

The 1553 Data Bus helped improve the physical integration of the avionics system by utilizing Multi-Functional Displays (MFDs) and reducing the number of cables required, which saved weight. This led to the concept called PAVE PILLAR, which was developed in the 1980s. PAVE PILLAR emphasizes the use of common modules or processors to perform different functions in order to maximize assets and streamline the BIT system. Using common modules increases the affordability of the system by reducing the number of components required as well as increasing the supportability of the system. In addition, these modules can be reconfigured in case of failure.

Another trend in military development programs that is recommended for adoption by the AHS lies in utilizing commercial off-the-shelf (COTS) hardware instead of developing special-purpose components. Commercial hardware yields cost savings, improved reliability, and minimal interactions (i.e., 386 and 486 chips, PROMs being used now). The downside is that these components are not tested to the internal and external environmental extremes that could occur in the AHS operating environment. Software would still be specialized for AHS purposes.

2.4. System Integration

Although the technology for an automatic highway control system is within the technical capabilities of the engineering community, the ability to get such a system in place is inhibited by the lack of a definitive direction in many subsystem design approaches. The principal element that is lacking is a system engineering approach that will select the group of subsystems that are most compatible and develop a complete AHS concept specification.

In September 1993, the Department of Transportation awarded contracts to teams led by Hughes Aircraft, Loral, Rockwell International, and Westinghouse

Electric to develop alternative architectures for an ITS. Each of the teams has a formidable supporting team of six to ten constituents including state departments of transportation, universities, and major manufacturers. The Phase I report was published in November 1994. This report emphasizes user services and communications, offering many options. The systems in the report call for a man-in-the-loop both for the infrastructure and for the vehicle. In addition, cost-controlling standards for all elements are needed. The report does not sort out system architecture, design requirements, and policy. ITS is relying on a Phase II architecture development program (led by Rockwell and Lockheed) and stakeholders' forums (such as regional forums, ITS America committees) to formulate a plan.

The design of the AHS must begin with a top-level concept for system integration from which requirements for each system element flow down. This top-level concept should minimize the interfaces required, and set common design standards for all control system, infrastructure, and vehicle elements. AHS development must be controlled by a single authority that will evaluate all system elements based on the top-level system model. This top-down approach is one of the strongest lessons learned from the development of complex integrated systems within the aerospace industry. What is often viewed as the easier approach, the bottoms-up integration of components optimized for their specific purpose, often leads to inefficiencies in the overall design and major problems arising out of the change process. Equally important is the involvement of the "ilities" (e.g., functionality, affordability, supportability, maintainability) with the design, development, and change process to be sure equal weight is given to all elements. Problems will be avoided late in a development program when all of the required area experts are included from day one. This can mean the difference between success and failure, and definitely prevents the need for large cost increases to fix problems discovered late. This approach is not new and is becoming institutionalized in businesses as Integrated Product Teams (IPT).

An unmanaged evolution toward an AHS and a lack of an integrated system approach raises questions on how the evolution of the AHS might occur and the transition requirements between the development of the components applicable to the system and the actual last step to a fully integrated system. Key elements needed for the AHS, ranging from ICC to active stability systems for trucks, will tend to be developed as independent products added as "features" to vehicles by manufacturers. This will by nature drive the AHS toward federated systems integrated on the vehicle by means of interface software and controllers. Additionally, there may be little standardization between the system designs used by different vehicle manufacturers, which will add another level of interface software and translators to interact with an AHS. Evolution does not easily lend itself to a top-level, cost-effective design approach. The only hope of countering this potential problem is to define the AHS approach near term and get agreement to a design approach and standards based on this approach from vehicle manufacturers and the component suppliers. The ITS Architecture program was developed to address such issues as strategies and standards as well as to provide a forum for stakeholders to discuss potential architectures. Stakeholders in the program consist of consumers, transportation infrastructure providers and planners, freight operations, public safety services, passenger operations, product and service providers. "These stakeholders have helped design a consensus building process to gain cooperation among many different stakeholders in achieving the goal of a nationally compatible intelligent transportation system."[2,p.8]

An additional problem faced on many recent projects, and one that is very likely to occur with the development of AHS if not monitored closely, is "requirements creep." Aerospace companies, not unlike other high-technology companies, are notorious for continuing to invent even when a system has entered engineering development for production. This leads to the desire to continually introduce "better" designs and components even when the baseline design meets all specifications. Requirements often have the tendency to follow capability and hence requirements creep occurs. This can obviously lead to increased cost to introduce a change and validate that it doesn't negatively impact any aspect of the system including O&S. Changes during the development process can potentially introduce technical risk, and definitely introduce schedule risk.

It is not the intent to deny all changes since problems will arise during development that require solutions. For example, the Tracking and Data Relay Satellite (TDRS) is a six-satellite system that saw three major design changes occur after the first and before the last spacecraft was launched. Parts unavailability, reliability improvements, selected mission improvements, and improved designs were major contributors to the changes. Management reserve in terms of both dollars and schedule is needed to account for these expected changes. However, it is imperative that a tightly controlled change process is implemented and owned by the top-level authority that evaluates every proposed change against a set of criteria such as:

- Is it an effective solution to an identified system deficiency?
- Does it reduce the risk for a high-risk development item in terms of meeting system specifications and/or schedule?
- Will it significantly reduce the operating and support cost of the system with a reasonable impact on acquisition cost and schedule?
- Does it significantly reduce the acquisition cost of an element of the system without unacceptably impacting the total system acquisition cost, the O&S cost, or the schedule?

This is not meant to be an all-inclusive list, and of course determining the meaning of "significant" is one of many challenges. The point is that all elements of the system development effort and long-term operations need to be evaluated before a change is accepted with a resulting development schedule disruption. Assuming the baseline system has a high probability of meeting all design requirements, change should be resisted in the development phase and potentially put into a planned multistage improvement program if future benefit is deemed high enough. Too many cost overruns and program delays have resulted from the disruption of a development program to introduce "better" approaches.

Two other common causes of requirements creep are customer technology personnel with no budget responsibility pushing their favorite technology and a customer base made up of multiple agencies that do not speak with one voice. The customer agent with budget responsibility needs to work closely with the integrated system developer to control the first. Companies are very sensitive to customer satisfaction and will tend to try to meet all perceived customer desires potentially resulting in negative impacts on the development effort both in terms of cost and schedule. This is especially true if an agency is felt to have strong influence on the program's continuation. The customer needs to effectively control the customer demands.

The single voice issue will potentially be much worse for the AHS than for a typical aerospace project. Generally a joint military program involves two services, three maximum. Commercial aircraft builders have a larger customer base, but all of the customers are working within a standardized system which limits the variables. Even with this limited and confined customer base, programs have been negatively impacted in terms of cost and schedule by the inability of the customers to speak with one voice when it comes to requirements (e.g., multiple avionics configurations on the same model aircraft). The AHS faces the possibility of having a customer community made up of a great number of agencies depending on the number of cities, counties, and states the highway passes through. Unless these multiple agencies can agree to a single set of requirements for the AHS program and empower a controlling AHS system development IPT, they will potentially be faced with meeting a widely diverse set of requirements that may even be conflicting in some cases. Of course, some variations from state to state, or even areas of a state, are to be expected depending on factors such as climate. But again, there needs to be one set of requirements for each set of environmental conditions that can be expected.

The change process itself must be set up at the system level. It should not be considered a software or a hardware change but a system change. Each problem should be recognized as a system problem that can impact a wide range of "ilities" each of which should be a participant in the change process. One aerospace lesson learned is that the statement "it's only a software change" is dangerous, especially if a federated system is used. The qualification tests required after each software test will need to involve all interfaces within the system, which is not a simple or inexpensive task. Additionally, configuration control, no matter what tool is being used to support it, is easily lost on a large complex integrated system. Again, change should be resisted unless it meets some stringent criteria.

Problems that can occur during the development effort of large programs are sometimes related to the fact that people do not take the time to truly understand the complexity of the problem they are addressing before proceeding. Perhaps this is a systems integrator's way of saying "look before you leap." Many changes have been vigorously pursued aimed at improving system effectiveness with a narrow or simplistic understanding of the total system. The risk of a detrimental impact on the total system performance is high and a detrimental impact on schedule and cost to remove or fix the improvement often results.

2.5. *Software*

"A systematic approach to the development, operation, maintenance, and retirement of software"(8,p.135) is needed from day one. There is no denying this statement, but experience shows that truly achieving this goal is critical, and the ability to successfully do it and maintain it is difficult. Key elements of a successful software management program include laboratory prototyping simulation before a new release of software, regression testing, and the use of quality control of blocks. Managing software documented by blocks with sufficient configuration release control can hold change problems to a minimum. The block cycle change rate is estimated to be from 3 to 5% per year, which includes the vehicle software/firmware in the vehicle. This is made more crucial by the fact that in the Department of Defense (DoD), software development is estimated to be 10% of the entire DoD budget with software evolution and maintenance costs accounting for more than 50–80% of the overall cost associated with a software system.

Computer Aided Software Engineering (CASE) tools were a giant step forward in this process, although the first-generation tools had limitations. For instance, they tended to only support a specific part of the software development process and failed to provide the necessary integration into the next step of the process. For example, many CASE tools support requirements analysis, but they do not support the translation of the requirements into a software specification and corresponding software design. It is anticipated that the second generation of CASE tools currently being researched and developed will provide much better support in these areas, but may still fall short of what is required.

Given the level of control that will ultimately be exercised by the AHS over the vehicles on the AHS, the level of safety inherent in the software design will have to be extremely high. Designing the software to account for all possible situations is not an easy task. Validating software for all possible contingencies is virtually impossible. The overall system software development control will be the ultimate driver determining when or if an AHS will exist.

The development of secure and reliable software will be a major problem for this system. There will be a need to prevent malicious entry into the computer system and to guarantee a level of assurance that software will operate correctly and safely. Also, predictable behavior is crucial, since failure to satisfy timing constraints may result in immediate catastrophic events, with little or no opportunity for human intervention. Software must satisfy complex timing properties, such as response deadlines, while interacting with external environmental phenomena, such as motion, temperature, pressure, and physical position.

2.6. *Affordability*

In order to be deemed affordable, the defense industry's goal is that "operations and support resource requirements be reduced by 50 percent without impairing combat capability."(8,p.3) Affordability requirements dominate all defense program thinking in these reduced budget times. In this case, affordability is defined as performance divided by cost. If both parts of the affordability equation are not met, no procurement will occur, and money will continue to be put into incremental improvements. New, advanced, integrated systems will not be developed. This has happened repeatedly in recent history within the aerospace industry. A recent program experienced continuous weight, and hence cost growth, while its ability to meet requirements eroded. It became unaffordable and was canceled.

Why is this important to AHS? AHS needs to meet the same strict affordability goals. It must provide the same flexibility as the present system at reduced cost or, perhaps more appropriately, provide a significant improvement at little or no increased life cycle cost (LCC). Unfortunately, the definition of "significant" is subjective. Improvement is defined as reduced travel time and increased safety, although convenience and quality of the travel experience are players as well. Reduced pollution probably will only be a factor in a very small fraction of the population. Some don't care and others care but not enough to pay the cost.

AHS-equipped vehicle cost is a critical consideration in AHS development. According to the U. S. Census Bureau, in 1993 the median household income was \$31,241 and 64% of U. S. households had incomes between \$15,000 and \$75,000. On the other hand, only 29% of the households had incomes in excess of \$50,000. The April 1995 issue of *Car and Driver Magazine* states that an average new automobile today costs \$20,045. Therefore, it is likely that the new-car buyers predominantly fall into 29% of the population who earn \$50,000 or more. The other 70% of the population will most likely have to retrofit their cars with AHS equipment. This puts tremendous pressure on developers to provide low-cost solutions for in-vehicle AHS systems.

Motivation must be very strong to get a significant portion of the population interested in buying AHS equipment. These motivations must overcome the fact that AHS implementation will be evolutionary. A potential user will be faced with the dilemma of buying an early evolution AHS-equipped automobile that will need to be fitted with new equipment as AHS evolution progresses. AHS motivations are not strong enough to support developing a fully capable AHS from the beginning. AHS evolution will be a function of the ability to introduce key components needed both in the infrastructure and in the vehicles based on other benefits, or through federal safety laws, that will make it an easy transition.

REFERENCES

1. Automated Highway System—System Objectives and Characteristics, 2nd Draft, The National AHS Consortium, Troy, MI, May 22, 1995.
2. ITS Architecture Development Program, Phase I, Department of Transportation, Washington, DC (Nov. 1994).
3. Y. ISHIWATARI, Mitsu Rolls Out a New Diamante, *Car and Driver* p. 42 (May 1995).
4. D. J. HOLT, Model Year 2013, *Automot. Eng.* p. 6 (June 1994).
5. P. RELMAN, Supersonic Sons and Daughters, *Air International* p. 332 (June 1995).
6. Inside IVHS, Jan. 16, 1995.
7. ITS CVO Update, Vol. 2.1, July 1995.
8. J. S. PRZEMIENIECKI (Ed.), *Critical Technologies for National Defense* (AIAA, Washington, DC, 1991).

CHAPTER 7

EVOLUTION TO AN AUTOMATED HIGHWAY SYSTEM

WILLIAM B. STEVENS

1. INTRODUCTION

In the Precursor Systems Analyses (PSA) sponsored by the Federal Highway Administration in 1993, several of the researchers addressed how AHS would be first introduced to the market. The researchers identified several problems to be overcome including driver acceptance and the ability of state and regional departments of transportation to operate and maintain an AHS.[1-3] The researchers concluded that one of the most serious issues is the "chicken-or-egg" problem—which comes first, the AHS highway lanes or the AHS-equipped vehicles? Minimum AHS vehicle penetration for a particular corridor or highway segment is reached when, after conversion of one of the existing highway lanes to AHS, the number of AHS lane users equals the number of manual users of the lane before conversion. It was estimated that this minimum AHS vehicle penetration ranged from 5 to 30% of the total vehicle population depending on many factors such as number of lanes already existing in the corridor, frequency of entry and exit lanes, and average trip distance.

Most researchers felt that a planned evolution path to AHS, rather than a revolutionary path, would be most desirable in meeting this problem of AHS

WILLIAM B. STEVENS • National AHS Consortium, Bethesda, Maryland 20817.

Automated Highway Systems, edited by Ioannou. Plenum Press, New York, 1997.

market introduction. By planned evolution it is meant that sufficient planning would occur to ensure that AHS would be introduced on a timely basis as one of the steps along the evolutionary path of introducing automated vehicle control (AVC) to the nation's vehicle–highway system.[4,5] It was argued that this kind of planned evolution would allow drivers to gradually accept various forms of AVC technology before the AHS introduction. It would also give engineers time to work out the details of the transfer of control between the driver and the AVC systems. Evolution would also support the development of vehicles that are far more suitable platforms for AHS than today's vehicles because many of the vehicle features required by some of the early AVC services may also be required by AHS and may be upward compatible. Finally, it was felt that evolution would provide time for the evolution of the state and local departments of transportation to the point that they will be able to manage their portion of a sophisticated, control system.

The logic of planned evolution appears to be both sound and desirable, but there are several potential issues to be resolved and concerns to be addressed as evolution planning proceeds. Below, a few of these are summarized.

1.1. Driver Acceptance

- How much driver appeal will the early partial AVC services have? Should AHS deployment be contingent on the success of those services?
- Driver motivations for using services with AVC are different than the societal motivations; if AVC evolves based purely on driver demand, will the big societal savings projected for AHS ever be realized?
- Transfer of control from the driver to the vehicle and back, is entirely different for the partial AVC services than for AHS; can one gracefully evolve into the other, or does it need to?
- Tied in with this concern is the issue of who has responsibility for the control of the vehicle under various postulated evolutionary steps.

1.2. Upward Compatible Vehicles

- Evolution assumes that some of the AVC-specific capabilities of partial AVC, such as sensors, actuators, and processors, will be upward compatible into AHS; that is, they will be usable as part of the AHS system. How reasonable is that assumption?
- Some feel that many of the vehicle features needed by AHS will be implemented regardless of the introduction of some of the partial AVC services; to what extent will the vehicle evolution precede and encourage the AVC service evolution?

1.3. Evolution of State DOTs

- Many of the partial AVC services are infrastructure-independent—only a few, such as lane keeping, may require infrastructure cooperation. How much preparation will these early partial AVC services really give to the state DOTs in increasing their ability to construct, operate, and maintain a sophisticated system?

Below we postulate a planned evolution, and address some of the potential issues to be resolved.

2. POTENTIAL EVOLUTIONARY STEPS

One evolutionary scenario assumes that partial AVC services will be introduced to the market before AHS with its full AVC, and that there will be evolutionary steps beyond AHS. Some of the specific services that have been mentioned as part of this postulated evolution are included in Table 1; below is a summary description of the services.

2.1. Adaptive Cruise Control (ACC)

This service is an enhanced cruise control; it incorporates forward-looking sensors so that when a driver turns the service on, the vehicle is continuously kept at a safe following distance from the vehicle in front. Automatic throttle control actuation is used to increase and decrease speed; in addition, some gentle actuation of the brakes may occur.

2.2. Frontal Collision Warning

This service requires logic and sensing that is in addition to the ACC to detect when the closing rate on the vehicle in front is high enough that a collision will occur if there is not additional braking; the driver will be warned, perhaps in stages, to brake to avoid the collision. At any time, the driver would be able to turn the service on or off as desired.

2.3. Frontal Collision Avoidance

This service will detect when the closing rate on the vehicle in front is high enough that a collision will occur if there is not additional braking, and will temporarily actuate the brakes in an attempt to avoid the collision. At any time, the driver would be able to turn the service on or off as desired.

TABLE 1
Potential Levels of Control

	Types of control									
								Transfer system control		
	Throttle	Throttle & Brake	Warning	Steering	Set speed & spacing	Dedicated lane?	Check-in and out?	To system	From system	Respon-sibility
Partial AVC										
ACC, col. warn.	X	X			Driver	No	No	Driver	Driver	Driver
Collision avoid.		X			Driver	No	No	Driver	Driver	Driver
Departure warn.			X		Driver	No	No	Driver	Driver	Driver
Lane keeping				X	Driver	No	No	Driver	Driver	Driver
ACC, lateral & longitudinal warning	X		X		Driver	No	No	Driver	Driver	Driver
Early full AVC										
Mixed auto. & manual		X		X	Driver	No	No	Driver	Driver	Driver
Dedicated to auto. control		X		X	System	Yes	Warning if failed	Driver	Driver	Driver
Full AVC										
AHS		X		X	System	Yes	Yes	System	System	System

2.4. *Side-Looking Collision Warning and Avoidance*

Sensors on the side of the vehicle will detect the presence of a vehicle in the adjacent lane; when the driver indicates intent to change lanes (e.g., actuates the turn indicator), the service may warn the driver or may temporarily actuate the steering to prevent the lane change and avoid the collision; the driver would be able to override the system and assume control at any time.

2.5. *Lane Keeping*

This service assumes that the vehicle is capable of sensing the center of the lane and is able to continuously control the steering to keep the vehicle in the center of the lane. Most of the approaches postulated for this service assume a "cooperative," passive, infrastructure such as accurately painted lane marker stripes, embedded magnetic nails, or radar-reflective stripes. The driver must know when the infrastructure is and is not supporting the service.

2.6. *Combined ACC, Frontal Collision Avoidance, and Lane Keeping*

By simultaneously turning on these services, the driver appears to be close to achieving full AVC; that is, drivers can take both hands off the steering wheel and both feet off the pedals, but this would need to be done cautiously since the vehicle is operating intermixed with manually driven vehicles. The vehicle could be enhanced with additional sensors and logic so that the vehicle is better prepared to accommodate the manually driven vehicles and to better protect the vehicle against the erratic actions of the manually driven vehicles. Again, the driver must know when lateral control is supported by the infrastructure.

2.7. *Combined with Dedicated Lane*

In theory, any of the services described above could be employed in a lane dedicated to equipped vehicles. For purposes of this analysis, the dedicated lane is defined only as a variant of the combined service described in Section 2.6.

In this potential evolutionary step, vehicles with ACC, frontal collision avoidance, and lane keeping services would be allowed to operate on a highway lane dedicated to those vehicles. The vehicles might also have the ability to communicate with roadside beacons that would advise the drivers of recommended basic operating parameters such as safe speed and spacing, or potential problems ahead[(2,6)]; also, it is assumed that the vehicles would have the ability to communicate among themselves to allow closer spacing and more efficient operation. The driver would be able to turn either service on or off as desired.

2.8. *Combined with Dedicated Lanes and Roadside Parameter Control*

This is a combined ACC/frontal collision avoidance/lane keeping service with a dedicated lane, where the operating parameters from the roadside actually set values in the longitudinal control circuitry of the ACC in the vehicles; that is, the driver chooses to turn control over to the system, but when he does, the system controls speed and spacing. The driver would still be able to turn the system on or off as desired.

2.9. *Fully Automated Vehicle Control on a Dedicated Lane (AHS)*

There is a basic operating difference between AHS and the service described above (combined partial AVC services on a dedicated lane); the driver is not able to turn the service off at will. The system assumes control of the vehicle after the driver has indicated a desire to enter the system and the vehicle and driver are accepted by the system. Once the system has assumed control, the system will move the vehicle into the dedicated lane and will retain control until the vehicle reaches the driver's desired exit. There, the system will move the vehicle into an exit area and verify that the driver is ready to assume control again; only then will the driver be allowed to resume control. This means that, with a few, infrequent exceptions (e.g., determined system intruder, or a total vehicle control failure where all vehicles are brought to a full stop), all vehicles on AHS will be system controlled at all times.

2.10. *Fully Autonomous Operation*

This more futuristic service would extend full AVC beyond the dedicated lanes for operation among manually driven vehicles. Drivers presumably would be able to enter their destination in their driveway, and the vehicle would take them to it under fully automated control; the driver could request to take control of the vehicle, but the vehicle would decide when the transfer of control is safe. Research and development work at the Robotics Institute at Carnegie Mellon University, as well as in the U. S. Army's Unmanned Ground Vehicle program, indicate that this vision may someday be technically and operationally feasible.

3. DRIVER ACCEPTANCE OF AVC SERVICES

Below, some of the advantages and disadvantages for each of the potential evolutionary steps mentioned in Section 2.

3.1. *ACC*

This service provides a nice convenience for users of cruise control. By tracking the vehicle in front up to the speed set by the driver, it keeps the driver from having to constantly reset the cruise control speed. The service may also allow the driver to adjust the following distance from the vehicle in front; in this way, drivers who prefer a large headway as well as drivers who prefer shorter headways can feel comfortable. The driver must remain fully alert.

3.2. Frontal Collision Warning

This service should provide an extra measure of safety to the ACC service. If the vehicle in front begins to brake hard, the system is to warn the driver, perhaps more quickly than the driver would have been able to judge the situation. It may also help detect some objects in the roadway and, again, give a warning to avoid a potential collision.

One of the concerns with this service is its level of sensitivity; if it frequently warns the driver of an impending collision when there is no serious problem, the driver will ignore the service and turn it off. But if the system does not warn the driver in a critical situation—and the driver expects the warning—then a collision could result. Another ramification is that if the driver travels for 2 years without having a system warning, when the warning does occur, will it be recognized?

The driver must maintain full awareness of the driving situation with this service, and must be capable of reacting immediately.

3.3. Frontal Collision Avoidance

This service goes a major step beyond the warning service by applying the brakes—perhaps hard—to avoid what the system believes to be an impending collision. It would probably be coupled with the warning service; that is, the system would warn the driver, but when the closing distance to the object ahead becomes critical, the system applies the brakes to avoid the collision or to lessen its impact.

The concerns about sensitivity expressed above for frontal collision warning also apply to this service. Frequent, or even occasional, application of the brakes when there is not a problem would not be tolerated for long by drivers—and if the brakes were not applied in time or sufficiently, then there might be a collision and the product supplier might be viewed as being partially responsible. Design issues include dealing with a multitude of situations caused by the other vehicles (i.e., those under manual control). For example, if a vehicle from the next lane pulls in front of a vehicle with the collision avoidance system (the "close cutoff" that has unfortunately become a typical occurrence on today's highways), can the system be designed to not cause a crash by immediately hard-braking? Even if the brakes are applied properly in all cases, if the driver of the vehicle behind is following too closely or not paying attention, and the braking vehicle is struck from the rear, will there be any attempt to blame the product manufacturer? As with any service in which the driver is able to turn the system on and off, the driver must maintain full awareness of the driving situation, and view the partial AVC services as driving aids.

Some feel that the problems of frontal collision avoidance systems warrant substantially more research before they become available in the market, particularly in the litigation-minded United States.

3.4. Side-Looking Collision Warning and Avoidance

Side-looking sensors detect vehicles in the adjacent lane so that when a driver indicates a lane change (e.g., puts on his turn indicator), the system warns the driver if there are any vehicles in that lane. There may be problems with sensitivity (false warnings, missed warnings), but there may be less of a problem than with frontal-looking sensor systems because of the relative closeness of the detected objects. The driver should view the system as supplemental only; that is, the driver should still check for the presence of vehicles since even if the system works perfectly, it may not detect all situations (e.g., a vehicle that is planning to move into the same lane from a different lane, or a vehicle that is rapidly gaining in that lane and is not yet detected by the sensors). A concern is that some drivers might come to rely solely on the warning system, and make driving errors because of that.

Temporary actuation of the steering wheel to prevent a side collision has been suggested. This could be a "stiffening" of the steering wheel movement that the driver could choose to override. The actuation could be coupled with the warning system so that if the driver does not listen to the warning, the system temporarily prevents the driver from moving into the collision situation. The driver must be able to immediately override the actuation in critical situations; for example, if the driver is making the lane change because there is a potential collision ahead, the system must not prevent the driver from making the lane change.

It has been postulated that if a vehicle has both frontal and side collision avoidance, then the collision avoidance maneuver could include a combination of temporary steering, braking, and throttle actuations to avoid a collision. The general assessment of this capability by the PSA researchers[5,7] was that the ability to successfully complete these complex maneuvers on a highway where the operation is intermixed with manual drivers would be very challenging and will require significant research. Successful return of control to the driver in these situations could also be a challenge.

3.5. Lane Keeping

This system might appeal to drivers who are covering long distances because it assumes steering of the vehicle to keep it in its lane of travel. This would appear to be a major advantage for long-haul truckers, for example. A disadvantage is that the driver could be lulled into inattentiveness and a false sense of safety. For this reason, one postulated lane keeping service would be as a warning service; it would warn the driver when his or her steering becomes too inaccurate and the vehicle is beginning to drift out of the lane and/or run off the road.

In the PSA studies, the most enthusiasm for lane keeping was in conjunction with transit bus operation.[2,8] The accurate lane keeping in a bus-only lane would allow buses to be operated in a manner similar to light rail, including the ability to approach entry-height loading platforms at tolerances close enough that wheelchair access would be possible.

3.6. Combined ACC, Frontal Collision Avoidance, and Lane Keeping

The driver should realize the levels of increased safety that would derive from both frontal collision avoidance and lane keeping separately. As noted above, the level of increased safety is somewhat limited because the road is shared with manual drivers; thus, it would not appear that there would be any additional, synergistic safety advantages above those mentioned above. However, if additional sensors and logic were added to help the vehicle accommodate the manually driven vehicles, then some incremental safety might be possible depending on the features added. Few if any reductions in trip time could be realized if most vehicles on the road are manually driven. However, once equipped vehicles are in the majority, then trip times may be shortened, particularly if the equipped vehicles can communicate.

In theory, drivers should be able to remove their hands from the steering wheel, and their feet from the pedals, but this could be dangerous because of the manual drivers sharing the road; particularly in dense traffic situations, the hands will need to hover near the wheel and the feet near the pedals—the driver will need "edge-of-the-seat" awareness of the traffic situation. Also, the driver's alertness is critical for obstacle detection.

Since the driver chooses to turn the systems on and off, the driver is responsible and must be prepared to assume control to avoid potentially dangerous situations when the system does not respond. As with lane keeping, maintaining driver awareness could be a problem, particularly on long trips where a false sense of safety might be assumed and/or where the driver's attention to the driving details would naturally tend to wander. As postulated here, the service could have significant safety implications. Additional research in the area is needed.

3.7. Combined with Dedicated Lane

Vehicle penetration with this combined service would build as part of the drivers' desire to operate on a dedicated lane with the ACC, collision avoidance, and lane keeping services. The drivers would be "in the loop"; that is, they would still be able to turn any one or all of these services on and off at their will. Driving responsibility would be with the drivers; they would be expected to maintain full awareness and control of the driving situation.

In order to increase the safety of the dedicated lanes and better ensure their dedicated use, barriers of some kind would probably be used. The safety of vehicle operation should be higher than operation with mixed traffic as long as everyone on the lane cooperates and does not drive manually. Throughput on the dedicated lane might be higher and trip times might be reduced, assuming that all of the drivers cooperate, particularly if the vehicles can communicate. Drivers who choose to assume manual control on the dedicated lane would decrease both the level of safety and level of efficiency of the other vehicles operating on the lane. Cooperation could be "encouraged" if driving on the dedicated lane were a privilege that could be withdrawn, or if fines could be imposed for manual operation. The feasibility of this kind of "encouragement" would need to be investigated.

Since any individual driver can choose to turn any one (or all) of the services off at any time, the operating parameters might need to be set assuming this will happen. This would mean that trip time reductions would be minimal. For example, if the posted speed limit on the highway lanes with manual drivers is 100 km/h, then the recommended speed on the lane dedicated to partial AVC vehicles might also be 100 km/h. If the dedicated lane is a single lane (as may be the case initially) with limited access, then drivers who choose to set their speed at 100 km/h (or less) will set the speed for the other drivers.

An additional problem arises regarding the recommended speed; if weather and/or road surface conditions degrade, will the highway operator be required to adjust the recommended speed based on the worsening conditions? If the highway operator does not adjust the recommended speed, is the operator possibly liable for resulting incidents? It would seem that the highway operator can choose to (1) give no recommended speed, other than the posted maximum speed, or (2) develop a system of sensors, processors, monitors, and traffic management personnel to ensure that the recommended speed for any given section of the highway is timely and conservative.

It is assumed that in this service the highway operator would not provide recommended spacing between the vehicles. The recommended spacing would be derived from the speed and the braking capabilities of the vehicles on the dedicated lane. A vehicle's braking capability can vary substantially based on, among other things, highway surface conditions, tire conditions, each vehicle's designed braking capacity, and the weight being carried in the vehicle. Unless the highway operator had knowledge about each individual vehicle (something intervehicle communications could provide to adjacent vehicles) it would need to give a very conservative recommended spacing based on the worst braking performance of the vehicles likely to enter the dedicated lane. This would reduce throughput of the dedicated lane to far below that of the manual lanes. Vehicle breakdowns in a limited access lane would create major problems. As the service is postulated here, there is no vehicle check-in; thus, there is no way of ensuring that most or all vehicles entering the lane are operationally sound. The dedicated lane would, therefore, need a shoulder or breakdown lane.

It is argued that this dedicated lane operation might be an interim step into AHS operation. It would seem that drivers would become accustomed to this "voluntary AHS"—certainly daily commuters would be likely to fall into a pattern; drivers would soon realize the value of cooperating with the other drivers in the lane. The drivers should become accustomed to the higher level of control by the system; and to the extent that everyone cooperates, the drivers could begin to enjoy some safety and, assuming communicating vehicles, reduced trip time benefits. But the driver would still need to remain alert for obstacles, malfunctions, or other unexpected events. The extent of the benefits and the feasibility of expecting the driver to remain alert have yet to be determined. One concern is how the drivers of non-equipped vehicles would react to having a lane dedicated to these specially equipped vehicles. Initially, the lane might need to be dedicated to transit and/or HOV use.

3.8. Combined with Dedicated Lanes and Roadside Parameter Control

In this variation of the previous combination of services, operating parameters from the roadside would actually set values in the longitudinal control circuitry of the ACC in the vehicles. The system controls speed and spacing; however, the driver would still be able to turn the system on or off at will.

The highway operator would be required to adjust the recommended speed and spacing based on the weather, traffic conditions, and knowledge of the vehicles on the roadway. As discussed above, this means that the highway operator must develop a system of sensors, processors, monitors, and traffic management personnel to ensure that the recommended speed for any given section of the highway is timely and conservative. It also means that the operator must develop a system for determining safe spacing among the vehicles. As explained above, to avoid having to recommend spacing based on the worst possible case of braking performance, the system operator would need to know the braking capabilities of each individual vehicle. If the vehicles were equipped to communicate among themselves, then each vehicle would know the approximate braking capability of the vehicle in front of it; in this way, it could adjust its spacing from the leading vehicle accordingly. This seems simpler than having infrastructure involvement.

As with the service described above, a shoulder or vehicle breakdown lane would be needed to avoid operational problems in cases of vehicle malfunctions.

Would roadside insertion of operating speed into each vehicle's ACC and lane keeping control circuitry improve the lane's operating efficiency? Efficiency should increase if the vehicles communicate and if the dedicated lane is allowed to operate at a speed higher than the manual lanes. This "waiver" for partially automated operation would be justified on the presumption that virtually all of the vehicles in the dedicated lane would be under automated control; therefore, the operator would need to impose tight restrictions regarding access to the dedicated lane (e.g., check-in) and operation within the lane. If automated operation is required within the lane then, it becomes a "mandated AHS" lane; that is, the requirement for automated operation is imposed through policies and fines rather than through system control, as in AHS. The level of increased efficiency and safety would depend on the extent to which the drivers follow the mandates; if all drivers follow the mandates, then the safety and efficiency of the lane would improve; but the improvements would probably not approach the projected improvements for AHS. There are two reasons for this: (1) as postulated, the drivers would still be responsible for driving onto and off of the dedicated lane; this means that drivers will be in control during the most critical maneuvers and would decrease system efficiency and safety; and (2) if drivers can assume control at any time, some of them will; again, this will subject the system to greater inefficiencies and greater safety risks.

If there is no legislation, the state and regional agencies may have no option but to put the system settings at the legal limit for the manual lanes. Traffic would move at that speed, or less if the system is slowed for adverse weather or traffic conditions. Meanwhile, on the manual lanes, if there is more than one lane, many drivers may choose to move at a speed somewhat greater than the speed limit. Only if the manual lanes were congested and the dedicated partial AVC lane was not could there definitely be a reduced travel time on the dedicated lane. In this case, it could reasonably be argued that the dedicated lanes would be safer; but would there be sufficient incentives for drivers to use the dedicated lane, possibly at an increased cost?

Responsibility for vehicle control while on the dedicated lane becomes less clear in this type of operation. The drivers would still have responsibility for the overall control of the vehicle, including detection of, and response to, obstacles and unexpected events. However, when the driver turns control over to the "system," then the roadway operator must assume responsibility for some of the operation. Responsibility for the on-vehicle portion of the system would need to be studied (see Section 3.2).

3.9. *Fully Automated Vehicle Control on a Dedicated Lane (AHS)*

Since the driver is "not in the loop" (that is, the driver cannot directly take control away from the system), responsibility for all of the driving functions must be reliably and safely assumed by the system. The system assumes responsibility for detecting and avoiding obstacles and unexpected events. In addition, the system's capability may increase to include traffic flow management for increased efficiency.

The driver should be able to sit back and relax, although it has been noted that the driver may want a "supervisory" role; that is, have the ability to enter data into the system that will affect its operation. At the least, the driver should be able to indicate a desire to exit, or stop in the breakdown lane (e.g., someone is carsick), or make an emergency stop—perhaps with an emergency stop button similar to those on elevators (e.g., the vehicle is on fire). The system should respond by exiting or even executing an emergency stop if necessary. The driver might also be encouraged to act as an observer; for example, drivers may see potential obstacles such as deer preparing to enter the roadway, before the system's sensors. Perhaps the driver could also enter his concerns about following a particular vehicle (exhaust stinks), or even enter a desire to increase the spacing between his vehicle and the vehicle in front. The system would need to build an "electronic cocoon" around the driver to allow him to interact with the system but not allow the driver to exert damaging control.

The system would verify each vehicle's AHS operating status before allowing it onto the dedicated lane; vehicle's not passing the test, including non-AHS vehicles, would be denied entry. The system would continue to monitor the condition of each vehicle as it is moved through the system as well as the condition of the roadside equipment to help ensure that no breakdowns on the AHS lane occur. Nevertheless, regardless of the extent of checking, there will be occasional vehicle malfunctions; for this reason, an AHS lane will need an occasional or continuous breakdown lane. This could serve as a second operating lane in normal operating conditions.

Obstacle avoidance will need to be accomplished using a combination of techniques; (1) obstacle exclusion from the dedicated lanes; (2) obstacle detection by both the vehicles and the infrastructure; and (3) communications about obstacles throughout the system.

In this approach, primary responsibility for operation of the vehicle on the dedicated lane will lie with the "system," both roadside and on board the vehicle. However, part of the responsibility may lie with the driver if he or she is responsible for some aspect of system verification.

With full AVC on a dedicated lane, most of the major causes of system unpredictability have been removed. A vehicle's speed on any section of an AHS lane will be the lowest speed derived from (1) the condition of the roadway, (2) the capabilities of the vehicles on the lane, or (3) the capability of the AHS system, including the humans in the operations center. Driver capability will not be a deciding factor. It is for this reason that the AHS will have dramatically improved safety and efficiency. And because the system is responsible for the operation, the driver will be able to relax, even in heavy traffic. The driver will be able to turn the travel time into additional leisure time (read a book or play a game) or work time (write a report). A new market for products to use during the in-vehicle travel time will develop.

3.10. Autonomous Control

This service postulates fully automated vehicle control in most if not all traffic conditions. There are several different postulated concepts for this futuristic service, all of which are preliminary. For this reason, a discussion of the service and its advantages and disadvantages is not given.

4. UPWARD COMPATIBILITY OF VEHICLES

The extent to which transition to AHS can be aided by, and actually proceed through, the levels of control described above is being addressed by the National AHS Consortium. The introduction of the partial AVC services, while potentially useful to AHS, will be driven by market demands; that is, they will be introduced as each individual vehicle manufacturer determines that there is sufficient demand to warrant the design, development, and marketing of a product.

There is no reason to believe that the initial introduction of these services by the various manufacturers will be standardized in any way. There could be some commonality between a manufacturer's components used for sensing and controlling vehicles for ACC, frontal collision avoidance, and/or lane keeping, and its components used for AHS. The extent of this commonality, however, will be driven by (1) the early knowledge that AHS is to be introduced in the United States with certain specific requirements and/or (2) the extent to which the AHS performance characteristics differ from the partial AVC services. Once the AHS control and sensing requirements are specified, then vehicle and vehicle electronics manufacturers may choose to produce sensors and control mechanisms for partial AVC services that are also capable of being used for AHS if economically justified. Alternatively, once manufacturers begin production of AHS systems, they may choose to design a modular system that is also capable of operating as ACC or lane keeping when not on a dedicated AHS lane.

Once the AHS service is sufficiently defined, it is possible that industry standards could be developed for some partial AVC services to help ensure consistent, safe performance of any upward compatibility of the human interface and operation with AHS.

Below, some of the sensing and control components in which there may be commonality are described.

4.1. Electronically Actuated Throttle, Brakes, and Steering

Most plans for AVC assume that computers will actuate these vehicle control functions. Today, electronically actuated throttles and steering units are available on some vehicles, and prototype electronically actuated braking units exist. The need for faster AHS reaction times (because of potentially higher speeds and closer headways) may work against commonality unless the manufacturers specifically have AHS in mind as they design the actuators. Attention will need to be paid to the detailed characteristics of the control algorithms and communications links between the control processors and the actuators to ensure compatibility with AHS.

4.2. Sensors

The sensors used for detecting vehicles or objects in front of the vehicle in an ACC system may possibly be used for AHS sensing as well, but this will depend on the eventual AHS operating characteristics. The AHS may be designed to operate at high speeds, at close headways, and/or at higher levels of reliability than ACC or lane keeping systems. Unless the ACC sensors are designed with AHS in mind, they may not be usable on AHS, or they may be supplemented with other sensors to enrich the signals and raise reliability. This may not be as true for the lane center sensors used for the lane keeping service.

4.3. Processors

The AHS control logic contained in an on-board processor will be significantly more complex than the logic required for ACC or lane keeping (one exception: a system designed for simultaneous ACC and lane keeping operation when intermixed with manually operated vehicles would be more complex than AHS operation on a dedicated lane). Again, if the ACC and lane keeping services are designed assuming a later addition of AHS logic, then the control logic could possibly be preserved, and a new processor could be added to operate on top of the existing logic. This means that the AHS may be an overlay on top of the ACC, collision avoidance, and lane keeping services that provides an added layer of functionality (including processing and sensing) and could use the partial AVC services as system backup. This kind of modular design would be useful in allowing the AHS retrofit of existing vehicles. Because of the expected continued dramatic increase in processing power per dollar, processing may not be a major AHS cost.

5. DRIVER MOTIVES AND SOCIETAL MOTIVES

In the PSA studies, some of the researchers[1,7] cautioned that inextricably tying AHS to some of the partial AVC services might be risky because those services may not have a broad appeal to drivers—certainly not the level of appeal that an AHS is expected to have. It was agreed, however, that even though partial AVC services could not match the benefits of an AHS, users would benefit from the convenience of ACC and/or lane keeping; and collision avoidance services should provide some increased in trip safety. It was also pointed out that a combination of these services could possibly enhance the level of benefits that could be realized by the drivers. The drivers will be attracted to the partial AVC services because of the convenience, the increased safety, and possibly some reduced travel time.

Even though it is expected that AHS benefits will eventually result in high appeal to drivers, some will not perceive how AHS can benefit them until they see an AHS lane efficiently moving traffic during periods of congestion. Even then, some drivers will wonder why automated vehicles cannot be just as effective on nondedicated roadways. Additionally, many individual drivers may not perceive the increased safety of AHS (just as many drivers do not perceive freeways as being substantially safer than local roads); they will recognize it when the increased safety is reflected in lower insurance costs. Until the drivers can directly see these benefits, they will need encouragement to convert. Incentives for drivers to initially upgrade their vehicles would speed conversion.[1,3]

The incentives for initial transitions to AHS will come from the state and regional transportation agencies that must meet their community transportation needs. The major throughput and safety gains that come with a dedicated AHS lane will more directly benefit the community and society as a whole. It is the transportation planners and community leaders who will project the impact of AHS on congestion and reduction in highway deaths and property losses, and who will be able to justify an AHS operation because of the major benefits. It is believed that once the AHS is in operation, then the highway users (e.g., truckers, commuters, vanpool riders, transit riders, vacationers, shoppers) will be able to see the benefits of reduced and dependable travel time, and greater user comfort. AHS conversions of existing highway lanes will be more easily accomplished as the user demand for AHS grows.

6. DEFINITION OF TERMS

- *Automated vehicle control* (AVC)—this is provided when a computer-based system temporarily assumes control of a vehicle's brakes, throttle, or steering.

- *Minimum AHS penetration*—is reached when, after conversion of one of the existing highway lanes to AHS, the number of AHS lane users equals the number of manual users of the lane before conversion.
- *Partial AVC services*—ITS services that incorporate automated vehicle control of brakes and/or steering and/or throttle, but not all three.
- *Planned evolution*—provide focus and timely evolution of partial AVC products so that they become stepping stones toward full AVC through guidance, spin-off technology, and standards.

SUMMARY

Partial automated vehicle control products will help build a base from which the transition to a fully automated AHS can be made. This base may include vehicles on the highways that are equipped with sensors and electronically actuated controls. The base will also include a segment of the driving population that has experience with some level of automated control.

The initial step into AHS will probably be made by a regional transportation agency which has determined that an AHS provides the needed increases in safety and capacity for the least cost.

REFERENCES

1. L. BONDERSON, Overview Report, Precursor Systems Analysis Studies, Delco/Hughes Electronics, Dec. 1994.
2. J. ELIAS, Overview Report, Precursor Systems Analysis Studies, Calspan Corp., Dec. 1994.
3. J. PITTENGER, Final Report, Precursor Systems Analysis Studies, Battelle, Dec. 1994.
4. P. IOANNOU, Vehicle Analysis Report, Precursor Systems Analysis Studies, Raytheon Corp., Dec. 1994.
5. W. STEVENS, *et al.*, Summary and Assessment of Findings from the Precursor Analyses of Automated Highway Systems, MITRE, July 1995.
6. D. BLANCETT, Final Report, Precursor Systems Analysis Studies, Dec. 1994.
7. S. SHLADOVER, Final Report, Precursor Systems Analysis Studies, Dec. 1994.
8. M. MARTIN, Final Report, Precursor Systems Analysis Studies, BDM, Dec. 1994.
9. R. BARBER, Final Report, Precursor Systems Analysis Studies, Rockwell, Dec. 1994.

CHAPTER 8

Spacing and Capacity Evaluations for Different AHS Concepts

Alexander Kanaris, Petros Ioannou, and Fu-Sheng Ho

1. INTRODUCTION

Urban highways in many major cities are congested and need additional capacity. Historically, capacity has been added by building additional lanes and new highways. Scarcity of land and escalating construction costs make it increasingly difficult to add capacity this way. One possible way to improve capacity is to use current highways more efficiently. The concept of Automated Highway Systems (AHS) was introduced to improve the capacity of the current transportation systems by using automation and intelligence.

Highway capacity depends on two variables: the velocity of the vehicles and the distance between them. Clearly, the higher the velocity of the vehicles, the higher the number of vehicles per lane per hour will be. But the vehicles need to maintain a certain amount of "safety distance" between them, to accommodate for the case that the flow of vehicles has to be slowed down or stopped, by applying the brakes. The moment that each vehicle starts applying its brakes typically involves a couple of seconds of delay in relation to the onset of braking of

Alexander Kanaris, Petros Ioannou, and Fu-Sheng Ho • Center for Advanced Transportation Technologies, University of Southern California, Los Angeles, California 90089-2562.

Automated Highway Systems, edited by Ioannou. Plenum Press, New York, 1997.

the vehicle in front, reflecting the fact that the human drivers need some time to process the information they perceive,[1] plus an additional time delay to react and a delay for the mechanical and hydraulic systems of the vehicle to respond. During this time, the vehicle continues moving forward at practically the same speed and if there is not sufficient space between the leading and the following vehicle at the moment the leading vehicle applies the brakes and begins to decelerate, a collision would be inevitable. Even if the follower begins to apply its brakes at exactly the same time as the leader, the deceleration of the leading and the following vehicle may not match[2,3] and this generates the need for additional intervehicle distance during the cruising stage in order to accommodate for the difference in braking performance.

Some of the earlier studies[5] made the assumption that the leading and trailing vehicles have the same deceleration characteristics. In more recent studies[20] these assumptions were removed.

Heavy vehicles travel a significantly longer distance from the moment they apply their brakes until they come to a complete stop. This has to be accommodated for by allowing a significantly larger intervehicle spacing. On the other hand, when a light vehicle follows a heavy vehicle, the braking distance is not the limiting factor because typically the light vehicle will be able to come to a stop in a much shorter time and distance. In this case, the limiting factors are the initial conditions and the total delay between the time that the leader starts decelerating and the time that the follower starts decelerating at the maximum possible deceleration.

The delay in detecting and in reacting to the leading vehicle's deceleration can be reduced significantly, by taking the human driver out of the "control loop."[4–7] With advances in technology and vehicle electronics, systems that were previously considered impossible to implement or too costly are becoming feasible and available. One such system is a functional extension of the classic cruise control.[5] The cruise control which is widely available on luxury cars today, is a controller that controls a throttle actuator in order to maintain constant vehicle speed. The next step in functionality is a controller that uses a sensor to measure the relative distance and the relative speed to any vehicle ahead and controls a throttle and a brake actuator in order to follow at the same speed and maintain a fixed relative distance.[3,8,9] Such vehicles can follow each other in the same lane automatically by relying on their own sensors and controls. Vehicles that rely on their own sensors, controls, and intelligence to operate in a highway environment are referred to as autonomous vehicles.

Advances in communications made it possible for vehicles to communicate with each other exchanging information about braking intentions and capabilities, acceleration, lane changing, and so forth. The infrastructure may also support vehicle following and maneuvers by providing desired speed and spacing commands in addition to traveler information. This distribution of intelligence gives rise to the operating concept referred to as infrastructure-supported free agent.

When the infrastructure becomes actively involved by sending braking commands for emergency stops and lane changing maneuvers, we have an operating concept referred to as infrastructure-managed free agent.

Another concept is to organize vehicles in platoons of a certain size where the intraplatoon spacing is very small and the interplatoon spacing could be larger for safety purposes. In this case each platoon appears to the infrastructure as a single unit and therefore can be managed more efficiently. Each platoon is now responsible for the control of its vehicles.

If the intervehicle separation becomes very small, the laws of physics dictate that collisions between vehicles may be inevitable. In the interest of safety and avoidance of vehicle damage it will be of paramount importance that the energy dissipated during the collision be constrained. Since safety cannot be compromised for the sake of capacity, it becomes a serious constraint in most AHS design decisions.[(6)]

In this study we consider a family of AHS platooning concepts. For each concept we calculate the minimum intervehicle spacing that would be required to guarantee either collision-free following or bounded energy dissipation in the event of a collision. We will be assuming that if the collisions are relatively rare events, are always very minor, and cause no permanent damage to the vehicles, the public might be willing to accept the fact that collisions may happen. Allowing for collisions to happen can reduce the minimum headway requirements for a platoon-based AHS architecture.

Finally, in a slightly different operational concept, a high level of synchronization is introduced where each vehicle is allocated a slot in time and space. The infrastructure manages the slot distribution by issuing the appropriate commands for each vehicle.

The degree of infrastructure involvement and distribution of intelligence lead to different operational concepts and architectures for AHS. The purpose of this section is to study the minimum safety spacing (MSS) for a number of different AHS concepts and architectures and to obtain capacity estimates.

2. SAFE INTERVEHICLE SPACING ANALYSIS

Intervehicle spacing during vehicle following is a very critical parameter of highway traffic. Insufficient spacing is usually the cause of rear-end collisions. In principle, the possibility of having a rear-end collision can be reduced by increasing the intervehicle spacing. However, the spacing that guarantees collision-free vehicle following can be characterized only when the braking scenario is known and well defined.

A braking scenario, which describes exactly how the vehicles brake, is usually specified by the deceleration profiles of the vehicles as a function of time. For each scenario there is a minimum spacing that must be maintained during steady state traffic flow, if collision-free vehicle following must be guaranteed. In this section we develop the basic equations that can be used to calculate the minimum spacing for

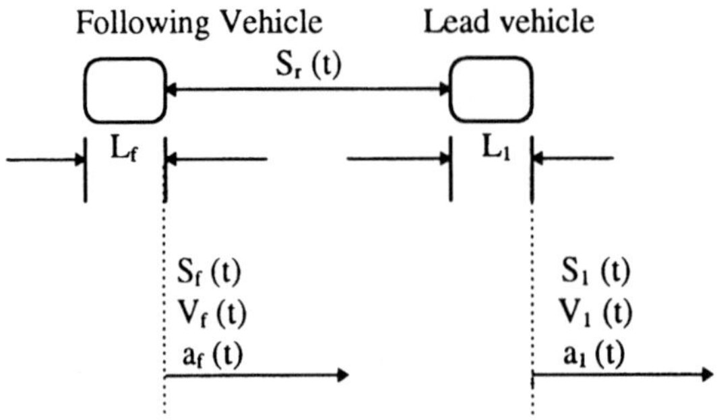

FIGURE 1. Vehicle following.

collision-free vehicle following, given the deceleration response information for both the leading and the following vehicle.

2.1. Minimum Spacing for Collision Avoidance

Consider two vehicles following each other, as shown in Figure 1. Assume that at $t = 0$ the leading vehicle begins to brake according to the deceleration profile defined by $a_l(t)$ and the following vehicle brakes according to the deceleration profile defined by $a_f(t)$. Assume that L_l and L_f are the lengths of the leading and following vehicles, respectively. At $t = 0$ the leading vehicle has a velocity $V_l(0) = V_{l0}$ and a position $S_l(0) = S_{l0}$ and the following vehicle has a velocity $V_f(0) = V_{f0}$ and a position $S_f(0) = S_{f0}$. If the spacing between the two vehicles at $t = 0$, $S_r(0) = S_{l0} - S_{f0} - L_l$, is large enough, then there would be no collision during braking maneuvers.

For a given braking scenario we would like to calculate the minimum value of the initial intervehicle spacing $S_r(0)$ for which there will be no collision. We refer to this value as the minimum safety spacing.(2)

The spacing between the two vehicles measured from the front of the following vehicle to the rear of the lead vehicle is given by

$$S_r(t) = S_l(t) - L_l - S_f(t) \tag{1}$$

where

$$S_l(t) = S_l(0) + \int_0^t V_l(\tau)\, d\tau \tag{2}$$

$$S_f(t) = S_f(0) + \int_0^t V_f(\tau)\, d\tau \tag{3}$$

and

$$V_l(t) = V_l(0) + \int_0^t a_l(\tau)\, d\tau \tag{4}$$

$$V_f(t) = V_f(0) + \int_0^t a_l(\tau)\, d\tau \tag{5}$$

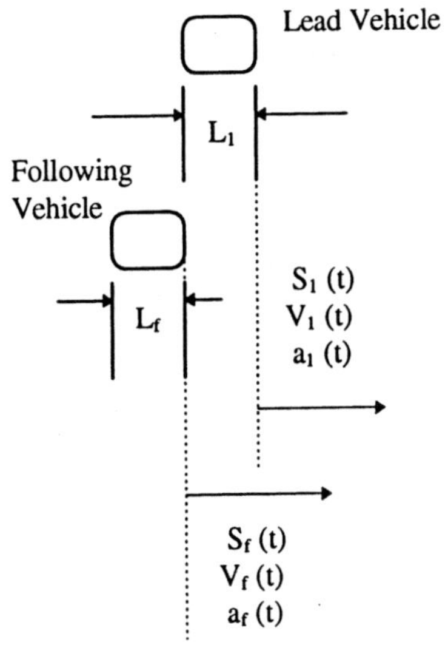

FIGURE 2. Hypothetical vehicle motion.

If the decelerations $a_l(t)$ and $a_f(t)$ and initial positions and velocities are specified, the MSS can be calculated as follows:

Assume that the two vehicles travel in the same direction but in two separate lanes. The position of the vehicles at time $t = 0$ is shown in Figure 2.

Let t_s be the stopping time of the following vehicle. Then

$$V_f(0)+\int_0^{t_s} a_f(\tau)d(\tau)=0 \tag{6}$$

$$S_f(t)=S_f(0)+\int_0^{t} V_f(\tau)d(\tau), \qquad \forall t \le t_s \tag{7}$$

and

$$S_f(t)=S_f(t_s), \qquad \forall t > t_s \tag{8}$$

The position of the leading vehicle at each time t is given by

$$S_l(t)=S_l(0)+\int_0^{t} V_l(\tau)d(\tau), \qquad \forall t \le t_s \tag{9}$$

The relative spacing at each time t is given by

$$S_r(t)=S_l(t)-L_l-S_f(t) \tag{10}$$

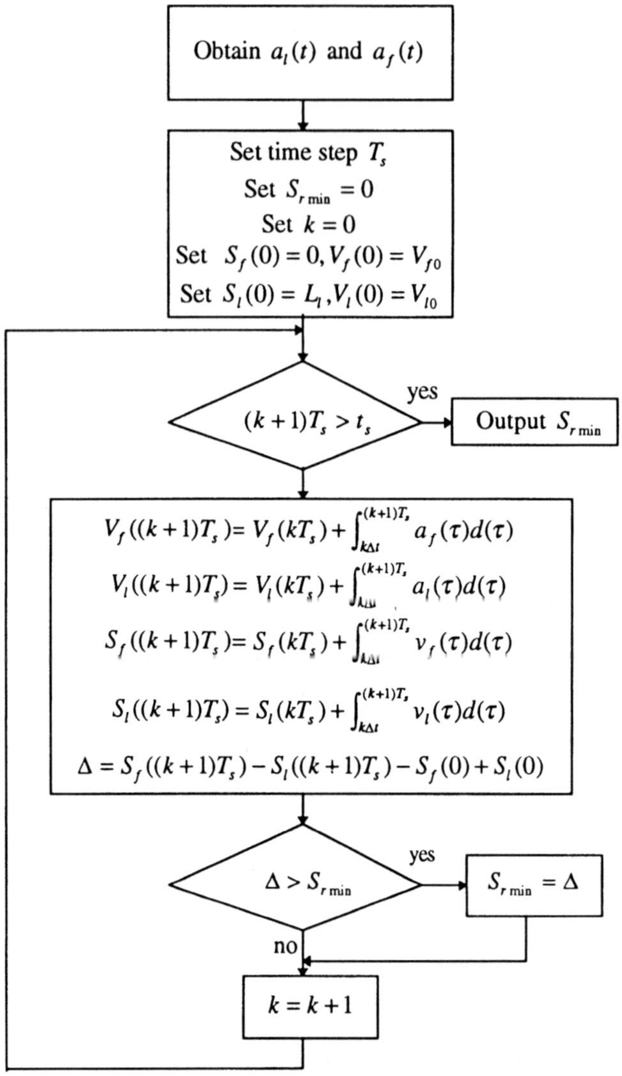

FIGURE 3. Flowchart for MSS calculation.

If both the leading and following vehicle are in the same lane, then $S_r(t) > 0$ for all $t \in (0, t_s]$ will imply no collision, whereas $S_r(t) < 0$ at some $t = t_c \in (0, t_s]$ will imply collision.

The MSS value denoted by $S_{\min}$ is given as $S_{\min} = -\min[S_r(t), 0]\ \forall\ t \in (0, t_s]$.

In other words, $S_{\min}$ is equal to the maximum distance by which the following vehicle would overtake the leading vehicle at any time t in the interval $[0, t_s]$ in the scenario shown in Figure 2.

Based on the above analysis, we adopt a numerical method to calculate S_{min}. Assume that the following vehicle brakes and it does so by following the given deceleration profile, and comes to a full stop at $t = t_s$. We divide the interval $[0, t_s]$ into small time steps and consider the time instants $t = 0, T_s, 2T_s, ..., kT_s$, where T_s is the length of the time step and k is an integer with the property $kT_s \leq (k + 1)T_s$. The method of calculation of S_{min} is shown in the flowchart of Figure 3.

2.2. *Minimum Spacing for Low-Impact Collisions*

The relative velocity at impact is the most significant factor determining the severity of the collision and the extent of property damage and the possibility of passenger injury.[10] In vehicle following situations, the relative velocity between the leader and the follower is determined by differences in deceleration rate and by the time differential of the onset of braking. Assuming the leader and the follower had been traveling at approximately the same speed, the intervehicle spacing becomes the critical parameter. In principle, the possibility of having a rear-end collision can be reduced by increasing the intervehicle spacing. However, the spacing that theoretically guarantees collision-free vehicle following can be characterized only when the braking scenario is known and well defined and the parameters are not subject to variations. Furthermore, the amount of spacing required in order to provide a guarantee at a 100% confidence level that collisions will never happen, might be surprisingly large, much larger than the spacing we are used to seeing with manual driving. Hence, it might be very hard or impossible to guarantee a collision-free environment. The dynamics and effects of intervehicle collisions should therefore be analyzed and understood.

Accepting the fact that intervehicle collisions may occasionally happen, requires that we carefully study the effects of such collisions to the vehicles involved. The conservation of momentum theorem states that after the collision of two objects the vector sum of the momentum before the collision will be equal to the vector sum of the momentum after the collision. If the two objects have mass m_1 and m_2 respectively and velocities u_1 and u_2 respectively before the collision, they will have velocities v_1 and v_2 respectively after the collision, such that:

$$m_1u_1 - m_2u_2 = m_1v_1 + m_2v_2 \tag{11}$$

The collision coefficient cc has been defined to be the scalar:

$$\text{cc} = \frac{v_2 - v_1}{u_1 - u_2} = -\frac{\Delta v}{\Delta u} \tag{12}$$

The collision coefficient is the ratio of the relative velocity at which the two objects separate after a collision over the relative velocity that the two objects approached each other before the collision. When cc = 1 we have what we call "elastic" impact. When cc = 0 we have what we call "plastic" impact. In the former case the two

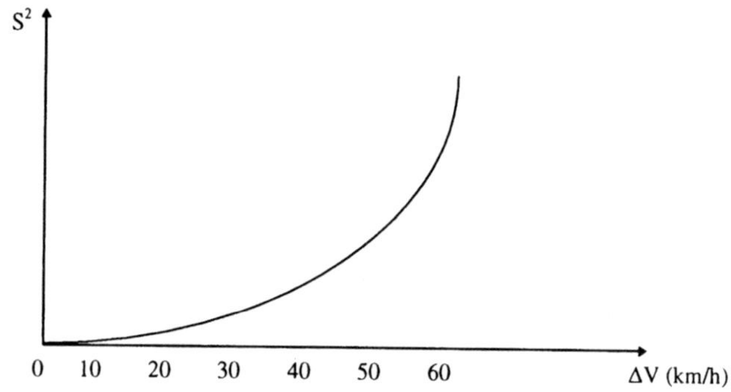

FIGURE 4. The severity (impact energy) versus relative velocity at impact.

objects bounce off each other at a relative velocity equal to their relative velocity before the impact. In the latter case the two objects essentially "stick" to each other and keep moving as one. Real-world objects rarely behave like any of these extremes, so the collision coefficient will be assuming values between 0 and 1.

In this section we develop the basic equations that can be used to calculate the minimum spacing for vehicle following, given the deceleration response information for both the leading and the following vehicle parameterized in terms of the value of the collision coefficient.

2.3. *Bounded Collision Energy Analysis*

In recent literature, Glimm and Fenton[11] expressed the accident severity index (S^2) for a platoon of (n + 1) vehicles that collide as

$$S^2 = \sum_{i=1}^{n} \Delta V_{i+1,i}^2(t_{ci})$$

where $\Delta V_{i+1,i}^2(t_{ci})$ denotes the relative speed at impact between vehicle (i) and (I + 1), at time t_{ci}, the moment of the collision. The nonlinear relation between relative velocity at collision and impact energy can be seen in Figure 4.

When only two vehicles are involved, the severity index is simply

$$S^2 = \Delta V^2(t_c) = [V_f(t_c) - V_l(t_c)]^2$$

where t_c is the time of the collision.

Consider two vehicles following each other, as shown in Figure 1. Assume that at $t = 0$ the leading vehicle begins to brake according to the deceleration profile defined by $a_l(t)$ and the following vehicle brakes according to the deceleration profile defined by $a_f(t)$. Assume that L_l and L_f are the lengths of the leading and following vehicles, respectively. At $t = 0$ the leading vehicle has a velocity $V_l(0)$

$= V_{l0}$ and a position $S_l(0) = S_{l0}$ and the following vehicle has a velocity $V_f(0) = V_{f0}$ and a position $S_f(0) = S_{f0}$. We want to determine the necessary spacing between the two vehicles at $t = 0$, $S_r(0) = S_{l0} - S_{f0} - L_l$ such that if there is a collision during braking maneuvers, the impact will happen at a relative velocity bounded by a preset upper limit, ΔV_s that gives a low accident severity index S^2.

For a given braking scenario we would like to calculate the minimum value of the initial intervehicle spacing $S_r(0)$ that will lead to collisions at relative velocities smaller than ΔV_s. We will refer to this value as the minimum impact spacing (MIS).

The spacing between the two vehicles measured from the front of the following vehicle to the rear of the lead vehicle is given by

$$S_r(t) = S_l(t) - L_l - S_f(t) \tag{13}$$

where

$$S_l(t) = S_l(0) + \int_0^t V_l(\tau)d\tau \tag{14}$$

$$S_f(t) = S_f(0) + \int_0^t V_f(\tau)d\tau \tag{15}$$

and

$$V_l(t) = V_l(0) + \int_0^t a_l(\tau)d\tau \tag{16}$$

$$V_f(t) = V_f(0) + \int_0^t a_f(\tau)d\tau \tag{17}$$

If the decelerations $a_l(t)$ and $a_f(t)$ and initial positions and velocities are specified, the MIS can be calculated in a way very similar to the method used earlier. Let's assume that we want to bound the energy of the collision by limiting the relative velocity just before the collision to less than ΔV. Let's also assume the existence

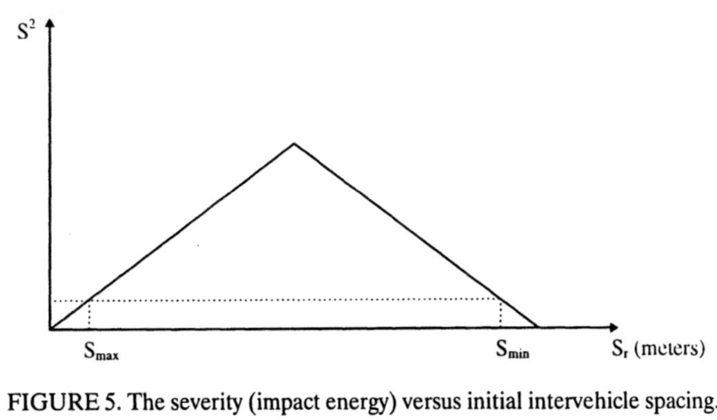

FIGURE 5. The severity (impact energy) versus initial intervehicle spacing.

of energy-absorbing bumpers that can absorb and dissipate the energy of the collision, thus guaranteeing a perfectly plastic collision (cc = 0). Figure 5 indicates that there are two ways to limit the relative velocity before the collision.

Assuming initial conditions where the leading and the following vehicle travel at approximately the same speed, we can guarantee that there is not enough time for a velocity differential to develop by limiting the relative spacing between vehicles to a very small distance. This leads to one possible vehicle following scenario, where in the event of an emergency the vehicles will always collide with each other and with the assumption of plastic collisions they will continue traveling as a single body until they come to a full stop.

The second likely braking scenario assumes that there is sufficient headway between vehicles but somewhat less than what would be required to guarantee no collisions in the event of emergency braking. We can apply the same methodology we used earlier to determine the minimum headway between vehicles that guarantees collisions with relative velocity less than a preselected ΔV. Assume that the two vehicles travel in the same direction but in two separate lanes. The position of the vehicles at time $t = 0$ is shown in Figure 2.

Let t_{sl} be the time needed by the leading and the following vehicle to slow down from their initial velocities V_{lo} and V_{fo} to velocities V_{lsl} and V_{fsl} such that $V_{\mathrm{fsl}} - V_{\mathrm{lsl}} < \Delta V$. This condition may occur more than once, from the moment the leading vehicle applied deceleration until the moment the following vehicle comes to a full stop. Therefore, we are interested in computing the headway for the two boundary cases: the case where the vehicles have first developed a sufficient ΔV and the case where the vehicles are at the end of the braking trajectory, the leader may have already stopped, but the follower is still moving and there is still a ΔV between them. The equations are practically the same as before. We have:

$$V_{\mathrm{f}}(0) + \int_0^{t_{\mathrm{s}}} a_{\mathrm{f}}(\tau)\,d\tau = 0 \tag{18}$$

$$S_{\mathrm{f}}(t) = S_{\mathrm{f}}(0) + \int_0^t V_{\mathrm{f}}(\tau)\,d\tau, \qquad \forall t \le t_{\mathrm{sl}} \tag{19}$$

and

$$S_{\mathrm{f}}(t) = S_{\mathrm{f}}(t_{\mathrm{s}}), \qquad \forall t > t_{\mathrm{sl}} \tag{20}$$

The position of the leading vehicle at each time t is given by

$$S_{\mathrm{l}}(t) = S_{\mathrm{l}}(0) + \int_0^t V_{\mathrm{l}}(\tau)\,d\tau, \qquad \forall t \le t_{\mathrm{sl}} \tag{21}$$

The relative spacing at each time t is given by

$$S_{\mathrm{r}}(t) = S_{\mathrm{l}}(t) - L_{\mathrm{l}} - S_{\mathrm{f}}(t) \tag{22}$$

and the relative speed at each time t is given by

$$-\Delta V(t) = V_1(0) + \int_0^t a_1(\tau)d\tau - V_f(0) + \int_0^t a_f(\tau)d\tau \tag{23}$$

In this case we have to determine the time instances t_{c1} and t_{c2} where the relative velocity is equal to the desired threshold. Having determined t_{c1} and t_{c2} we can then determine the relative spacing between the two vehicles. Therefore, the MIS has a minimum value and a maximum value. To limit the impact of the first collision at t_{c1}, we must allow for a maximum headway of $S_{max} = -\max S_r(t), \quad \forall t \in (0, t_{c1}]$.

To limit the impact of the last collision at t_{c2}, we must allow for a minimum headway of $S_{min} = -\min[S_r(t), 0], \quad \forall t \in [t_{c1}, t_{c2}]$.

From this description it becomes clear that the required headway must be either less than S_{max} or greater than S_{min} (see Figure 5).

The two limits, S_{max} and S_{min}, are equal to the distance by which the following vehicle would have overtaken the leading vehicle at the time instances t_{c1} and t_{c2}, respectively, which corresponds to the time instances when their relative velocity is equal to ΔV, assuming the initial conditions shown in Figure 2. Based on the above analysis, we use numerical methods to calculate S_{max} and S_{min}.

3. VEHICLE FOLLOWING CONCEPTS

With advances in technology and in particular in vehicle electronics, systems that were previously considered impossible or too costly to implement are becoming feasible and available. One such system is a functional extension of the classic cruise control. It consists of a controller that uses a sensor to measure the relative distance and the relative speed to any vehicle ahead and controls a throttle and a brake actuator in order to follow at the same speed and maintain a desired relative distance. The relative distance may be characterized in terms of a constant length or it may be a function of the speed. If the majority of vehicles have such a controller on board, we can have an environment where vehicles follow each other automatically, in the same highway lane, without any other kind of interaction such as communication between them. The highway may provide a level of support to the vehicles by transmitting information about road conditions, congestion, routing suggestions, and possibly recommended speeds. If the vehicles do not communicate and do not require any infrastructure support, they are said to operate autonomously. A system like that may provide a capacity increase by smoothing out traffic flow and eliminating the mistake that human drivers tend to make, that is, to follow at short and unsafe distances and then overcorrecting by slowing down too much when a vehicle ahead starts to decelerate.

A further functionality enhancement comes by allowing the vehicles to communicate and notify each other about their braking intentions. Also the infrastructure may become involved in setting the desired velocity for each section of the highway, communicating to vehicles about the need for emergency braking and coordinating the flow of the traffic. Such systems may achieve significant improvements in flow rates and capacity increases of the existing highways. By adding more equipment and intelligence to the vehicle–infrastructure system we can come up with more advanced concepts that have the potential for bigger benefits. In this section we describe a number of operating AHS concepts for automatic vehicle following.

3.1. Autonomous Vehicles

A possible AHS concept is one where the vehicles operate independently, i.e., autonomously, using their own sensors. Each vehicle senses its environment, including lane position, adjacent vehicles, and obstacles. The infrastructure may provide basic traveler information services, i.e., road conditions and routing information. The infrastructure may also provide some means to assist the vehicle in sensing its lane position. Many different systems have been proposed to help the vehicle sense its position, such as implanted magnetic nails, magnetic stripes, radar reflective stripes, radio frequency cables, or GPS satellites.[12]

In an autonomous environment, the vehicle does not rely on communications with other vehicles or the infrastructure in order to make vehicle following decisions. Each autonomous vehicle maintains a safe distance from the vehicle it is following or if a vehicle is not present within the sensing distance it travels at a constant speed in accordance with the posted speed limits and regional safety regulations and of course road conditions. In other words, if there is no vehicle ahead within the maximum safety distance, the vehicle travels at the speed limit or at a lower speed depending on the road conditions.

Since there is no communication between vehicles, each vehicle senses the relative spacing and speed to the vehicle ahead and selects a headway based on its own braking capabilities and by assuming that the vehicle in front may brake with the 'worst' possible deceleration. The technology that allows the vehicle to sense the relative position and speed to the vehicle ahead can also be adapted to allow the vehicle to estimate the size and indirectly the vehicle class and braking capabilities of the vehicle ahead. This knowledge will allow a less conservative assumption about the braking capabilities of the leading vehicle that will lead to a more accurate selection of intervehicle spacing. In the case where mixing of vehicle classes in the same lane is allowed, distinguishing whether the vehicle ahead is a truck, bus, or a passenger vehicle will have a significant effect on the selection of spacing and therefore on capacity.

3.2. Free Agent Vehicles—Infrastructure Supported

A vehicle is considered a "free agent" if it has the capability to operate autonomously but it is also able to receive communications from other vehicles and from the infrastructure. This implies that the infrastructure may get involved in a supporting role, by issuing warnings and recommendations for desired speed and headways but the infrastructure will not have the authority to issue direct control commands. Therefore, this concept has been referred to as "infrastructure supported." The fundamental difference between this concept and that described in Subsection 3.1 is that there is vehicle-to-vehicle and vehicle-to-infrastructure communication. Each vehicle communicates to the vehicle behind its braking capabilities and its braking intentions. This allows the vehicle behind to choose its headway. For example, a shorter headway can be selected by a passenger vehicle if the vehicle ahead is a heavy truck or a bus. A larger headway must be selected by a heavy vehicle if the vehicle ahead is a passenger vehicle. A free agent vehicle uses its own sensors to sense its position and environment, including lane position, adjacent vehicles, and obstacles.

With this concept the MSS between vehicles is expected to be smaller than that on conventional highways because of the intelligent longitudinal control system and vehicle-to-vehicle and infrastructure-to-vehicle communications. Each vehicle senses the relative spacing and speed to the vehicle ahead and decides and selects a headway based on its own braking capability, the braking capability of the vehicle ahead, and the road surface conditions which are either sensed by the vehicle or are broadcasted from the infrastructure. When a vehicle starts to brake, it notifies the vehicle behind about the magnitude of its braking force. Even if we assumed a relatively primitive form of communication between vehicles like a line-of-sight communication that transmits the applied braking force, we can achieve better separation control as we eliminate the delay in deciding if the vehicle ahead is performing emergency braking or routine braking.

3.3. *Free Agent Vehicles—Infrastructure Managed*

The concept of free agent vehicles with infrastructure management is based on the assumption that the traffic is composed of vehicles acting as free agents while the infrastructure assumes a more active and more complex role in the coordination of the traffic flow and control of vehicles. Each vehicle is able to operate autonomously and uses its own sensors to sense its position and environment, including lane position, adjacent vehicles, and obstacles. The difference in this centrally managed architecture is that the infrastructure has the ability to send commands to individual vehicles.

This is envisioned to be a "request–response"-type architecture, in which individual vehicles ask permission from the infrastructure to perform certain activities and the infrastructure responds by sending commands back to the requesting vehicle and to other vehicles in the neighborhood.

It is expected and assumed that the infrastructure is able to detect emergency situations and whenever it detects such emergency, the infrastructure will have the responsibility to send an emergency braking command to all vehicles affected. This concept minimizes the delay in performing emergency braking. This allows for some further reduction of the minimum headway, compared to the concepts presented so far. On the other side, the accurate timing of the emergency and stopping commands for each vehicle that must be issued by the infrastructure, requires accurate tracking of individual vehicles as well as extensive and frequent communications between individual vehicles and the infrastructure.

3.4. Platooning without Coordinated Braking

This concept represents the possibility that the safest and possibly most cost-effective way of achieving maximum capacity is by making platoons of vehicles the basic controlling unit. This will boost road capacity by expanding on the concept of infrastructure managed control.[13–15]

Platoons are clusters of vehicles with short spacing between individual vehicles in each group and longer spacing between platoons. The characterizing differentiation is that the platoon is to be treated by the infrastructure as an "entity" thereby minimizing some of the need for communicating with and coordinating individual vehicles. The infrastructure does not attempt to control any individual vehicle under normal circumstances, keeping the cost and necessary bandwidth low. The infrastructure is expected to be an intelligent agent that monitors and coordinates the operation of the platoons.

Tight coordination is required within the platoon in order to maintain a close spacing and this requires that the vehicles must be communicating with each other, constantly. The significantly longer interplatoon spacing is required to guarantee no interplatoon collisions.

Each vehicle is expected to be equipped with the sensors and intelligence to maintain its lane position, sense its immediate surroundings, and perform the functions of merging into and splitting off a platoon. It is not expected to accomplish lane changes, or merging and splitting without the infrastructure's or the platoon entity's help.

The main mode of operation of the infrastructure would be of a request–response type. Each platoon's and/or vehicle's request is processed and appropriate commands are sent to the appropriate vehicles/platoons to respond to that request. The infrastructure takes a more proactive role in monitoring traffic flow, broadcasting traffic flow messages, advising lane changes to individual vehicles and platoons in addition to the usual information provider functions.

Once a vehicle has merged into a platoon, the headway maintenance controller must take into account the braking capabilities of the vehicle ahead in the platoon in order to set an appropriate separation distance that minimizes the possibility of collision. The platoon leader may also provide corrections to the individual intraplatoon headways in order to reduce the possibility of a rear-end collision between two vehicles propagating to the other members of the platoon.

Mixing of vehicle classes, although an implicit feature of the present highway system, creates a major complication because of the dissimilar braking characteristics of each vehicle class. Therefore, it makes sense to form platoons of vehicles belonging to the same class, exclusively. In this concept we assume that no coordination of the braking sequence takes place within a platoon in order to distinguish it from the next one where coordinated braking is employed.

3.5. Platooning with Coordinated Braking

The platooning concept with coordinated braking is based on the concept of maximizing capacity by carefully coordinating the timing and degree of braking among the vehicles participating in a platoon entity. This allows the minimization of the spacing between vehicles without compromising safety. For example, during a braking maneuver the platoon leader may dictate a braking sequence to be followed by each vehicle so that the maneuver is performed without any intraplatoon collision. Such a sequence may require the last vehicle to brake first followed by the second last vehicle, etc. The distinguishing feature of this concept is the minimization of intraplatoon spacing and the promise of higher capacity.

3.6. *Infrastructure-Managed Slotting*

Under the infrastructure-managed slotting concept, an infrastructure-based control system creates and maintains vehicle "slots" in space and time. Slots can be thought of as moving roadway segments, each of which holds at most one vehicle at any time. The vehicles are identified and managed only by association with these slots. For simplicity in management, i.e., to achieve slots of uniform length, vehicles that need more space may be assigned multiple slots. Heavy-loaded light trucks may be assigned two slots, unloaded semis may be assigned three slots, loaded heavy trucks may be assigned four slots, and so forth.

In the basic slotting concept the slots should be of fixed length. The virtual leading edge of each slot can be thought of as a moving point that the vehicle assigned to the slot has to follow. Thus, the controller on the vehicle is assigned to follow this virtual moving point, not another vehicle. In essence this relieves the requirement of using headway sensors on the vehicle and of sensing the relative distance and speed to any other vehicle. Under no circumstances is a vehicle allowed to violate the edges of its assigned slot.

The distinguishing feature of this concept is that the sensing requirements are theoretically simplified. At least, the vehicle does not need to sense the relative position and speed of other vehicles. Yet the vehicle must be able to sense its position relative to the edge of the slot and the virtual point it tries to follow. A global and accurate longitudinal position sensing system is required.

In terms of separation policy, the slotting method is bounded by the limitations of the inherently "synchronous" architecture. This means that the size of each slot must be sufficient such that the spacing between individual vehicles occupying a single slot is sufficient to avoid collisions under the worst-case scenario. Thus, the weakest link in the chain is the vehicle with the worst braking performance that the system tries to accommodate in a single slot. Once the spacing is set to accommodate such a vehicle, every other vehicle that has better braking performance will not be able to utilize this capability to shorten the spacing to the vehicle in front. There will be "dead space" in between them. Similarly, a vehicle that does not meet the minimum braking requirement to occupy a single slot will be assigned two (or more) consecutive slots, with the resulting inefficiency of wasting even more space than is really needed.

By comparison, an architecture where each vehicle optimizes the headway between itself and the vehicle in front based only on the braking capabilities of the two vehicles involved is inherently an "asynchronous" architecture, which results in true minimization of the unused space between vehicles.

The relative merits of a "synchronous" versus "asynchronous" architecture have been intriguing the designers of computers and communications systems ever since digital systems became a reality. The typical trade-off is complexity versus performance. It has been well established through extensive research in other fields that asynchronous architectures provide the potential for maximizing performance at the cost of increased complexity.[16] It is almost obvious that the same is true on the subject of the AHS separation policy architecture.

4. SPACING AND CAPACITY EVALUATIONS

In this section we present briefly the fundamental factors that affect traction during vehicle acceleration and braking. Traction is what ultimately defines the braking capabilities of any kind of vehicle, under any kind of weather and road conditions. Then we develop likely emergency stopping scenarios for each AHS concept under consideration which we then use to calculate intervehicle spacing and capacity.

4.1. Adhesion and Friction

The friction force between two surfaces is defined as the force opposing the relative displacement of the two surfaces when a force is applied as shown in Figure 6. In the context of vehicle traction this force is referred to as adhesion. Adhesion (attraction between two surfaces) and friction (resistance to relative motion of adjacent surfaces) are very complex physical phenomena. But for practical purposes it is common to use the approximation that the magnitude of the friction force F depends on two factors only, namely, the normal force G between the two surfaces and a dimensionless coefficient of friction μ, such that:

$$F = \mu G \tag{24}$$

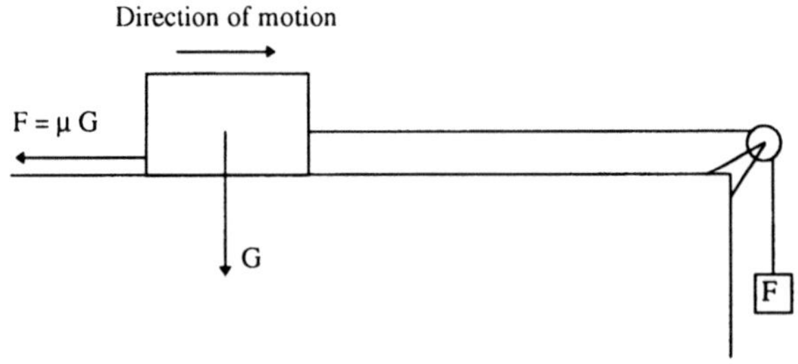

FIGURE 6. Physical representation of friction force *F*.

The value of the coefficient of friction μ depends on the characteristics of the two surfaces, primarily their smoothness and their hardness, and on the relative speed V_r between them. For most surfaces, as V_r increases, μ decreases. When the two surfaces do not move, μ assumes a considerably higher value, referred to as the static friction coefficient.

Applying the general concept to the problem of vehicle traction, it is clear that the maximum tractive or braking effort (TE_{max}) that can be utilized is limited by the tire-to-road surface adhesion:

$$TE_{max} = \mu G_a \tag{25}$$

where G_a is the weight on the wheels that apply the force. For propulsion G_a is the weight on the powered axle while for braking G_a represents the total vehicle weight G since the brakes act on all wheels. The actual weight distribution between front and rear axles depends on vehicle design and furthermore varies as a function of the actual deceleration as a result of the mass transfer phenomenon.

The change of μ with speed is very important in traction and friction. It makes braking at high speeds more difficult than at low speeds because it increases the possibility of skidding. Any spinning or skidding of the wheels results in a rapid increase of the relative speed V_r between the wheels and the road surface and therefore a sudden reduction of μ. As a result, traction is lost. To restore the friction coefficient, spinning or skidding must be terminated by reducing the tractive or braking effort. This is the principle of operation of the so-called antilock braking systems (ABS). The value of μ for vehicles depends on the type and condition of the road surface, the vehicle speed, and the condition of the tires. A range of values of μ for most types of vehicles is shown in Figure 7.[17]

The braking ability of all vehicles is best on dry pavement. It degrades substantially on wet pavement and braking ability is virtually lost on snow.

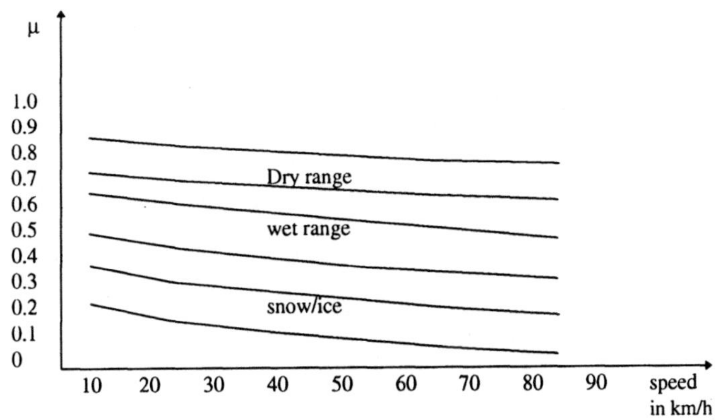

FIGURE 7. Friction coefficient of vehicles with rubber tires.

In our analysis, we use data from vehicle tests performed by established authorities. For passenger vehicles, we use information from *Consumer Reports*[2] and the consumer-oriented *Road and Track* magazine.[3] For heavy vehicles like buses and trucks, we obtained information from vehicle braking tests.[18] Based on these data, we have estimated the braking capabilities of a range of passenger and heavy vehicles on dry, wet, and snowed road pavement. In a more or less expected fashion, we found that sports cars can achieve the best braking distances (highest deceleration), followed by middle- and upper-class medium-size vehicles (such as in the "sports sedan" category), followed by small or economy-class vehicles. The last finding is a little counterintuitive, based on the fact that small vehicles are lightweight and thus require less energy dissipation to achieve braking and are less demanding of good tire performance. Yet there is an obvious trend for auto manufacturers to try to match the braking capabilities with the acceleration capabilities of a given vehicle. We found that the trend is to offer approximately double the deceleration (in g's) to the available acceleration (also in g's) in low gear. This is a ballpark figure, of course, and deviations do exist.

The braking capability of any vehicle degrades on wet pavement by a factor determined by the texture of the pavement and the type of tires used. We represent that as a change in the friction coefficient μ. The data collected give a quantitative estimate of the friction coefficient on dry, wet, and snowed pavement. The numbers of course vary depending on the vehicle, its tires, and the presence of ABS. A typical vehicle that can achieve $0.8g$ deceleration on dry pavement can go down to $0.55g$ in wet conditions and to as low as $0.15g$ in snow conditions. The collected braking test results are presented in Appendix A.

In our study, we simplified somewhat our assumptions regarding the friction coefficient μ. Instead of assuming a maximum deceleration of $1g$ and scaling it by the typical value of μ, i.e., 0.8 for passenger vehicles, we used the value $0.8g$ for maximum deceleration and assumed that μ is 1.0. This does not affect the results for braking on dry road pavement. Then for wet road conditions we assumed a worst-case scenario where the friction coefficient becomes half, i.e., μ becomes 0.5 while the maximum deceleration remains at $0.8g$ for passenger vehicles. Similarly, instead of assuming different values of μ for buses and for heavy trucks, we used the same value for all of them, but we used a different value of maximum deceleration for each class. We used $0.4g$ maximum deceleration for buses and $0.3g$ maximum deceleration for heavy trucks. These numbers are based on measurements on actual vehicles, and the data can be found in Appendix A.

The maximum deceleration that each vehicle can achieve depends on many factors and therefore it cannot be predicted exactly. It depends mostly on the tires of course, like the quality and type of tread, hardness, temperature, inflation pressure, and the age of the tire. It also depends on the size and type of friction materials in the brakes, the mass distribution of the vehicle, the presence of ABS and many other factors. In our analysis we simplify these complex dependences by using the abstraction of uniform value of m and assuming appropriate values for maximum deceleration for different classes of vehicles, without affecting the accuracy of the results.

During the emergency braking phase the jerk is not intentionally limited and the maximum deceleration is allowed to be as large as the vehicle can achieve. The jerk clearly depends on the mass of the vehicle first and on the hydraulic brake system second. It clearly depends on the rate of change of the force that the driver applies on the brake pedal in the case of manually driven vehicles. For automated vehicles it will depend on the dynamics of the brake actuator. It would simply be inversely proportional to the mass of the vehicle if all of the vehicles had exactly the same actuators and hydraulic systems, but this is certainly not going to be the case.

Based on our experience with an actual brake system that is in use in a prototype automated passenger-class vehicle, we made an educated guess for other classes of vehicles. We assumed that the maximum jerk is limited to 50 m/s^3 for passenger vehicles, 40 m/s^3 for buses, and 30 m/s^3 for heavy-loaded trucks.

4.2. Uniform versus Nonuniform Braking

For a realistic estimation of the theoretical capacity, we have assumed a "typical" maximum deceleration level for each class of vehicles, based on actual test data. Since discrepancies of 10% or more can be clearly seen in the braking capabilities among vehicles of the same class, we have made the assumption of a 10% discrepancy in maximum deceleration between the leader and the follower in the sense that the follower has inferior maximum deceleration capability, an assumption that inevitably generates the need for more spacing.

To be realistic, this discrepancy exists mostly at the limit of the braking capability of the vehicles, when braking occurs in the unstable region where the slope of μ versus wheel slip is negative as seen in Figure 8.[19] At that point, demanding slightly higher deceleration results in skidding of the tires and in a sharp reduction in the μ and in overall deceleration. In our effort to represent a realistic worst-case scenario, we assumed 10% deviation from the maximum braking capability for the following vehicle in all cases of unrestricted braking, i.e., when the traction of the tires is pushed to the limits. On the other hand, braking by applying less than the

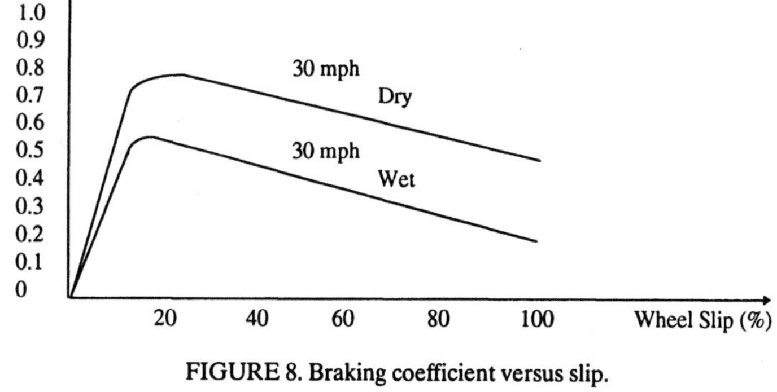

FIGURE 8. Braking coefficient versus slip.

maximum deceleration is easier because we can stay away from the unstable region of the μ curve. This can be used to our benefit if we impose a limit in deceleration for all vehicles. This limit is a common denominator that all vehicles should be able to meet by a proper design of their control system. This is the definition of the concept we will henceforth call "uniform braking." By staying away from the unstable braking region we can almost guarantee a better control of the magnitude of the deceleration. This justifies using only 5% deviation from the nominal braking capability for the follower in the case of uniform braking. Uniform braking is more crucial in platooning where, in the interest of efficiency, vehicles within each platoon have to have similar performance. For completeness and for the sake of comparison, we analyzed the effects of uniform braking in both platooning and nonplatooning environments.

The concept that all vehicles should be restricted to a closely matched (i.e., uniform) degree of deceleration is clearly an architectural decision. We assumed that the braking deceleration on a dry road can be restricted to $0.5g$ for all passenger vehicles, $0.3g$ for all buses, and $0.2g$ for all heavy trucks. The idea here is to use a number that every vehicle in its respective class can comfortably achieve. This helps guarantee that the deviation from one vehicle to another will be less than 5% in the worst case. So we used a 5% discrepancy in the deceleration of the leading and following vehicle to represent the worst-case mismatch in the case of uniform braking.

4.3. Mixing of Vehicle Classes

The mixing of different classes of vehicles on the same AHS will affect capacity because of the different braking capabilities of the different classes of vehicles. In our analysis we consider three different vehicle classes, possessing fundamentally different characteristics: passenger vehicles (P), buses (B), and heavy trucks (T).

This leads to the following possible combinations:

PP: A passenger vehicle leading a passenger vehicle
PB: A passenger vehicle leading a bus
PT: A passenger vehicle leading a truck
BP: A bus leading a passenger vehicle
BB: A bus leading a bus
BT: A bus leading a truck
TP: A truck leading a passenger vehicle
TB: A truck leading a bus
TT: A truck leading a truck

We made the following distinctions in mixing possibilities:

a. *No mixing.* Traffic consists of passenger vehicles only, i.e., we have 0% mixing. In this case, the passenger vehicle-to-passenger vehicle (PP) minimum headway was assumed between all vehicles.

b. *Allowed mixing of vehicle classes.* All cases of mixing assume uniform mixing, i.e., the minority vehicles are uniformly distributed among the population of passenger cars. This is a realistic assumption as long as the percentage of mixing is fairly low.

Case 1:
Traffic consisting of passenger vehicles with 5% mixing of buses. In this case, the PP minimum headway was assumed between 90% of the vehicles; PB, between 5% of the vehicles; and BP, between 5% of the vehicles.

Case 2:
Traffic consisting of passenger vehicles with 5% mixing of trucks. In this case, the PP minimum headway was assumed between 90% of the vehicles; PT, between 5% of the vehicles; and TP, between 5% of the vehicles.

Case 3:
Traffic consisting of passenger vehicles with 2.5% mixing of buses and 2.5% mixing of trucks. In this case, the PP minimum headway was assumed between 90% of the vehicles; PB, between 2.5% of the vehicles; PT, between 2.5% of the vehicles; BP, between 2.5% of the vehicles; and TP, between 2.5% of the vehicles.

Case 4:
Traffic consisting of passenger vehicles with 5% mixing of buses and 5% mixing of trucks. In this case, the PP minimum headway was assumed between 80% of the vehicles; PB, between 5% of the vehicles; PT, between 5% of the vehicles; BP, between 5% of the vehicles; and TP, between 5% of the vehicles.

4.4. Autonomous Vehicles

In the case of autonomous vehicles, each vehicle relies on its own sensors to determine the motion intentions of the leading vehicle. Since there is no vehicle-to-vehicle communication, each vehicle has to use relative speed and spacing measurements to determine the intentions of the vehicle ahead. Therefore, in calculating a safe intervehicle spacing we consider the following worst-case stopping scenario.

The acceleration (actually deceleration) profile of the leading and following vehicles involved in a braking maneuver is assumed to follow the trajectories shown in Figure 9.

The leading vehicle performs emergency braking at time $t = 0$, at a maximum rate of change (jerk) equal to J_{lmax} until it reaches a maximum deceleration of a_{lm}. The follower, which might have been accelerating initially, at a_{fac} starts decelerating after a detection and brake actuation delay equal to t_{fa} in an effort to maintain the desired spacing. Since initially the follower is not aware that the leader is performing emergency braking, it limits its jerk and deceleration to J_{fc} and a_{fauto} respectively, in an effort to meet the vehicle control objective and at the same time maintain passenger comfort. The follower detects and initiates emergency braking at $t = t_{fc}$.

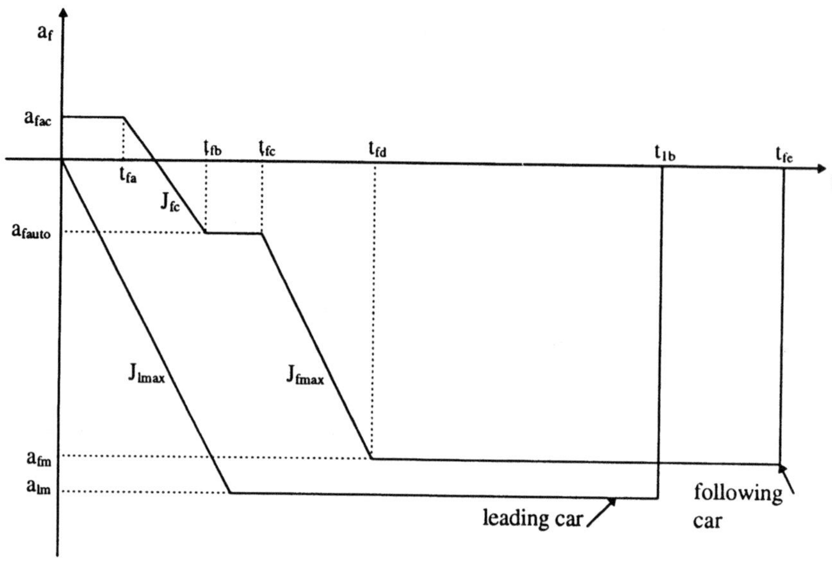

FIGURE 9. Autonomous vehicles.

At this time passenger comfort is no longer a crucial issue and braking is done with maximum jerk J_{fmax} and maximum deceleration a_{fm}.

In this section we use the above stopping scenario to calculate the minimum time headway for collision-free vehicle following by substituting appropriate numerical values for all of the above parameters.

In evaluating the above scenario we adopted a set of likely initial conditions at the onset of braking. The assumptions regarding the initial conditions are the following: The leader has been traveling at a speed of 60 mph while the follower has an instantaneous velocity 5% higher, i.e., 63 mph, and an instantaneous acceleration $a_{fac} = 0.15g$. These conditions represent the realistic scenario that the follower had been performing a position adjustment as in trying to catch up with the leader. Therefore, the vehicle is accelerating just before it has to start braking. When the vehicle detects that the leader is braking (which involves a 0.1-s delay for detection and a 0.1-s delay in the actuator) it starts braking until it reaches the maximum allowable deceleration $a_{fauto} = -0.1g$ for passenger comfort.

The vehicle initially applies a limited amount of braking because at the onset of braking it is not known if the leader is simply slowing down or performing emergency braking. If the follower applies emergency braking every time it detects the leader slowing down, it would be detrimental to the stability of the traffic flow. Therefore, the follower applies limited braking at first, with the objective of not upsetting the quality of the ride of the passengers or the position and velocity error of any vehicles behind. For this reason, the jerk is limited to 5 m/s^3 during this phase. Eventually, the follower will detect that the headway is diminishing rapidly and

TABLE 1
Autonomous Vehicles: Dry Road Surface

		PP	PB	PT	BP	BB	BT	TP	TB	TT
V_{lo}	mph	60	60	60	60	60	60	60	60	60
V_{fo}	mph	63	63	63	63	63	63	63	63	63
A_{lmax}	g	0.8	0.8	0.8	0.4	0.4	0.4	0.3	0.3	0.3
A_{fmax}	g	0.72	0.36	0.27	0.72	0.36	0.27	0.72	0.36	0.27
J_{lmax}	m/s^3	50	50	50	40	40	40	30	30	30
J_{fmax}	m/s^3	50	40	30	50	40	30	50	40	30
μ_{lmax}		1	1	1	1	1	1	1	1	1
μ_{fmax}		1	1	1	1	1	1	1	1	1
A_{fauto}	g	0.1	0.1	0.1	0.1	0.1	0.1	0.1	0.1	0.1
A_{fac}	g	0.15	0.15	0.15	0.15	0.15	0.15	0.15	0.15	0.15
J_{fc}	m/s^3	5	5	5	5	5	5	5	5	5
t_{fa}	s	0.2	0.2	0.2	0.2	0.2	0.2	0.2	0.2	0.2
t_{fc}	s	0.3	0.3	0.3	0.3	0.3	0.3	0.3	0.3	0.3
min headway	s	**0.66**	**2.63**	**3.97**	**0.063**	**1.04**	**2.37**	**0.045**	**0.171**	**1.28**
	m	18.71	74.2	111.7	1.79	29.15	66.63	1.29	4.81	36.07

therefore the leader is performing an emergency braking maneuver. We assumed that the detection of emergency braking involves 0.3 s of delay.

Using these parameter values, we computed the necessary headways for different road conditions and levels of mixing of classes of vehicles using the algorithm presented in Section 2.1. The spacing results are presented in Table 1 for the case of dry road surface. The spacing results for the case of wet road surface are presented in Table 2.

TABLE 2
Autonomous Vehicles: Wet Road Surface

		PP	PB	PT	BP	BB	BT	TP	TB	TT
V_{lo}	mph	60	60	60	60	60	60	60	60	60
V_{fo}	mph	63	63	63	63	63	63	63	63	63
A_{lmax}	g	0.8	0.8	0.8	0.4	0.4	0.4	0.3	0.3	0.3
A_{fmax}	g	0.72	0.36	0.27	0.72	0.36	0.27	0.72	0.36	0.27
J_{lmax}	m/s^3	50	50	50	40	40	40	30	30	30
J_{fmax}	m/s^3	50	40	30	50	40	30	50	40	30
μ_{lmax}		0.5	0.5	0.5	0.5	0.5	0.5	0.5	0.5	0.5
μ_{fmax}		0.5	0.5	0.5	0.5	0.5	0.5	0.5	0.5	0.5
A_{fauto}	g	0.1	0.1	0.1	0.1	0.1	0.1	0.1	0.1	0.1
A_{fac}	g	0.15	0.15	0.15	0.15	0.15	0.15	0.15	0.15	0.15
J_{fc}	m/s^3	5	5	5	5	5	5	5	5	5
t_{fa}	s	0.2	0.2	0.2	0.2	0.2	0.2	0.2	0.2	0.2
t_{fc}	s	0.3	0.3	0.3	0.3	0.3	0.3	0.3	0.3	0.3
min headway	s	**1.03**	**4.99**	**7.65**	**0.065**	**1.77**	**4.43**	**0.049**	**0.211**	**2.26**
	m	29.01	140.7	215.6	1.847	49.77	124.7	1.379	5.937	63.57

TABLE 3
Autonomous Vehicles: Uniform Braking, Dry Road

		PP	PB	PT	BP	BB	BT	TP	TB	TT
V_{lo}	mph	60	60	60	60	60	60	60	60	60
V_{fo}	mph	63	63	63	63	63	63	63	63	63
A_{lmax}	g	0.5	0.5	0.5	0.3	0.3	0.3	0.2	0.2	0.2
A_{fmax}	g	0.475	0.285	0.19	0.475	0.285	0.19	0.475	0.285	0.19
J_{lmax}	m/s^3	50	50	50	40	40	40	30	30	30
J_{fmax}	m/s^3	50	40	30	50	40	30	50	40	30
μ_{lmax}		1	1	1	1	1	1	1	1	1
μ_{fmax}		1	1	1	1	1	1	1	1	1
A_{fauto}	g	0.1	0.1	0.1	0.1	0.1	0.1	0.1	0.1	0.1
A_{fac}	g	0.15	0.15	0.15	0.15	0.15	0.15	0.15	0.15	0.15
J_{fc}	m/s^3	5	5	5	5	5	5	5	5	5
t_{fa}	s	0.2	0.2	0.2	0.2	0.2	0.2	0.2	0.2	0.2
t_{fc}	s	0.3	0.3	0.3	0.3	0.3	0.3	0.3	0.3	0.3
min headway	s	**0.72**	**2.73**	**5.25**	**0.075**	**1.00**	**3.52**	**0.045**	**0.100**	**1.36**
	m	20.33	76.83	147.7	2.112	28.27	99.15	1.290	2.833	38.19

The spacing calculations in Tables 1 and 2 are based on the assumption that vehicles can brake with maximum possible deceleration depending on their capabilities. Another possible scenario is to use the concept of uniform braking that limits the maximum deceleration and maximum jerk to values that could be met and used by all vehicles of the same class. These limits will make the braking performance of the vehicles very similar. Using this scenario we calculated spacings based on the vehicle values shown in Table 3. In this case due to uniformity we assume 5% deviation between decelerations of vehicles of the same class. This 5% deviation accounts for inaccuracies in measuring acceleration/deceleration and maintaining the desired one using the on-board vehicle controller.

Based on the above spacings the maximum possible throughput referred to as the capacity C measured as the number of vehicles per hour per lane is given by the formula

$$C = (360000V)[(100-2W_T-W_B)(L_P+h_{PP}V)+W_T(L_P+h_{PT}V+h_{TP}V+L_T) \\ +W_B(L_P+h_{PB}V+h_{BP}V+L_B)]^{-1} \tag{26}$$

where V is the speed of flow measured in meters per second, L_p is the length of passenger cars, L_B is the length of buses, and L_T is the length of trucks with trailers, in meters. We assumed $L_p = 4.75$ m, $L_B = 12$ m, and $L_T = 18$ m. The parameter h_{PP} is the minimum time headway between passenger cars, h_{PT} is the minimum time headway between a passenger car and a truck that follows it, h_{TP} is the minimum time headway between a truck and a passenger car that follows it, h_{PB} is the minimum time headway between a passenger car and a bus that follows it, and h_{BP} is the

TABLE 4
Autonomous Vehicles: Capacity Estimates under Different Road Condition Assumptions

	Dry road surface	Wet road surface	Uniform braking
A. With identification of different vehicle classes			
0% mixing	4116	2860	3850
5% buses	3746	2516	3525
5% trucks	3458	2278	3096
2.5% buses + 2.5% trucks	3596	2391	3297
5% buses + 5% trucks	3193	2054	2882
B. No identification of different vehicle classes			
0% mixing	4116	2860	3850
5% buses	3631	2432	3416
5% trucks	3356	2207	3007
2.5% buses + 2.5% trucks	3488	2314	3198
5% buses + 5% trucks	3026	1943	2735

minimum time headway between a bus and a passenger car that follows it, in seconds. W_B is the percentage of buses and W_T is the percentage of trucks in the mix. We use equation (26) and the numerical results of Tables 1, 2, and 3 to calculate the capacity values that are presented in Table 4A.

In equation (26) we assumed that a bus or a truck is always between two passenger vehicles and the passenger vehicle recognizes when its leader is a truck or a bus. This is a reasonable assumption because the radar sensors used for ranging measurements can be equipped with the feature of being able to distinguish different classes of vehicles. Without this assumption each vehicle has to assume the worst possible situation, which is the one where each vehicle treats its leader as a passenger vehicle, i.e., a vehicle with the highest possible braking capability. In this case equation (26) is modified to

$$C = (360000V)[(100 - 2W_T - 2W_B)(L_P + h_{PP}V) + W_T(L_P + h_{PT}V + h_{PP}V + L_T) + W_B(L_P + h_{PB}V + h_{PP}V + L_B)]^{-1} \quad (27)$$

The capacity results for this case are listed in Table 4B.

4.5. *Free Agent Vehicles—Infrastructure Supported*

In the case of free agent vehicles we assumed the braking scenario shown in Figure 10. The use of vehicle-to-vehicle communication simplifies the task of deter-

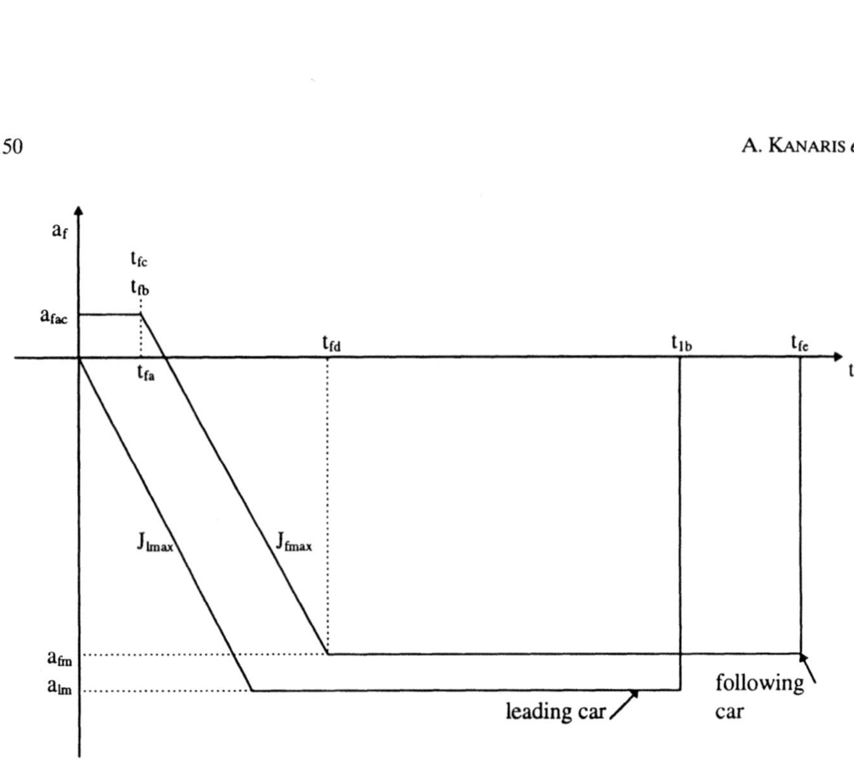

FIGURE 10. Infrastructure-supported free agent vehicles.

mining when the leading vehicle is performing emergency braking. The leader at $t = 0$ starts performing emergency braking. At $t = 0$ it communicates its intention to the following vehicle. The following vehicle receives the information from the leader and verifies using its own sensors that it has to perform an emergency braking as well.

TABLE 5
Free Agent Vehicles: Infrastructure Supported, Dry Road

		PP	PB	PT	BP	BB	BT	TP	TB	TT
V_{lo}	mph	60	60	60	60	60	60	60	60	60
V_{fo}	mph	63	63	63	63	63	63	63	63	63
A_{lmax}	g	0.8	0.8	0.8	0.4	0.4	0.4	0.3	0.3	0.3
A_{fmax}	g	0.72	0.36	0.27	0.72	0.36	0.27	0.72	0.36	0.27
J_{lmax}	m/s^3	50	50	50	40	40	40	30	30	30
J_{fmax}	m/s^3	50	40	30	50	40	30	50	40	30
μ_{lmax}		1	1	1	1	1	1	1	1	1
μ_{fmax}		1	1	1	1	1	1	1	1	1
A_{fauto}	g	0.1	0.1	0.1	0.1	0.1	0.1	0.1	0.1	0.1
A_{fac}	g	0.15	0.15	0.15	0.15	0.15	0.15	0.15	0.15	0.15
J_{fc}	m/s^3	10	10	10	10	10	10	10	10	10
t_{fa}	s	0.1	0.1	0.1	0.1	0.1	0.1	0.1	0.1	0.1
t_{fc}	s	0.1	0.1	0.1	0.1	0.1	0.1	0.1	0.1	0.1
min headway	s	**0.463**	**2.432**	**3.762**	**0.027**	**0.832**	**2.162**	**0.021**	**0.088**	**1.077**
	m	13.03	68.50	106.0	0.784	23.45	60.91	0.600	2.466	30.35

TABLE 6
Free Agent Vehicles: Infrastructure Supported, Wet Road

		PP	PB	PT	BP	BB	BT	TP	TB	TT
V_{lo}	mph	60	60	60	60	60	60	60	60	60
V_{fo}	mph	63	63	63	63	63	63	63	63	63
A_{lmax}	g	0.8	0.8	0.8	0.4	0.4	0.4	0.3	0.3	0.3
A_{fmax}	g	0.72	0.36	0.27	0.72	0.36	0.27	0.72	0.36	0.27
J_{lmax}	m/s^3	50	50	50	40	40	40	30	30	30
J_{fmax}	m/s^3	50	40	30	50	40	30	50	40	30
μ_{lmax}		0.5	0.5	0.5	0.5	0.5	0.5	0.5	0.5	0.5
μ_{fmax}		0.5	0.5	0.5	0.5	0.5	0.5	0.5	0.5	0.5
A_{fauto}	g	0.1	0.1	0.1	0.1	0.1	0.1	0.1	0.1	0.1
A_{fac}	g	0.15	0.15	0.15	0.15	0.15	0.15	0.15	0.15	0.15
J_{fc}	m/s^3	10	10	10	10	10	10	10	10	10
t_{fa}	s	0.1	0.1	0.1	0.1	0.1	0.1	0.1	0.1	0.1
t_{fc}	s	0.1	0.1	0.1	0.1	0.1	0.1	0.1	0.1	0.1
min headway	s	**0.828**	**4.792**	**7.451**	**0.037**	**1.564**	**4.224**	**0.028**	**0.140**	**2.054**
	m	23.34	135.0	210.0	1.039	44.07	119.0	0.800	3.951	57.85

The assumptions regarding the initial conditions are the same as in the previous case: We assume the leader has been traveling at a speed of 60 mph while the follower has an instantaneous velocity of 63 mph and an instantaneous acceleration of $0.15g$, as if the follower had been trying to catch up with the leader.

When the vehicle detects the leader is braking and at the same time receives the information that this is emergency braking, it bypasses the limited jerk/limited

TABLE 7
Free Agent Vehicles: Infrastructure Supported, Uniform Braking, Dry Road

		PP	PB	PT	BP	BB	BT	TP	TB	TT
V_{lo}	mph	60	60	60	60	60	60	60	60	60
V_{fo}	mph	63	63	63	63	63	63	63	63	63
A_{lmax}	g	0.5	0.5	0.5	0.3	0.3	0.3	0.2	0.2	0.2
A_{fmax}	g	0.475	0.285	0.19	0.475	0.285	0.19	0.475	0.285	0.19
J_{lmax}	m/s^3	50	50	50	40	40	40	30	30	30
J_{fmax}	m/s^3	50	40	30	50	40	30	50	40	30
μ_{lmax}		1	1	1	1	1	1	1	1	1
μ_{fmax}		1	1	1	1	1	1	1	1	1
A_{fauto}	g	0.1	0.1	0.1	0.1	0.1	0.1	0.1	0.1	0.1
A_{fac}	g	0.15	0.15	0.15	0.15	0.15	0.15	0.15	0.15	0.15
J_{fc}	m/s^3	10	10	10	10	10	10	10	10	10
t_{fa}	s	0.1	0.1	0.1	0.1	0.1	0.1	0.1	0.1	0.1
t_{fc}	s	0.1	0.1	0.1	0.1	0.1	0.1	0.1	0.1	0.1
min headway	s	**0.519**	2.525	**5.041**	**0.036**	**0.800**	**3.317**	**0.023**	**0.058**	**1.151**
	m	14.64	71.11	142.0	1.030	22.55	93.44	0.668	1.638	32.43

TABLE 8
Free Agent Vehicles: Infrastructure Supported, Capacity Estimates

	Dry road surface	Wet road surface	Uniform braking
0% mixing	5400	3425	4942
5% buses	4730	2923	4377
5% trucks	4276	2605	3730
2.5% buses + 2.5% trucks	4492	2755	4025
5% buses + 5% trucks	3845	2304	3400

braking stage shown in Figure 9 in the previous section. In Figure 10, we have clustered the detection and the actuation delay into a single 0.1-s delay before the follower applies emergency braking. In effect, the actuation delay is compensated for by the fact that the vehicle knows in advance it will have to apply the brakes, and the brake actuator may be preloaded. Therefore, in Figure 10 we assume $t_{fa} = t_{fc} = 0.1$ s. The minimum headway results together with the numerical values of the variables shown in Figure 10 are presented in Tables 5, 6, and 7. Equation (27) is used to calculate capacity for different levels of mixing of different classes of vehicles. The results are shown in Table 8.

4.6. *Free Agent Vehicles—Infrastructure Managed*

In the case of free agent vehicles with infrastructure management we have assumed that the infrastructure has the primary responsibility of detecting the pres-

TABLE 9
Free Agent Vehicles: Infrastructure Managed, Dry Road

		PP	PB	PT	BP	BB	BT	TP	TB	TT
V_{lo}	mph	60	60	60	60	60	60	60	60	60
V_{fo}	mph	63	63	63	63	63	63	63	63	63
A_{lmax}	g	0.8	0.8	0.8	0.4	0.4	0.4	0.3	0.3	0.3
A_{fmax}	g	0.72	0.36	0.27	0.72	0.36	0.27	0.72	0.36	0.27
J_{lmax}	m/s^3	50	50	50	40	40	40	30	30	30
J_{fmax}	m/s^3	50	40	30	50	40	30	50	40	30
μ_{lmax}		1	1	1	1	1	1	1	1	1
μ_{fmax}		1	1	1	1	1	1	1	1	1
A_{fauto}	g	0.1	0.1	0.1	0.1	0.1	0.1	0.1	0.1	0.1
A_{fac}	g	0.15	0.15	0.15	0.15	0.15	0.15	0.15	0.15	0.15
J_{fc}	m/s^3	20	20	20	20	20	20	20	20	20
t_{fa}	s	0	0	0	0	0	0	0	0	0
t_{fc}	s	0	0	0	0	0	0	0	0	0
min headway	s	**0.36**	**2.327**	**3.655**	**0.014**	**0.73**	**2.056**	**0.012**	**0.054**	**0.971**
	m	10.15	65.55	103.0	0.409	20.5	57.91	0.326	1.538	27.35

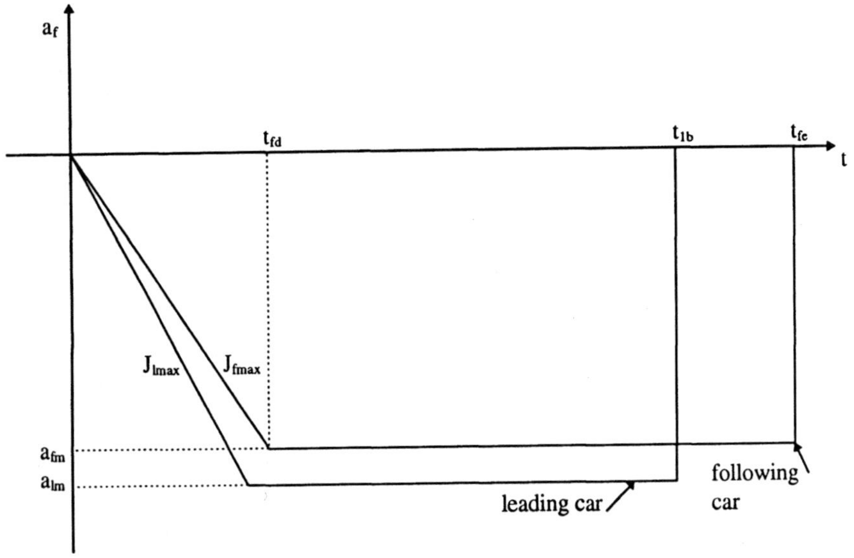

FIGURE 11. Infrastructure-managed free agent vehicles.

ence of emergencies and synchronizing the onset of emergency braking of all vehicles involved. This results in the most favorable timing for braking delays.

The infrastructure may simply issue the command "Begin emergency braking now" and all vehicles receiving this will have to apply maximum braking without further delay. This not only simplifies the task of determining when the leading

TABLE 10
Free Agent Vehicles: Infrastructure Managed, Wet Road

		PP	PB	PT	BP	BB	BT	TP	TB	TT
V_{lo}	mph	60	60	60	60	60	60	60	60	60
V_{fo}	mph	63	63	63	63	63	63	63	63	63
A_{lmax}	g	0.8	0.8	0.8	0.4	0.4	0.4	0.3	0.3	0.3
A_{fmax}	g	0.72	0.36	0.27	0.72	0.36	0.27	0.72	0.36	0.27
J_{lmax}	m/s^3	50	50	50	40	40	40	30	30	30
J_{fmax}	m/s^3	50	40	30	50	40	30	50	40	30
μ_{lmax}		0.5	0.5	0.5	0.5	0.5	0.5	0.5	0.5	0.5
μ_{fmax}		0.5	0.5	0.5	0.5	0.5	0.5	0.5	0.5	0.5
A_{fauto}	g	0.1	0.1	0.1	0.1	0.1	0.1	0.1	0.1	0.1
A_{fac}	g	0.15	0.15	0.15	0.15	0.15	0.15	0.15	0.15	0.15
J_{fc}	m/s^3	20	20	20	20	20	20	20	20	20
t_{fa}	s	0	0	0	0	0	0	0	0	0
t_{fc}	s	0	0	0	0	0	0	0	0	0
min headway	s	**0.726**	**4.687**	**7.344**	**0.025**	**1.460**	**4.117**	**0.019**	**0.109**	**1.947**
	m	20.46	132.0	206.9	0.697	41.12	116.0	0.546	3.066	54.85

TABLE 11
Free Agent Vehicles: Infrastructure Managed, Uniform Braking, Dry Road

		PP	PB	PT	BP	BB	BT	TP	TB	TT
V_{lo}	mph	60	60	60	60	60	60	60	60	60
V_{fo}	mph	63	63	63	63	63	63	63	63	63
A_{lmax}	g	0.5	0.5	0.5	0.3	0.3	0.3	0.2	0.2	0.2
A_{fmax}	g	0.475	0.285	0.19	0.475	0.285	0.19	0.475	0.285	0.19
J_{lmax}	m/s^3	50	50	50	40	40	40	30	30	30
J_{fmax}	m/s^3	50	40	30	50	40	30	50	40	30
μ_{lmax}		1	1	1	1	1	1	1	1	1
μ_{fmax}		1	1	1	1	1	1	1	1	1
A_{fauto}	g	0.1	0.1	0.1	0.1	0.1	0.1	0.1	0.1	0.1
A_{fac}	g	0.15	0.15	0.15	0.15	0.15	0.15	0.15	0.15	0.15
J_{fc}	m/s^3	20	20	20	20	20	20	20	20	20
t_{fa}	s	0	0	0	0	0	0	0	0	0
t_{fc}	s	0	0	0	0	0	0	0	0	0
min headway	s	**0.416**	**2.43**	**4.93**	**0.021**	**0.695**	**3.21**	**0.014**	**0.040**	**1.042**
	m	11.72	68.13	138.9	0.602	19.56	90.36	0.404	1.123	29.36

vehicle is performing emergency braking but also minimizes the relative delay in propagating the onset of emergency braking from each vehicle to the vehicle behind, effectively down to zero.

We have listed the actuation delay as a single 0.1-s delay before each vehicle applies emergency braking, but since all of the vehicles receive the command at the same time the relative delay is zero and this is reflected in the value of the parameter t_{fc}. The time t_{fc} represents the total delay between the onset of emergency braking between the leader and the follower and in this case $t_{fc} = 0$.

The assumptions regarding the initial conditions are the same as before: The leader has been traveling at a speed of 60 mph while the follower has an instantaneous velocity of 63 mph and an instantaneous acceleration of $0.15g$, as if the follower had been trying to catch up with the leader. The minimum headway results together with the numerical values of the variables shown in Figure 11 are presented in Tables 9, 10, and 11. Equation (27) is used to calculate capacity for different levels of mixing of different classes of vehicles. The results are shown in Table 12.

TABLE 12
Free Agent Vehicles: Infrastructure Managed, Capacity Estimates

	Dry road surface	Wet road surface	Uniform braking
0% mixing	6437	3823	5810
5% buses	5472	3197	5018
5% trucks	4873	2820	4184
2.5% buses + 2.5% trucks	5155	2997	4563
5% buses + 5% trucks	4299	2464	3756

4.7. Vehicle Platoons without Coordinated Braking

In the platooning without coordinated braking case, we have assumed that each vehicle notifies the vehicle behind about its braking capabilities and the magnitude and timing of the braking force used.

When the platoon leader detects an emergency, it immediately notifies the vehicle that follows. There will be a delay while the message propagates from each vehicle to the vehicle behind, as well as an actuation delay. But the actuation delay is not affecting the scenario as long as it is approximately the same for each vehicle. We have assumed that the total delay is 0.1 s for every vehicle and it is represented by the parameter t_{fa}. Therefore, we have accounted for only a 0.1-s total delay in propagating the message from each vehicle to the vehicle behind and this becomes the value of the parameter t_{fc}, which represents the delay of the onset of emergency braking. This deceleration profile is shown in Figure 12.

The assumptions regarding initial conditions are as follows: The leader has been traveling at a speed of 60 mph while the follower has an instantaneous velocity of 61.5 mph. Since the platoon protocol involves a much tighter control of individual vehicle velocity than in the case of free agents, only a 2.5% difference is assumed in the initial vehicle velocities. The instantaneous acceleration was also taken to be 0*g* as it would be impossible for a vehicle in a platoon to be accelerating while the vehicle ahead is maintaining constant speed. Both the velocities and the accelerations of vehicles in platoons are expected to be closely coordinated. In addition, for reasons explained earlier we assumed no mixing of vehicle classes.

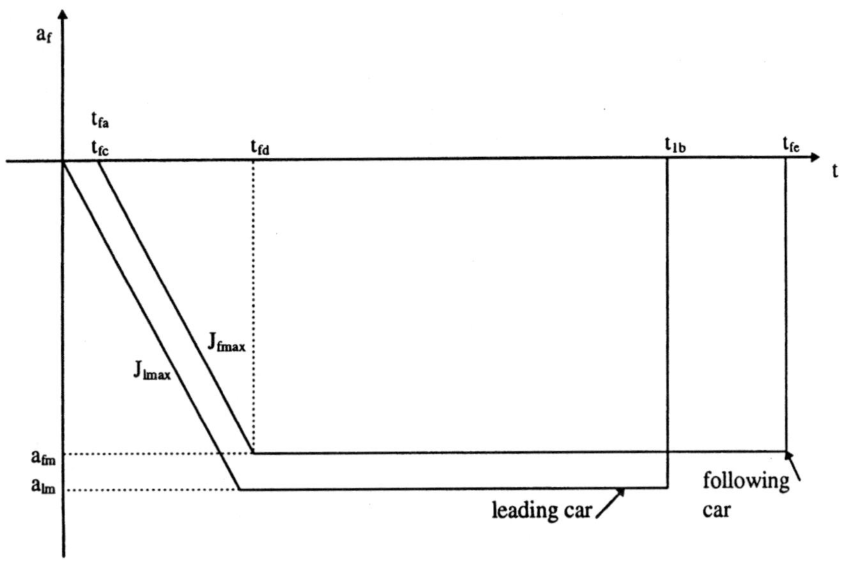

FIGURE 12. Platoons without coordinated braking.

The interplatoon spacing depends on the concept used for platoon following. We compared three different concepts.

1. Autonomous platoons, where platoons do not communicate with each other and each platoon relies on its own sensors to detect the motion of a leading platoon. In this case, the interplatoon spacing is calculated as in the case of autonomous vehicles. Therefore, each vehicle assumes t_{fc} = 0.1 s and each platoon entity assumes the parameters of autonomous vehicles: t_{fc} = 0.3 s both for 10-car platoons and for 20-car platoons.
2. Free agent platoons supported by the infrastructure where the interplatoon spacing is calculated as in the case of free agent vehicles with infrastructure support. Each vehicle in the platoon assumes t_{fc} = 0.1 s. Each platoon entity assumes the parameters of free agent infrastructure-supported vehicles: t_{fc} = 0.1 s both for 10-car platoons and for 20-car platoons.
3. Free agent platoons managed by the infrastructure where the interplatoon spacing is calculated as in the case of free agent vehicles with infrastructure management. Each vehicle in the platoon assumes t_{fc} = 0.1 s. Each platoon entity assumes the parameters of free agent infrastructure-managed vehicles: t_{fc} = 0 s both for 10-car platoons and for 20-car platoons.

The capacity is calculated in each case using the equation:

$$C = (3600VN)/[(h_{PP}V + L_P)(N-1) + H_{PP}V + L_P] \tag{28}$$

where L_p is the length of each vehicle in the platoon (we have assumed vehicles of the same length), h_{pp} is the intraplatoon time headway, H_{pp} is the interplatoon time headway, and N is the number of vehicles in the platoon. The resulting intraplatoon spacing for platoons without coordinated braking can be found in Table 13A. Allowing intervehicle collisions at up to 5 mph yields the results of Table 13B. The capacity results with and without intervehicle collisions are presented in Table 14.

4.8. *Vehicle Platoons with Coordinated Braking and No Delay*

In platooning with coordinated braking we assume that the vehicle in the platoon leader position assumes the primary responsibility of detecting emergencies and notifying each and every vehicle in the platoon. This notification takes place through a network-style vehicle-to-vehicle communications system that minimizes the communication delays. The platoon leader notifies all of the vehicles in the platoon about the magnitude of the braking force that is to be applied and also the exact time this is to be applied. This architecture not only eliminates the need for each vehicle to detect the magnitude of braking and if the braking should be limited or emergency braking, but also can adjust the onset of emergency braking for an effective 0 s relative delay, or even to an artificial negative relative delay.

TABLE 13
A. Platoons without Coordinated Braking

		Dry	Wet	Uniform
V_{lo}	mph	60	60	60
V_{fo}	mph	61.5	61.5	61.5
A_{lmax}	g	0.8	0.8	0.5
A_{fmax}	g	0.72	0.72	0.475
J_{lmax}	m/s^3	50	50	50
J_{fmax}	m/s^3	50	50	50
μ_{lmax}		1	0.5	1
μ_{fmax}		1	0.5	1
A_{fauto}	g	0	0	0
A_{fac}	g	0	0	0
J_{fc}	m/s^3	20	20	20
t_{fa}	s	0.1	0.1	0.1
t_{fc}	s	0.1	0.1	0.1
min headway	s	**0.37**	**0.65**	**0.38**
	m	10.26	17.93	10.48

B. Platoons without Coordinated Braking, Allowing 5-mph Collisions

		Dry	Wet	Uniform
V_{lo}	mph	60	60	60
V_{fo}	mph	61.5	61.5	61.5
A_{lmax}	g	0.8	0.8	0.5
A_{fmax}	g	0.72	0.72	0.475
J_{lmax}	m/s^3	50	50	50
J_{fmax}	m/s^3	50	50	50
μ_{lmax}		1	0.5	1
μ_{fmax}		1	0.5	1
A_{fauto}	g	0	0	0
A_{fac}	g	0	0	0
J_{fc}	m/s^3	20	20	20
t_{fa}	s	0.1	0.1	0.1
t_{fc}	s	0.1	0.1	0.1
min headway	s	**0.36**	**0.63**	**0.36**
	m	9.90	17.22	9.94
max headway	s	**0.076**	**0.186**	**0.277**
	m	2.09	5.14	7.61

The brake actuation delay can be completely compensated for and it is not affecting the scenario as long as it is approximately the same for each vehicle. We have assumed it is 0.1 s on every vehicle. Therefore, we have made the assumption of exactly 0-s total delay for the onset of braking for each vehicle in the platoon and this is the value of the parameter t_{fc} which represents this delay. This deceleration profile is shown in Figure 13.

TABLE 14
Platoons of Passenger Vehicles without Coordinated Braking (t_{fc} = 0.1 s): Capacity Estimates with/without 5-mph Collisions

	Dry road surface	Wet road surface	Uniform braking
A. Autonomous platoons			
10-car platoons	6217/6090	4171/4059	6139/5955
20-car platoons	6399/6257	4280/4156	6349/6142
B. Free agent infrastructure supported platoons			
10-car platoons	6453/6317	4276/4158	6369/6172
20-car platoons	6522/6374	4335/4207	6470/6255
C. Free agent infrastructure managed platoons			
10-car platoons	6580/6438	4331/4211	6495/6289
20-car platoons	6586/6435	4363/4234	6534/6314

The other assumptions regarding the initial conditions are the same as in all architectures involving platoons. The leader has been traveling at a speed of 60 mph while the follower has an instantaneous velocity of 61.5 mph. The instantaneous acceleration was also taken to be 0*g* as it would be impossible for a vehicle in a platoon to be accelerating while the vehicle ahead is maintaining constant speed.

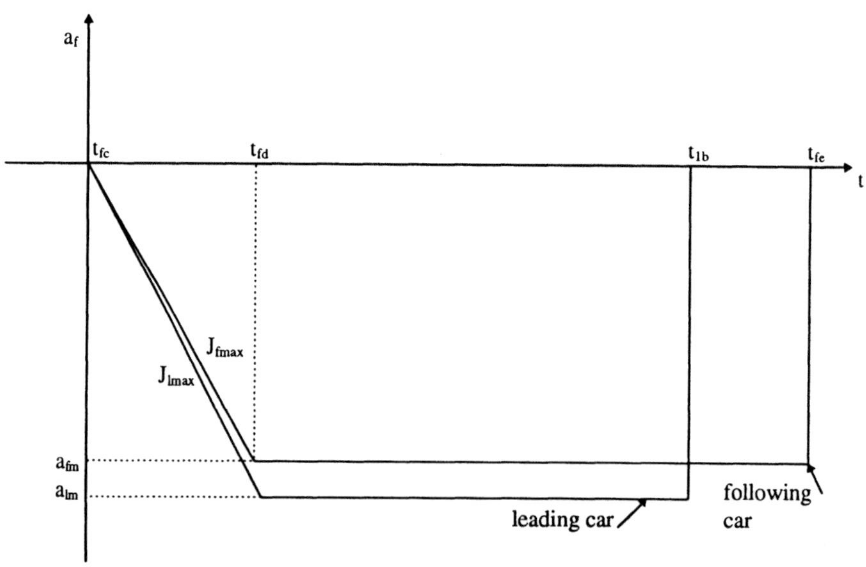

FIGURE 13. Platoons with coordinated braking and no delay.

Both the velocities and the accelerations of vehicles in platoons are expected to be closely coordinated.

For the interplatoon spacing we used and compared three different concepts.

1. Autonomous platoons where the interplatoon spacing is calculated as in the case of autonomous vehicles. Therefore, each vehicle assumes $t_{fc} = 0$ s and each platoon entity assumes the parameters of autonomous vehicles: $t_{fc} = 0.3$ s for both 10-car platoons and for 20-car platoons.
2. Free agent platoons supported by the infrastructure where the interplatoon spacing is calculated as in the case of free agent vehicles with infrastructure support. Each vehicle in the platoon assumes $t_{fc} = 0$ s. Each platoon entity assumes the parameters of free agent infrastructure-supported vehicles: $t_{fc} = 0.1$ s for both 10-car platoons and for 20-car platoons.
3. Free agent platoons managed by the infrastructure where the interplatoon spacing is calculated as in the case of free agent vehicles with infrastructure management. Each vehicle in the platoon assumes $t_{fc} = 0$ s. Each platoon entity assumes the parameters of free agent infrastructure-managed vehicles: $t_{fc} = 0$ seconds for both 10-car platoons and for 20-car platoons.

The interplatoon spacing results for platoons with coordinated braking are calculated using equation (28), based on the intraplatoon spacings presented in Table 15A. Allowing intervehicle collisions at up to 5 mph yields the results of Table 15B. The capacity results with and without intervehicle collisions are presented in Table 16.

4.9. Vehicle Platoons with Coordinated Braking and Staggered Timing

This case is identical to the previous one except for the purposeful timing of the onset of emergency braking. In the platooning with coordinated braking case we have assumed the vehicle in the platoon leader position assumes the primary responsibility of detecting emergencies and notifying each and every vehicle in the platoon. This notification takes place through a network-style vehicle-to-vehicle communications system that minimizes the communication delays. The platoon leader notifies all of the vehicles in the platoon about the magnitude of the braking force that is to be applied and also the exact time this is to be applied. This architecture not only eliminates the need for each vehicle to detect the magnitude of braking and if the braking should be limited or emergency braking, but also can adjust the onset of emergency braking to an artificial negative relative delay.

Therefore, we have made the choice of using a 0.1-s total delay for the onset of braking for each vehicle in the platoon going from the tail to the head, in the sense that the tail of the platoon is requested to brake first, then the vehicle ahead after a delay of 0.1 s, until the command to begin braking becomes effective for the platoon leader. Therefore, we used a negative value, –0.1 s, as the value of the parameter t_{fc} which represents the relative delay for two consecutive vehicles within the platoon. This deceleration profile is shown in Figure 14.

TABLE 15
A. Platoons with Coordinated Braking, No Delay

		Dry	Wet	Uniform
V_{lo}	mph	60	60	60
V_{fo}	mph	61.5	61.5	61.5
A_{lmax}	g	0.8	0.8	0.5
A_{fmax}	g	0.72	0.72	0.475
J_{lmax}	m/s^3	50	50	50
J_{fmax}	m/s^3	50	50	50
μ_{lmax}		1	0.5	1
μ_{fmax}		1	0.5	1
A_{fauto}	g	0	0	0
A_{fac}	g	0	0	0
J_{fc}	m/s^3	20	20	20
t_{fa}	s	0	0	0
t_{fc}	s	0	0	0
min headway	s	**0.27**	**0.55**	**0.28**
	m	7.51	15.18	7.73

B. Platoons with Coordinated Braking, No Delay, Allowing 5-mph Collisions

		Dry	Wet	Uniform
V_{lo}	mph	60	60	60
V_{fo}	mph	61.5	61.5	61.5
A_{lmax}	g	0.8	0.8	0.5
A_{fmax}	g	0.72	0.72	0.475
J_{lmax}	m/s^3	50	50	50
J_{fmax}	m/s^3	50	50	50
μ_{lmax}		1	0.5	1
μ_{fmax}		1	0.5	1
A_{fauto}	g	0	0	0
A_{fac}	g	0	0	0
J_{fc}	m/s^3	20	20	20
t_{fa}	s	0	0	0
t_{fc}	s	0	0	0
min headway	s	**0.26**	**0.52**	**0.26**
	m	7.16	14.47	7.20
max headway	s	**0.109**	**0.214**	**0.26**
	m	3.00	5.89	7.20

We cannot omit mentioning the fact that the platoon leader which detects the presence of emergency is subsequently restrained from braking until every other vehicle in the platoon has begun braking. Therefore, while this architecture allows us to minimize the necessary spacing between vehicles in the platoon, it increases the interplatoon spacing requirement.

TABLE 16
Platoons of Passenger Vehicles with Coordinated Braking (t_{fc} = 0 s): Capacity Estimates with/without 5-mph Collisions

	Dry road surface	Wet road surface	Uniform braking
A. Autonomous platoons			
10-car platoons	7391/7217	4671/4531	7280/7028
20-car platoons	7733/7532	4841/4683	7660/7365
B. Free agent infrastructure supported platoons			
10-car platoons	7727/7537	4802/4654	7607/7331
20-car platoons	7913/7703	4911/4748	7836/7529
C. Free agent infrastructure managed platoons			
10-car platoons	7909/7710	4872/4720	7786/7498
20-car platoons	8007/7792	4947/4782	7930/7615

The other assumptions regarding the initial conditions are the same for all architectures involving platoons. For the interplatoon spacing we used and compared several different concepts.

1. Autonomous platoons where the interplatoon spacing is calculated as the sum of the intervehicle spacing used in the case of autonomous vehicles

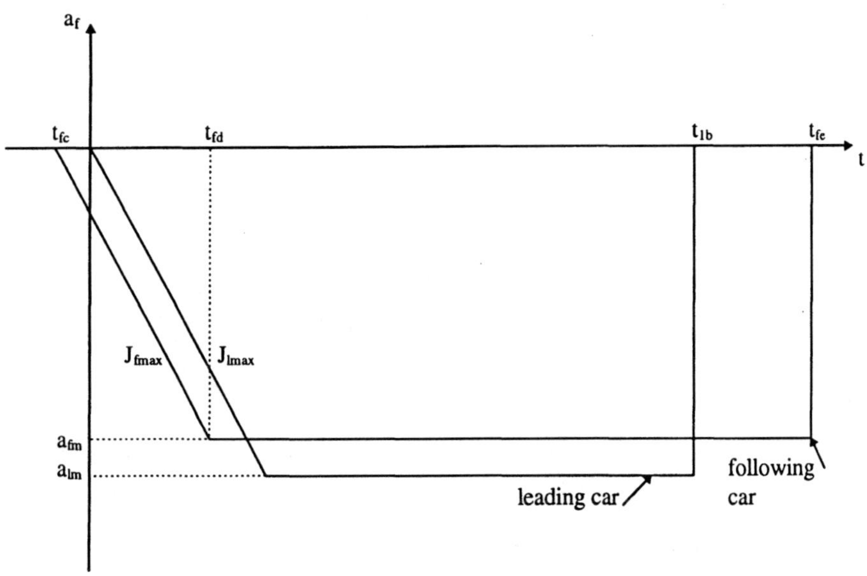

FIGURE 14. Platoons with coordinated braking with staggered delay.

TABLE 17
A. Platoons with Coordinated Braking: Delay of 0.1 s From Tail to Head

		Dry	Wet	Uniform
V_{lo}	mph	60	60	60
V_{fo}, mph	61.5	61.5	61.5	61.5
A_{lmax}	g	0.8	0.8	0.5
A_{fmax}	g	0.72	0.72	0.475
J_{lmax}	m/s^3	50	50	50
J_{fmax}	m/s^3	50	50	50
μ_{lmax}		1	0.5	1
μ_{fmax}		1	0.5	1
A_{fauto}	g	0	0	0
A_{fac}	g	0	0	0
J_{fc}	m/s^3	20	20	20
t_{fa}	s	0	0	0
t_{fc}	s	–0.1	–0.1	–20.1
min headway	s	**0.173**	**0.452**	**0.18**
	m	4.76	12.431	4.98

B. Platoons with Coordinated Braking: Delay of 0.1 s From Tail to Head, Allowing 5-mph Collisions

		Dry	Wet	Uniform
V_{lo}	mph	60	60	60
V_{fo}	mph	61.5	61.5	61.5
A_{lmax}	g	0.8	0.8	0.5
A_{fmax}	g	0.72	0.72	0.475
J_{lmax}	m/s^3	50	50	50
J_{fmax}	m/s^3	50	50	50
μ_{lmax}		1	0.5	1
μ_{fmax}		1	0.5	1
A_{fauto}	g	0	0	0
A_{fac}	g	0	0	0
J_{fc}	m/s^3	20	20	20
t_{fa}	s	0	0	0
t_{fc}	s	–0.1	–0.1	–0.1
min headway	s	**0.160**	**0.426**	**0.164**
	m	4.41	11.72	4.50
max headway	s	**0.116**	**0.229**	**0.164**
	m	3.19	6.30	4.50

and the product of the coordinated braking delay with the number of vehicles in a platoon. Each vehicle in the platoon assumes t_{fc} = –0.1 s. Each platoon entity assumes t_{fc} = 1.3 s for 10-car platoons and t_{fc} = 2.3 s for 20-car platoons.

2. Free agent platoons supported by the infrastructure where the interplatoon spacing is calculated as the sum of the intervehicle spacing used in the case

TABLE 18
Platoons of Passenger Vehicles with Coordinated Braking (t_{fc} = –0.1 s): Capacity Estimates with/without 5-mph Collisions

	Dry road	Wet road	Uniform braking
A. Autonomous platoons			
10-car platoons	7226/7060	4604/4468	7108/6889
20-car platoons	7637/7442	4802/4646	7551/7291
B. Free agent infrastructure-supported platoons			
10-car platoons	7540/7359	4729/4586	7408/7171
20-car platoons	7808/7604	4870/4709	7716/7445
C. Free agent infrastructure-managed platoons			
10-car platoons	7714/7525	4797/4649	7579/7330
20-car platoons	7901/7692	4905/4743	7808/7530

of free agent vehicles with infrastructure support and the product of the coordinated braking delay with the number of vehicles in a platoon. Each vehicle in the platoon assumes t_{fc} = –0.1 s. Each platoon entity assumes t_{fc} = 1.1 s for 10-car platoons and t_{fc} = 2.1 s for 20-car platoons.

3. Free agent platoons managed by the infrastructure where the interplatoon spacing is calculated as the sum of the intervehicle spacing used in the case of free agent vehicles with infrastructure management and the product of the coordinated braking delay with the number of vehicles in a platoon. Each vehicle in the platoon assumes t_{fc} = –0.1 s.

Each platoon entity assumes t_{fc} = 1.0 s for 10-car platoons and t_{fc} = 2.0 s for 20-car platoons.

The capacity is calculated using the formula:

$$C = (3600VN) / [(h_{PP}V + L_P)(N-1) + L_P + (H_{PP} + Nt_b)V] \tag{29}$$

where L_p is the length of each vehicle in the platoon (we have assumed vehicles of the same length), h_{pp} is the intraplatoon time headway, H_{pp} is the interplatoon time headway, N is the number of vehicles in the platoon, and t_b is the coordinated braking delay. The spacing is calculated using equation (29) based on the intraplatoon spacings given in Table 17A. Allowing intervehicle collisions at up to 5 mph yields the results of Table 17B. The capacity results with and without intervehicle collisions are presented in Table 18.

TABLE 19
Infrastructure-Managed Slotting: Capacity Estimates

	Dry road surface	Wet road surface	Uniform braking
0% mixing	4047	2826	3773

TABLE 20
Capacity Comparisons

	0% mixing of vehicles			5% mixing of buses			5% mixing of trucks		
Capacity without platooning	Dry	Wet	Uni-form	Dry	Wet	Uni-form	Dry	Wet	Uni-form
Autonomous vehicles with class identification	4116	2860	3850	3746	2516	3525	3458	2278	3096
Autonomous vehicles without class identification	4116	2860	3850	3631	2432	3416	3356	2207	3007
Free agents — infrastructure supported with class identification	5400	3425	4942	4730	2923	4377	4276	2605	3730
Free agents — infrastructure managed with class identification	6437	3823	5810	5472	3197	5018	4873	2820	4184
Infrastructure-managed slotting	4047	2826	3773						

	2.5% buses + 2.5% trucks			5% buses + 5% trucks		
	Dry	Wet	Uni-form	Dry	Wet	Uni-form
Autonomous vehicles with class identification	3596	2391	3297	3193	2054	2882
Autonomous vehicles without class identification	3488	2314	3198	3026	1943	2735
Free agents — infrastructure supported with class identification	4492	2755	4025	3845	2304	3400
Free agents — infrastructure managed with class identification	5155	2997	4563	4299	2464	3756

(*continued*)

TABLE 20 (*continued*)

	10-car platoons			20-car platoons		
Capacity with platooning	Dry	Wet	Uni-form	Dry	Wet	Uni-form
Autonomous platoons without coordinated braking	6090	5652	5955	6257	5977	6142
Infrastructure-supported platoons without coordinated braking	6312	5843	6166	6372	6081	6252
Infrastructure-managed platoons without coordinated braking	6434	5947	6283	6433	6137	6311
Autonomous platoons with coordinated braking	7217	4531	7028	7532	4683	7365
Infrastructure-supported platoons with coordinated braking	7531	4652	7323	7700	4747	7524
Infrastructure-managed platoons with coordinated braking	7704	4718	7489	7789	4780	7611
Autonomous platoons with delayed braking	7060	4468	6889	7442	4646	7291
Infrastructure-supported platoons with delayed braking	7359	4586	7171	7604	4709	7445
Infrastructure-managed platoons with delayed braking	7525	4649	7330	7692	4743	7530

4.10. Infrastructure-Managed Slotting

The infrastructure-managed slotting concept involves a different set of assumptions and parameters. We have not presented it in detail in the tables, except one table that shows capacity estimates under this architecture concept. We used the spacing data for passenger cars by assuming a doubling of all communication delays with an additional 3 m to account for position inaccuracy, related to the inability to utilize space effectively by using the exact slot size for each vehicle. We also assumed that the follower has no initial acceleration. The capacities computed under these assumptions can be found in Table 19.

5. DISCUSSION AND CONCLUSIONS

The capacity estimates for each concept considered are summarized in Table 20. These results indicate that the capacity is reduced by 30 to 40% by going from dry road to wet road conditions under each concept. The capacity is also reduced by about 10% if all vehicles are required to use lower but similar braking force during emergency stopping. Mixing of different classes of vehicles reduces capacity by about 11% in the case of mixing 2.5% buses and 2.5% trucks with passenger vehicles and by about 23% for 5% buses and 5% trucks. Platooning with coordinated braking gives the highest capacity. Infrastructure-managed slotting gives the lowest. The use of vehicle-to-vehicle communication for notifying vehicles about the onset of braking used in the free agent and platooning-based concepts helps increase capacity considerably.

The results developed are based on several assumptions regarding braking capabilities, worst-case stopping scenarios, and the like. We tried to make these assumptions as realistic as possible, by using braking data from actual experiments and by considering a wide class of concepts that cover a wide range of AHS configurations. Despite this effort there are still a lot of uncertainties in the choice of intervehicle spacing that need to be addressed. The level of conservatism is one of them and is related to the trade-off between safety and capacity. The frequency of failures on AHS operations that lead to the need for emergency braking is another uncertainty that depends on how AHS will be designed and operated. The results of this chapter are therefore qualitative in nature and can be used to compare the requirements and benefits of different AHS concepts.

ACKNOWLEDGMENTS. This work was performed as part of the California PATH Program of the University of California, in cooperation with the State of California Business, Transportation, and Housing Agency, Department of Transportation; and the U. S. Department of Transportation, Federal Highway Administration.

The contents of this chapter reflect the views of the authors who are responsible for the facts and the accuracy of the data presented herein. The contents do not necessarily reflect the official views or policies of the State of California. This chapter does not constitute a standard, specification, or regulation.

The authors would like to thank Steve Shladover, Jacob Tsao, Datta Godbole, and Jim Misener of PATH for many useful discussions on the subject of intervehicle spacing.

REFERENCES

1. M. SIVAK, Driver reaction times in car-following situations, Invited Presentation at the International Workshop on Driver Reaction Time, Jerusalem, Israel, January 5–6, 1987.
2. *Consumer Reports,* March to September 1995.
2. Y. SUN AND P. IOANNOU, A Handbook for InterVehicle Spacing in Vehicle Following, California PATH Research Report UCB-ITS-PRR-95-1, University of Southern California (1995).
3. *Road and Track,* April 1989 and October 1995.
4. C. C. CHIEN, Intelligent Vehicle Highway System (IVHS): Advanced Vehicle Control Systems, Doctoral dissertation, University of Southern California (1994).
5. M. LENARD, Safety Consideration for a High Density Automated Vehicle System (General Electric Co., Erie, PA, 1969).
5. C. C. CHIEN AND P. IOANNOU, Automatic vehicle following, *Proc. American Control Conference,* 1992.
6. A. RAHIMI *et al.*, Safety Analysis for Automated Transportation Systems, *Trans. Sci.* 7, 1 (1973).

6. J. K. HEDRIC, D. MCMAHON, V. NARENDRAN, AND D. SWAROOP, Longitudinal vehicle controller design for IVHS systems, *Proc. American Control Conference,* 1991, pp. 3107–3112,
7. U. KARASLAN, P. VARAIYA, AND J. WALRAND, Two proposals to improve freeway traffic flow, Preprint (1990).
8. P. IOANNOU, F. AHMED-ZAID, AND D. WUH, A time headway autonomous intelligent cruise controller: Design and simulation, Technical Report USC-SCT 92-11-01 (1992).
9. P. IOANNOU AND C. C. CHIEN, Autonomous intelligent cruise control, *IEEE Trans. Veh. Technol.* **42,** 657–672 (1993).
10. W. L. CALSON, Crash Injury Loss: The Effect of Speed, Weight and Crash Configuration. Accident Analysis and Prevention (1977).
11. J. GLIMM AND R. FENTON, Dynamic Behavior of Strings of Automated Transit Vehicles, *SAE Paper Series,* No. 770288 (1977).
12. S. E. SHLADOVER, C. A. DESOER, J. K. HEDRICK, M. TOMIZUKA, J. WALRAND, W. B. ZHANG, D. MCMAHON, and S. SHEIKHOLESLAM, Automatic Vehicle Control Developments in the PATH Program, *IEEE Trans. Veh. Technol.* **40,** 114–130 (1991).
13. B. S. Y. RAO, P. VARAIYA, AND F. ESKAFI, Investigations into achievable capacities and stream stability with coordinated intelligent vehicles, PATH Technical Report (1992).
13. A. SAGE (Ed.), *System Design for Human Interaction* (IEEE Press, New York, 1987).
14. S. SHEIKHOLESLAM AND C. A. DESOER, A system level study of the longitudinal control of a platoon of vehicles, *ASME J. Dyn. Sys. Meas. Contr.* **114,** 286–292 (1991).
15. S. E. SHLADOVER, Longitudinal control of automotive vehicles in close-formation platoons, *ASME J. Dyn. Syst. Meas. Contr.* **113,** 231–241 (1991).
16. B. W. STUCK AND E. ARTHURS, *A Computer Communications Network Performance Analysis Primer* (Prentice–Hall, Englewood Cliffs, NJ, 1985).
17. V. R. VUCHIC, *Urban Public Transportation, Systems and Technology* (Prentice–Hall, Englewood Cliffs, NJ, 1981).
18. NHTSA, Heavy duty vehicle brake research program report No. 1. Stopping capability of air braked vehicles volume 1, Technical Report No. DOT HS 806 738 (1985).
19. T. D. GILLESPIE, *Fundamentals of Vehicle Dynamics* (SAE, Warrendale, PA, 1992).
20. S. SWAROOP, C. C. CHIEN, P. IOANNOU, and J. K. HEDRICK, A comparison of spacing and headway control laws for automatically controlled vehicles, *J. Veh. Sys. Dyn.* **23,** 597–625 (1994).

APPENDIX A: VEHICULAR DATA REFERENCES

TABLE A.1
Braking Performance Comparisons of Popular Passenger Vehicles on Dry and Wet Roads

	Dry			Wet		
	Initial velocity	Stopping distance	Deceleration (av. g)	Initial velocity	Stopping distance	Deceleration (av. g)
			Family sedans[a]			
Chrysler Cirrus Lxi	60 mph	145 ft	0.83g	60 mph	167 ft	0.72g
Mercury Mystique LS	60 mph	140 ft	0.86g	60 mph	165 ft	0.73g
Ford Contour GL	60 mph	148 ft	0.81g	60 mph	158 ft	0.76g
Honda Accord LX	60 mph	143 ft	0.84g	60 mph	175 ft	0.69g
			Upscale sedans[b]			
Toyota Avalon XLS	60 mph	129 ft	0.93g	60 mph	146 ft	0.82g
Mazda Millenia S	60 mph	136 ft	0.88g	60 mph	157 ft	0.77g
Lexus ES300	60 mph	133 ft	0.90g	60 mph	167 ft	0.72g
Oldsmobile Aurora	60 mph	136 ft	0.88g	60 mph	155 ft	0.78g
			Low-priced sedans[c]			
Mazda Protege ES	60 mph	135 ft	0.89g	60 mph	167 ft	0.72g
Chevrolet Cavalier LS	60 mph	133 ft	0.90g	60 mph	165 ft	0.73g
Nissan Sentra GXE	60 mph	142 ft	0.85g	60 mph	158 ft	0.76g
Saturn SL2	60 mph	138 ft	0.87g	60 mph	157 ft	0.77g

[a]From *Consumer Reports*, March 1995.
[b]From *Consumer Reports*, May 1995.
[c]From *Consumer Reports*, June 1995.

TABLE A.2
Braking Performance Comparisons on Dry and Wet Roads of Popular Passenger Vehicles

	Dry			Wet		
	Initial velocity	Stopping distance	Deceleration (av. g)	Initial velocity	Stopping distance	Deceleration (av. g)
		Mid-sized coupes[a]				
Dodge Avenger ES	60 mph	129 ft	0.93*g*	60 mph	157 ft	0.77*g*
Ford Thunderbird LX	60 mph	131 ft	0.92*g*	60 mph	153 ft	0.79*g*
Chevrolet Monte Carlo Z34	60 mph	139 ft	0.87*g*	60 mph	165 ft	0.73*g*
Buick Riviera	60 mph	133 ft	0.90*g*	60 mph	147 ft	0.82*g*
		Sport-utility vehicles[b]				
Ford Explorer	60 mph	148 ft	0.81*g*	60 mph	181 ft	0.66*g*
Jeep Grand Cherokee	60 mph	144 ft	0.84*g*	60 mph	159 ft	0.76*g*
Chevrolet Blazer	60 mph	156 ft	0.77*g*	60 mph	172 ft	0.70*g*
Land Rover Discovery	60 mph	143 ft	0.84*g*	60 mph	202 ft	0.60*g*
		Small, cheap cars[c]				
Hyundai Accent 4-door	60 mph	137 ft	0.88*g*	60 mph	172 ft	0.70*g*
Hyundai Accent 2-door L	60 mph	145 ft	0.83*g*	60 mph	204 ft	0.59*g*
Toyota Tercel 4-door DX	60 mph	156 ft	0.77*g*	60 mph	195 ft	0.62*g*
Toyota Tercel 2-door base	60 mph	153 ft	0.79*g*	60 mph	193 ft	0.62*g*
Geo Metro 4-door LSi	60 mph	151 ft	0.80*g*	60 mph	172 ft	0.70*g*
Geo Metro 2-door LSi	60 mph	152 ft	0.79*g*	60 mph	199 ft	0.60*g*

[a]From *Consumer Reports*, July 1995.
[b]From *Consumer Reports*, August 1995.
[c]From *Consumer Reports*, September 1995.

TABLE A.3
Braking Performance Comparisons of Seven 4-Wheel Drive Vehicles on Dry Roads and on Snow[a]

	Dry			Snow		
	Initial velocity	Stopping distance	Deceleration (av. g)	Initial velocity	Stopping distance	Deceleration (av. g)
BMW 325iX	60 mph	142 ft	0.85*g*	20 mph	75 ft	0.18*g*
Audi 90 Quatro	60 mph	143 ft	0.84*g*	20 mph	99 ft	0.14*g*
VW Quantum GL5	60 mph	145 ft	0.83*g*	20 mph	59 ft	0.23*g*
Toyota Celica All-Trac	60 mph	146 ft	0.82*g*	20 mph	80 ft	0.17*g*
Subaru Justy 4WD GL	60 mph	151 ft	0.80*g*	20 mph	63 ft	0.21*g*
Subaru XT6 4WD	60 mph	153 ft	0.79*g*	20 mph	49 ft	0.27*g*
Pontiac 6000 STE 4WD	60 mph	N/A	N/A	20 mph	56 ft	0.24*g*

[a]From Road and Track, April 1989.

Braking Performance Comparisons on Dry Roads of Passenger Vehicles Representing Extremes[a]

	Dry			Dry		
	Initial velocity	Stopping distance	Deceleration (av. g)	Initial velocity	Stopping distance	Deceleration (av. g)
BMW 325i	60 mph	126 ft	0.95*g*	80 mph	212 ft	1.01*g*
Chevrolet Corvette LT1	60 mph	123 ft	0.98*g*	80 mph	225 ft	0.95*g*
Ford Mustang Cobra	60 mph	123 ft	0.98*g*	80 mph	214 ft	1.00*g*
Toyota Supra Turbo	60 mph	122 ft	0.99*g*	80 mph	208 ft	1.03*g*
Porsche 911 Turbo	60 mph	116 ft	1.04*g*	80 mph	199 ft	1.07*g*
BMW 740i	60 mph	144 ft	0.84*g*	80 mph	255 ft	0.84*g*
Chevrolet Camaro V6	60 mph	162 ft	0.74*g*	80 mph	282 ft	0.76*g*
Mercury Villager	60 mph	178 ft	0.68*g*	80 mph	293 ft	0.73*g*
Toyota Corolla DX	60 mph	186 ft	0.65*g*	80 mph	319 ft	0.67*g*
VW Golf III GL	60 mph	175 ft	0.69*g*	80 mph	301 ft	0.71*g*

[a]From Road and Track, October 1995.

Braking Performance Comparisons on Dry Roads of Air-Braked Heavy-Duty Vehicles[a]

	Initial velocity	Stopping distance	Deceleration (av. g)	Initial velocity	Stopping distance	Deceleration (av. g)
IH School Bus	20 mph	28 ft	0.48 *g*	60 mph	310 ft	0.34 *g*
Ford/IH Short School Bus	20 mph	36 ft	0.37 *g*	60 mph	375 ft	0.32 *g*
Thomas Transit Bus	20 mph	36 ft	0.37 *g*	60 mph	292 ft	0.41 *g*
Ford 4 by 2 Truck	20 mph	36 ft	0.37 *g*	60 mph	331 ft	0.36 *g*
GMC 6 by 4 Truck	20 mph	54 ft	0.25 *g*	60 mph	528 ft	0.23 *g*
Mack 6 by 4 Truck	20 mph	44 ft	0.30 *g*	60 mph	363 ft	0.33 *g*
Peterbilt 4 by 2 Tractor	20 mph	39 ft	0.34 *g*	60 mph	407 ft	0.30 *g*
Ford 4 by 2 Tractor	20 mph	30 ft	0.45 *g*	60 mph	289 ft	0.42 *g*
White 4 by 2 Tractor	20 mph	42 ft	0.32 *g*	60 mph	366 ft	0.33 *g*
IH 6 by 4 Tractor	20 mph	51 ft	0.26 *g*	60 mph	475 ft	0.25 *g*
Western Star 6 by 4 tractor	20 mph	46 ft	0.29 *g*	60 mph	431 ft	0.28 *g*
Stuart Conv. auto hauler	20 mph	43 ft	0.31 *g*	60 mph	434 ft	0.28 *g*
Stuart Stringer auto hauler	20 mph	39 ft	0.34 *g*	60 mph	354 ft	0.34 *g*

[a]From NHTSA test data.

APPENDIX B: TABLES OF RESULTS

TABLE B.1
Symbols and Notation

Symbol	Meaning
PP	Passenger car leader, passenger car follower
PB	Passenger car leader, bus follower
PT	Passenger car leader, truck follower
BP	Bus leader, passenger car follower
BB	Bus leader, bus follower
BT	Bus leader, truck follower
TP	Truck leader, passenger car follower
TB	Truck leader, bus follower
TT	Truck leader, truck follower
L_P	Length of a passenger vehicle, in meters (assumed equal to 4.75 m)
L_B	Length of a bus, in meters (assumed equal to 12 m)
L_T	Length of a truck with trailer, in meters (assumed equal to 18 m)
h_{PP}	Minimum time headway between passenger car leader passenger car follower, in seconds
h_{PB}	Minimum time headway between passenger car leader, bus follower, in seconds
h_{PT}	Minimum time headway between passenger car leader, truck follower, in seconds
h_{BP}	Minimum time headway between bus leader, passenger car follower, in seconds
h_{BB}	Minimum time headway between bus leader, bus follower, in seconds
h_{BT}	Minimum time headway between bus leader, truck follower, in seconds
h_{TP}	Minimum time headway between truck leader, passenger car follower, in seconds
h_{TB}	Minimum time headway between truck leader, bus follower, in seconds
h_{TT}	Minimum time headway between truck leader, truck follower, in seconds
V_{lo}	Leading vehicle initial velocity, in miles per hour
V_{fo}	Following vehicle initial velocity, in miles per hour
A_{lm}	The maximum achievable deceleration of the leading vehicle in g
A_{fm}	The maximum achievable deceleration of the following vehicle in g
J_{lmax}	The maximum achievable jerk of the leading vehicle in meters/sec^3
J_{fmax}	The maximum achievable jerk of the following vehicle in meters/sec^3
μ_{lmax}	The maximum road–tire friction coefficient (dimensionless)
μ_{fmax}	The maximum road–tire friction coefficient (dimensionless)
A_{fauto}	The acceleration value under automatic brake control during soft braking, in g
A_{fac}	The initial acceleration value during vehicle following, in g
J_{fc}	The jerk value under automatic brake control during soft braking, in meters/sec^3
t_{fa}	Detection and brake actuation delay applicable to the following vehicle, in seconds
t_{fc}	The time at which the following vehicle starts the emergency braking maneuver, in seconds

CHAPTER 9

COMMUNICATION TECHNOLOGIES FOR AHS

ANDREAS POLYDOROS AND PROKOPIOS PANAGIOTOU

1. INTRODUCTION

The world is going digital, exclusively, and wireless—to a good degree. Anyone with a digital wristwatch, a computer, or a fax machine is well aware of the first truth, and anyone who owns a cordless phone is familiar with the second. Digital cellular technology will soon be used in all forms of broadcasting, from satellite Direct Audio Broadcasting (DAB) of high-quality FM directly to home or car, to High-Definition TV (HDTV), interactive TV (pay-per-view movies and assorted transactions from the home), wireless Personal Communications Services (PCS), including Low-Earth Orbiting (LEO) satellite systems now being designed, and multimedia machines for education and entertainment. A huge eager market already exists for these technologies, or is fast being created, the industrial players have the resources to design and deploy them, and the legal/regulatory framework is being readjusted to accommodate them.

What does this digital communication revolution have to do with Automated Highway Systems (AHS)? Plenty, as it turns out. AHS is an emerging application area that exemplifies the transition between the challenges of the second half of the 20th century and those at the start of the third millennium. It

ANDREAS POLYDOROS AND PROKOPIOS PANAGIOTOU • Communication Sciences Institute, University of Southern California, Los Angeles, California 90089-2565.

Automated Highway Systems, edited by Ioannou. Plenum Press, New York, 1997.

aims to utilize, among other things (such as control and transportation technologies), the vast telecommunication knowledge base developed over the last 40 years in order to develop and deploy solutions that can alleviate the existing transportation problems, and therefore enhance the efficiency, quality, and safety of an activity that is integral to daily life, namely, the use of highways and streets. Efficient use of the existing transportation facilities and energy resources promises to reduce the current levels of congestion, thus improving the air quality and protecting the environment.[1]

The purpose of this chapter is to present candidate communication technologies (existing, emerging, and future) that can provide real-time, reliable data communications between vehicles and roadway infrastructure as well as between vehicles. Significant attention is paid to the emerging technology of packet radio, as it is envisioned that it will play a major role in forthcoming solutions to the communication subsystem design problem. The chapter is organized as follows: In Section 2, we discuss the AHS functional requirements, as well as the information flow requirements for two major AHS applications: Advanced Traffic Management and Information Systems (ATMIS) and Advanced Vehicle Control Systems (AVCS). The suitability of several existing and emerging communication technologies for AHS applications is examined in Section 3, whereas Section 4 presents a survey of AHS-related worldwide efforts. Future technologies (in particular packet radio technologies) are presented in Section 5. Finally, Section 6 summarizes the conclusions of this chapter.

2. AHS FUNCTIONAL REQUIREMENTS

It is expected that various types of applications/services will be provided to meet the goals of AHS and satisfy customer needs. The right choice of communication technology for each application depends heavily on the type of *functional requirements* arising from that application. We review here the most prominent of such requirements as they arise within the AHS framework. A partial list of AHS user services that are envisioned is as follows[2,3]:

- *Traveler information*—such as traveler advisory, traveler information services (value-added), trip planning, position location systems, navigation, route guidance, weather information, visitor attractions, restaurants, events, hotels, gas stations.
- *Traffic management*—such as incident detection and management, demand management, traffic network monitoring, traffic control, parking management, workzone management, and automatic toll collection.
- *Vehicle control systems*—such as adaptive cruise control, collision warning, collision avoidance (automatic braking and lane departure warning systems), longitudinal (speed and spacing) and lateral control (steering).

- *Freight and fleet management*—such as intermodal transport planning, route planning and scheduling, hazardous material monitoring and tracking, vehicle and cargo monitoring, law enforcement and regulatory support.
- *Public transportation and emergency vehicle management*—such as planning and scheduling systems, signal pre-emption traffic control, automatic payment, dynamic ride sharing, prediction of arrivals, and emergency services systems management.
- *Additional services*—such as traveler safety and security, Mayday transmissions, and hazard warning.

Naturally, it would be desirable to have a single communications infrastructure to support all of the applications envisioned within AHS. It is conceivable, however, that more than one communications interface will end up being necessary within a vehicle because of the differing requirements of different applications. Each of the envisioned applications has its detailed requirements such as area of coverage (local, regional), directionality (one-way, two-way, vehicle-to-infrastructure, vehicle-to-vehicle), addressability (broadcast, multicast, point-to-point), voice and/or data communications, message size, message frequency (periodic, nonperiodic), latency, reliability, and prioritization of messages.[4] Clearly, a wide range of communication requirements emerges from such a diverse landscape of application prerequisites.

Route guidance and automatic toll collection provide a good example of services that have completely different requirements. For route guidance, the contents of the messages from and to the roadside infrastructure do not directly depend on each other. Thus, a half-duplex communication system ensures that a vehicle transmits its request and then receives the necessary information. Automatic toll collection, however, requires full-duplex communications between vehicles and roadside infrastructure. To illustrate further the disparity of requirements, we present below the information flow requirements of two AHS applications, namely, ATMIS and AVCS.

2.1. Advanced Traffic Management and Information Systems (ATMIS)

ATMIS combines the services offered by Advanced Traffic Management Systems (ATMS) and Advanced Traveler Information Systems (ATIS). The integration of these two applications promises to improve the traffic flow by detecting accidents and bottlenecks and provide drivers with valuable guidance information.[1] In ATMS, the traffic conditions are monitored by a surveillance system and the collected data are processed in a Traffic Management Center (TMC). These data are used to inform drivers about traffic conditions, accidents, and alternative routes. ATMS services make possible the real-time adjustment of traffic control systems, which in conjunction with Changeable Message Signs (CMS) for driver advice can reduce delays, expected travel time, and accidents, thus optimizing the traffic flow on highways and surface streets.

In ATIS, communication between vehicles and a TMC provides the drivers with information about traffic conditions and information about services such as hotels, gas stations, restaurants, etc. In-vehicle display units provide electronic maps, route selection, and real-time route guidance. In addition to trip-related information, ATIS services offer electronic vehicle identification for toll debiting, safety advisory and warning messages, Mayday signaling and capabilities for immediate response in emergencies.

2.2. Advanced Vehicle Control Systems (AVCS)

AVCS technologies have the potential to increase substantially the transportation safety and capacity of the highways, and to provide drivers with electronic equipment capable of performing driving functions. AVCS will combine current and future technologies (intelligent cruise control, vision enhancement systems, collision warning and avoidance systems, lane departure warning systems, automatic braking systems) to provide full vehicle and highway automation.[1]

The first partially automated vehicles will be equipped with sensors and detection radars, making them capable of obstacle detection and avoidance. Later, the development of lateral (steering) and longitudinal (speed and spacing) control systems will enable platooning of vehicles and automatic lane-changing maneuvers. The full automatic steering of the vehicles will be the last step towards a completely automated vehicle control system, which will virtually eliminate the human driving functions and help to reduce accidents and increase traffic flow.[1] Simulation results have shown that the goals of AVCS can be achieved by organizing the vehicles in platoons.[5] According to a glossary of AVCS terminology,[6] a platoon is defined as "a group of vehicles operating under closely-coupled vehicle follower longitudinal control. The close coupling is achieved by communication among the member vehicles of information about vehicle movements and potential anomalies. Platooning can not be achieved with autonomous vehicles or point-follower longitudinal control."

The two applications have been thoroughly examined in Ref. 7 in order to extract the communication needs and the data requirements of an integrated ATMIS/AVCS system. The information flow requirements associated with the various communication needs are summarized in Table 1.[7] We notice that uplink vehicle-to-infrastructure and vehicle-to-vehicle communications require the exchange of short messages (64–100 bits) for both ATMIS and AVCS transactions, since they contain only control information (velocity and acceleration), or simple requests for information.[8,9] The message size for the downlink (infrastructure to vehicle), however, depends on the application. AVCS services require transmission of short messages, whereas ATMIS services such as traveler advisory and information (including the transmission of digitized maps) require the transmission of long messages which can be broken up into several packets.[7] The study of the data requirements for the integrated system has concluded that the AVCS communication needs require periodic, high-data-rate, real-time data traffic, whereas the ATMIS data traffic is nonperiodic, and low data rate. Although some ATMIS services require real-time data traffic, the overall ATMIS data requirements are not as stringent as the ones for AVCS.

TABLE 1
ATMIS/AVCS Information Flow Requirements

			Vehicle-to-Vehicle	
	Uplink	Downlink	Intra-platoon	Inter-platoon
ATMIS	Info requests Traffic probe (short messages)	Traveler info & advisory (long messages)	—	—
AVCS	Platoon status information Requests for lane changes (short messages)	Target velocity & acceleration Handling of lane changes (short messages)	Lane keeping maneuvers Lane changing maneuvers (short messages)	Lane changing maneuvers (short messages)

3. TECHNOLOGY AVAILABILITY

There arises next the natural question of what available technologies exist today for accommodating the above AHS requirements, and what needs to be developed further for such a need. Several existing and emerging technologies have been suggested as candidates for providing communications for AHS applications, with each technology having its advantages and disadvantages. The evaluation of their suitability for certain applications depends on the communication requirements and spectrum needs of each application, which in turn depend on the application's data requirements.(10) The task of determining the detailed data requirements of the various applications and proposing standards is particularly difficult because of the great variety of applications under consideration and of their evolutionary nature. Furthermore, the specification of appropriate communication protocols depends on the spatial and temporal aspects of the users' distribution in the area of coverage and on the way that intelligence is distributed between the infrastructure and the vehicles. In the following subsection, we discuss the main features of some of the candidate technologies.

3.1. Candidate Technologies

The following technologies are included here: Highway Advisory Radio (HAR) and Advanced Highway Advisory Radio (AHAR), Frequency Modulation (FM) Subcarriers, Radio Data System (RDS), High Speed FM Subcarrier Data System (HSDS), Vertical Blanking Interval (VBI) and Secondary Audio Programming (SAP), Roadside Beacons, Infrared Technologies, Digital Cellular Systems, Personal Communications Services (PCS) Systems, Mobile Satellite Systems, Satellite-based Position Location Systems, and Meteor Burst Systems. We briefly discuss the salient features of each technology and comment on their suitability within the AHS framework.

3.2. Highway Advisory Radio (HAR) and Advanced Highway Advisory Radio (AHAR)

These two systems operate at low power levels in the AM band (530 and 1610 kHz).(11) In the future, HAR and AHAR stations will operate anywhere in the available portion of the AM band. They have been used extensively for providing voice services like broadcasting of traveler information (e.g., traffic information, parking availability). However, they offer limited capacity for data communications because the maximum modulating frequency in the AM band is 5 kHz. Furthermore, since HAR stations operate in a limited geographical area, their use for AHS applications requires the installation of a large number of stations.

3.3. Frequency Modulation/Subsidiary Communications Authorization (FM/SCA)

The Federal Communications Commission (FCC) has given authorization to FM stations to use the spectrum that is not occupied by the FM channel for broadcasting of other programs (e.g., sports information, paging data, teleprinter newscasting) on SCA subcarriers, provided that there is no interference with the audio channel.[11-13] The North American standard has allocated 200 kHz for FM stations and allowed 53 to 99 kHz for a subcarrier frequency. The FCC has not imposed any restrictions on the choice of modulation. Depending on the modulation and coding schemes, the data rate of the SCA channels can be more than 16 kbps. Since SCA transmitters and receivers are relatively inexpensive and FM broadcast stations cover more than 90% of the area of the United States, an installed infrastructure that can support some AHS applications requiring the broadcasting of information (e.g., traffic information) already exists. This technology, however, is not suitable for providing two-way communications between vehicles and roadway infrastructure.

3.4. *Radio Data System (RDS)*

RDS uses an AM subcarrier located at 57 kHz on the FM baseband.[11,14] The total baseband bandwidth of RDS is approximately 7 kHz (53–60 kHz) and the data rate is 1187.5 bps. In Europe, RDS has been used extensively for a variety of services (including traffic information). There are plans for installation of the system in the United States as well, but the only envisioned service so far is program type identification (e.g., news, sports, kind of music). RDS is a very robust system with relatively low cost. However, the fact that it can only provide one-way communications and the low data rate make RDS prohibitive for certain AHS services (e.g., route guidance).

3.5. *High-Speed FM Subcarrier Data System (HSDS)*

HSDS is a time division multiplexed system that has been developed by Seiko Telecommunication Systems, Inc.[14] It is a one-way broadcasting system that can provide services such as personal messaging, traffic information, news, and weather information. HSDS uses Double-Sideband Suppressed-Carrier Amplitude Modulation (DSB-SC AM) with duobinary encoding. The HSDS signal is modulated as a subcarrier (typically 10%), and the data rate is 19 kbps in a 19-kHz bandwidth centered at 66.5 kHz. The system has been designed so that the interference to the main audio channel, or to any existing RDS subcarrier, is negligible. HSDS appears to be a more promising technology than FM/SCA and RDS. Both the infrastructure and the receivers required are relatively inexpensive. It is, however, a one-way broadcast system and is not interoperable with other systems.

3.6. *Vertical Blanking Interval (VBI) and Secondary Audio Programming (SAP.)*

Every TV station uses two separate transmitters for the video and audio signals.[11] VBI is used to modulate the first 20 picture lines of each frame, which do not appear on the screen because they occur during the blanking time interval. A single VBI line (Line 4) is already used extensively to provide closed captioning.

Existing technology can combine all 20 lines into a single data channel, with a resultant data rate of about 56 kbps.

Besides some different technical characteristics, the TV audio transmitter operates like an FM station and, therefore, subcarriers can also be supported. One of the standard TV audio subcarriers is Secondary Audio Programming (SAP). It is being used to broadcast a running commentary of TV programs (usually news). The wider bandwidth of the SAP channel allows for more than one wideband subcarrier, which can be used for data traffic. TV stations can provide extra capacity in major metropolitan areas, where there are more TV stations than FM stations.

3.7. Roadside Beacons

Many existing systems (see Section 4) offer data services by using beacons installed at the roadside. These beacons provide a short-range, wireless communication link for services such as toll collection, roadside safety inspection, and in-vehicle signing.[14] The three candidate technologies for this type of link are active radio frequencies [e.g., the Vehicle to Roadside Communication (VRC) system proposed by Hughes[2]], passive (modulated backscatter) radio frequencies, and infrared. A widely deployed beacon-based communication system would be technically capable of accommodating the communication needs of most of the envisioned AHS applications. However, the deployment of such a system faces two obstacles: delay (transmission delays which occur when a vehicle is located in a zone that is not covered by any beacon) and complexity.

3.8. Infrared Communication Links

Infrared technologies have received considerable attention for both short-range, vehicle-to-infrastructure (via infrared beacons) and vehicle-to-vehicle communications.[8] Under ideal conditions, an infrared communication link can support data rates up to 1 Mbps. The data rate, however, reduces as the distance between the transmitter and the receiver increases. Furthermore, the reliability of such links is severely affected by weather conditions. For distances more than 5 m, attenuation varies from 0.01 dB (clear conditions) to 10 dB (heavy fog).[15] Reliability and alignment sensitivity problems are the major drawbacks in the deployment of infrared technologies for vehicle-to-vehicle communication links.

3.9. Digital Cellular Systems

The increased demand for cellular telephony services all over the world has led to the development of new digital cellular systems that have the capability of increasing the user capacity of analog systems. In the United States, two standards have been proposed to replace the existing analog cellular system AMPS (Advanced Mobile Phone System): IS-95 (developed by Qualcomm Inc.) and IS-54.[16] IS-95 uses Direct Sequence Code Division Multiple Access (DS-CDMA) as a multiple access mechanism. It has been designed to replace a number of existing AMPS channels with a single DS-CDMA channel. The other standard, IS-54, has already been deployed commercially in several areas of the United States. It employs a 3 slot/frame Time Division Multiple Access (TDMA) system and each channel replaces one existing 30-kHz AMPS channel. In Japan, a similar TDMA system has been designed to operate within the narrow channel bandwidth (25 kHz) of the existing analog system. The Pan European digital cellular system is called GSM[(17)] (Global System for Mobile Communications—previously called Groupe Spécial Mobile) and uses an 8 slot/frame TDMA system at a data rate of 270.83 kbps.

Although these systems have been designed primarily for voice services, there are plans to utilize them for a variety of data services as well.[(18)] In the United States, three primary data services have been considered: asynchronous data (e.g., file transfers), facsimile transmissions, and Short Message Services (SMS). In Europe, the interest in providing mobile data communications using GSM has led to the development of a family of standards, called TETRA (Trans-European Trunked Radio).[(18)]

All of these systems operate in a circuit-switched mode, where a channel is assigned to a user for the duration of the call. This mode is very efficient for providing voice services and for transfers of large files. It is, however, inefficient for data communications because of the bursty nature of data messages, which occupy a channel for a short period of time.[(18)] Cellular operators have developed a new standard, called Cellular Digital Packet Data (CDPD),[(14,18)] as an overlay on the existing analog cellular telephony system AMPS in order to allow users to transmit data over a cellular channel. CDPD users can transmit digital data messages at a burst rate of 19.2 kbps over a channel that has been assigned either permanently or dynamically to them. There is no need for dedicated CDPD channels (although some service providers may choose to do so), because CDPD has been designed to use a channel with no voice activity and then switch to another channel when there is voice activity in the channel being used.

Since packet data services do not require the setup of a virtual circuit between the transmitter and the receiver and the overhead associated with it (as opposed to voice services), CDPD can increase the user capacity of a cellular system. Furthermore, in CDPD there is no modem synchronization delay, unlike circuit-switched mode, where modem synchronization can take as long as 20 s when a circuit is established. CDPD has the potential to provide many services (e.g., electronic mail, vehicle theft recovery, database services) including a variety of AHS services (in particular, those that require vehicle-to-infrastructure communications).[(14)]

3.10. *Personal Communications Services (PCS) Systems*

The FCC has defined PCS as "a family of mobile or portable radio communications services which could provide services to individuals and business and be integrated with a variety of competing networks. . . . The primary focus of PCS will be to meet communications requirements of people on the move."[(19)] Recently, the FCC allocated spectrum in the 1.85–1.97- and 2.13–2.20-GHz bands for use by PCS systems.[(20)] Both voice and data services are envisioned within PCS; they include mobile microcellular services, microcellular cordless telephone services (for pedestrians and offices), facsimile transmissions, and others.

PCS systems have the potential to provide some AHS services as well. One form of PCS that has this potential is a digital cellular system operating in the assigned bands.[14] Both IS-95 and GSM (with some modifications) have been proposed for use in the PCS bands. An extension of the GSM standard (called DCS 1800) has been already deployed in Europe in the 1.8-GHz band.[16,17]

3.11. *Mobile Satellite Systems*

Although Geosynchronous Earth Orbit (GEO) satellites are being used extensively for providing worldwide telecommunication services, attention has been shifted in the past few years to the use of LEO satellite systems for wireless communications between mobile users anywhere in the world.[14] LEO systems have low power requirements, thus allowing for smaller user terminals, and short time delays, because they are closer to earth than GEO systems. There are two types of LEO systems: little LEO and big LEO systems.[14] Little LEO systems operate in the VHF band and have been designed for data communications only. Big LEO systems operate above 1 GHz and offer both voice and data communications. They also use more satellites to achieve higher capacity and reliability. We note that all of these systems provide position location information as an alternative to the system described below.

3.12. *Satellite-Based Position Location Systems*

Position location information is a crucial element for the deployment of many AHS applications. One way that this information can be provided to mobile users is triangulation by waveforms from satellite transmitters. Transit was the first satellite-based system used for position location.[21] The most popular system currently in use is GPS (Global Positioning System),[21] which uses 24 satellites in 12-h orbits. The satellites are spaced uniformly around the earth and their positions are known precisely at all times. Each satellite transmits two spread spectrum signals at 1575.42 and 1227.60 MHz which also convey low-data-rate information about the satellite. A mobile user can determine its position by acquiring the codes of at least four different satellites (in its field of view) and solving simultaneously the corresponding delay equations. The resulting position accuracy is of the order of tens of feet, which can be reduced to about a foot by using signal processing techniques (differential GPS).

3.13. *Meteor Burst*

A meteor burst[14] (or meteor scatter) communication system can provide long-range packet data communications using the ionized trails of meteors as reflectors. Meteor burst systems can be used for communication services in rural areas (provided that there is no need for emergency messages), where conventional cellular systems cannot offer cost-effective coverage. They can also be a cheap alternative to expensive satellite systems, because of their low cost and the small number of base stations needed for their deployment. They have the potential to support AHS services such as wide-area Automatic Vehicle Location (AVL) and Fleet Management in rural areas. In urban areas, however, the noise levels are too high for meteor burst systems and therefore, a combination with cellular systems should be considered for seamless coverage.

3.14. Technology Applicability

The applicability of existing and emerging communication technologies to the various AHS applications is shown in Table 2.[(4)] It is seen that no single technology can accommodate the communication needs of all applications. Therefore, it is necessary to identify new technologies that can meet the remaining needs. Furthermore, it is interesting to explore technologies that may be capable of providing reliable communications for integrated systems (e.g., an integrated ATMIS/AVCS system). The difference between existing and new technologies is the degree of interactivity (i.e., one-way versus two-way communications) between the users and the infrastructure. Several systems that have already been deployed use sensors for collection of information. This information is transmitted to a TMC, which in turn regulates Changeable Message Signs (CMS), or broadcasts this information to all vehicles. This is an example of the crudest form of interactivity that already exists. In a higher level of interactivity, information is exchanged between users and infrastructure, and action is taken by the user in an automated way. The focus of current recent efforts is on the use of technologies that can provide real-time, reliable data communications between vehicles and roadway infrastructure as well as between vehicles. A critical participant in this new list of technologies—packet radio—is discussed in Section 5.

4. AHS-RELATED WORLDWIDE PROGRAMS

A lot of effort has been undertaken worldwide, by governments and private industry alike, to explore this envisioned growth area. The following subsections present a survey of North American, European, and Japanese AHS-related programs. This is by no means an exhaustive list of all worldwide efforts; that would be beyond the scope of this chapter. Our main purpose here is to present the communication technologies that have been used in the most prominent programs.

4.1. North American Programs

North American programs include Pathfinder, SMART Corridor (it is also the test site for Pathfinder), and PATH (Partners for Advanced Transit and Highways) in California, TravTel (Travel Technology) in Orlando, Florida, Advance in the Chicago area, and GuideStar in the Minneapolis/Saint Paul area.[(1,2,22)] All of these

TABLE 2
Applicability of Candidate Technologies to AHS Applications

Technologies	ATMIS	AVCS	Public Transit	Fleet Management	Truck Monitoring	Rural Systems
Short Range Radio (HAR, AHAR)	Some voice	No	No	No	No	Some voice
Subcarriers (FM/SCA, RDS, HSDS, VBI, SAP)	Some voice All 1-way data	No	Some voice and data (1-way)	Niche applications	No	Some voice and data (1-way)
Roadside Beacons	All data	Some	All data	Some data	Some	Some data
Cellular	All	Some	All (voice and data)	All (voice and data)	Some	All (voice and data)
PCS	Maybe all	Maybe Some	All (voice and data)	All local (voice and data)	All local (voice and data)	No
Satellite	All (voice and data)	No	Wide area only	All (voice and data)	Some	All
Meteor Burst	No	No	No	No	Some data	No

programs have focused on the integration of traffic sensors, computers, and various communication technologies in order to coordinate traffic (both in highways and surface streets) and disseminate traffic conditions to drivers. Traffic information is collected from various sources, processed at a TMC, and then broadcasted to all vehicles using a broad range of methods (e.g., CMS, HAR). In addition to providing traffic information, some of these programs have tested and implemented route guidance systems using dead-reckoning equipment with map-matching and GPS receivers for position calibration.

The PATH program is also investigating the use of current and future technologies (e.g., collision warning and avoidance systems, lane departure warning systems) for the development of fully automated vehicles and highways, which will increase the safety and mobility of the existing highway system.[23,24] PATH research also includes studies of the performance and impact of route guidance systems, as well as studies on driver behavior.

4.2. European Programs

Most of the numerous European projects have been fully coordinated elements of major programs that transcend the borders of individual countries. The two major European programs were DRIVE (Dedicated Road Infrastructure for Vehicle Safety in Europe) and PROMETHEUS (Program for European Traffic with Higher Efficiency and Unprecedented Safety).[1,2,22]

The principal objective of DRIVE was to develop standards and specifications for a coherent Integrated Road Transport Environment (IRTE) in order to provide services such as traffic information, route guidance, advanced traffic management and control, collision avoidance, and so on. SOCRATES (System of Cellular Radio for Traffic Efficiency and Safety) and TARDIS (Traffic and Roads-Drive Integrated Systems) were the two major DRIVE projects. SOCRATES has investigated the use of dedicated channels from the European cellular standard GSM for broadcasting information to all properly equipped vehicles in a cell. TARDIS has tried to establish common functional specifications for systems that depend on communication between vehicles and a roadside infrastructure. It has also investigated the feasibility of a single communication system that can support both route guidance and automatic toll collection.

Whereas DRIVE was infrastructure-oriented, PROMETHEUS was a vehicle-oriented project. An important PROMETHEUS concept was that of the *intelligent copilot*; an integrated set of vehicular systems linked by communication networks with the roadside infrastructure and with other vehicles, whose purpose was to aid and enhance the capabilities of the driver.

Several navigation and route guidance systems [e.g., Ali-Scout (now renamed Euro-Scout), Trafficmaster, Autoguide, LISB (Leit und Information System Berlin)] have already been demonstrated in Europe.[1,2] All of them are real-time, traffic information systems that provide drivers with up-to-the-minute information on the speed, direction, and length of any traffic backup. They enable drivers to assess the situation and take avoiding action, if desired, but do not suggest what evasive action to take. A network of infrared roadside beacons has been utilized to provide two-way communications between vehicles and a TMC.

4.3. Japanese Programs

The four major Japanese programs are AACS (Automobile–Automobile Communication System), RACS (Road/Automobile Communication System), AMTICS (Advanced Mobile Traffic Information and Communication System), and VICS (Vehicle Information and Communication System).[1,2,22,25]

The AACS program has investigated the area of autonomous vehicle-to-vehicle communications that can provide services such as relay of traffic information, warning messages (accidents, head-on collision), and emergency messages. Communication technologies that have been used include inductive radio, low-power UHF, and packet radio.

In the RACS design, on-board navigation equipment is incorporated in the vehicle and the optimal route is selected using dead-reckoning with map-matching (the maps are stored in CD ROMs). The RACS program has employed roadside beacons that were connected to a TMC, so that traffic information could be continuously updated. Besides providing information services, they were also used for transmission of location data so that accumulated errors (inherent in dead-reckoning systems) could be corrected. Two communication systems have been tested: an inductive radio system and a microwave system in the 2.5-GHz band.

The in-vehicle equipment that has been used in the AMTICS program is similar to that of RACS. The communication interface, however, was different. AMTICS used a one-way communication system for broadcasting traffic data and other information (e.g., traffic conditions, weather, available parking spaces) from the Teleterminal System (a cellular digital packet radio system with 3-km-radius areas, operating at 800 MHz). An uplink capability was also available through the Teleterminal System, but no vehicle-to-roadside infrastructure communication was required in the basic AMTICS design.

The VICS program has been created in order to unite the AMTICS and RACS approaches, so that a consensus on the communication architecture and protocols can be established. Possible communication technologies that have been considered include the 800-MHz Teleterminal System, FM multiplex, microwave beacons, and signposts (13-GHz microwave radio units).

The above is but a sample of the host of activities taking place everywhere, and is indicative of the intense interest of all major industrial players in AHS.

5. FUTURE TECHNOLOGIES: PACKET RADIO

Packet Radio Architectures (PRA) are a set of design options around the notion of radio transmissions within a dense network of small distributed antennas installed in cars, as well as in a future terrestrial infrastructure. Radio is the medium of choice for the wireless data transport since it is currently used more extensively than either optical (laser) or sound waves, and it promises to lead to inexpensive and widely available components. Packetized communication, the other main feature of these architectures, pertains to the process of "chopping" communication into small pieces or "packets" which are then exchanged between transmitter and receiver by following specific rules (called link-access protocols).

Packet radio technology has been experimented with over the last 15 years or so for large, complex, highly mobile networks such as those found in the defense world, especially for rapidly deployable forces. It has also been implemented commercially in a variety of systems, ranging from stationary, point-to-point systems [e.g., High-Speed Wireless Local Area Networks (LANs)] to Low-Speed Wide Area systems providing mobile data services. Existing mobile data networks (EMBARC, Skytel, ARDIS, RAM Mobile Data Network) use packet radio technology to provide services like one-way paging, two-way transfer of data files, and computer-aided dispatching with store-and-forward and transparent roaming capabilities.[18] Tracking companies, emergency vehicles and other fleet management operations have capitalized on these recent systems to improve their efficiency (Nextel Communications). The development of CDPD (see Section 3.9) has enabled users to utilize the existing analog cellular system for data services. The allocation of the Industrial–Scientific–Medical (ISM) band for unlicensed operation (provided that it complies with the FCC restrictions) has also provided fertile ground for the development of low-cost spread spectrum packet radio systems providing services such as automated utility meter reading (e.g., Utilicom Inc.). The following subsection identifies the design issues of packet radio systems for AHS applications.

It is expected that at least some of the AHS communication needs (most likely those without time criticality as an element) will be satisfied by these new entrants in the packet radio world. Many of the envisioned AHS applications require the exchange of messages that are short in duration. Thus, it proves advantageous from the standpoint of channel utilization to use such packet switching techniques. On the other hand, some applications imply large messages, high data rates, and heavy usage of the channel (e.g., transfers of large files). Whether these two classes of applications should use separate frequency bands or can be merged and designed as a composite packetized system is one of the many architectural questions.

The previous experience is helpful in visualizing packet radio as a suitable vehicle for AHS, but it cannot be adopted outright because of the idiosyncrasies of the envisioned AHS applications. For example, there will be hundreds or even thousands of users on a segment of a highway, each generating message traffic in a highly dynamic fashion. Many of these messages will demand extremely high reliability and almost instantaneous success during transmission, in particular for automated highway maneuvers. And, finally, it is expected that information will flow back and forth between the vehicles and the infrastructure, or from vehicle to vehicle, in a possibly unpredictable way.

Research efforts have focused on architectures that include fixed, distributed, dedicated communication nodes (called base stations), to which all of the mobiles are to be connected (in a wireless fashion, of course) via a cellular structure much like the one envisioned for PCS, except geared specifically to AHS and centered around the highway system. These base stations will transmit and receive information to and from the vehicles using these packet radio techniques. For some applications (e.g., ATMIS), this information will then be relayed to a TMC; information will also flow from the TMC to these base stations. The two-way links for this latter part of the overall communication process can be either through dedicated ground networks (e.g., fiber, wire) or through additional point-to-point wireless links. For other applications (e.g., AVCS), the base stations can also serve as intermediary nodes for exchange of information between moving vehicles.

Designing and building a packet radio system for AHS is a multi-faceted and challenging endeavor. Not only should the details of the information flow be worked out in a flexible and efficient way, but many other aspects of the problem must be addressed as well. The most prominent of these are:

a. Specification of the data requirements for a diverse menu of applications and services
b. Identification of the communication requirements associated with these specifications
c. Identification of design options capable of accommodating communication requirements
d. Legal/regulatory issues, including the crucial aspect of spectrum allocation
e. Financial, marketing, and consumer issues
f. Interface, coexistence, and coordination with various other communication systems

In order to develop technical guidelines for ultimate system composition, design options must be analyzed. Some of the areas in which design options exist are:

1. Choice of the physical layout of the network, including the density of the base stations and their location in relation to the highway
2. Spectral allocation of the system, an issue closely related to spectrum availability,[26] external interference, and propagation characteristics in environment
3. Signal design, which includes the composite choice of the modulation, error correction coding, spectral spreading (if any), and multiple-access protocol

A generic conceptual model that can be used to study and quantify the design trade-offs of the Physical and the Link-Access layers of a packet radio system that can accommodate real-time, interactive communications between vehicles and infrastructure as well as between vehicles is shown in Figure 1.[7] The model includes two modules for the performance evaluation of the two layers and identifies the

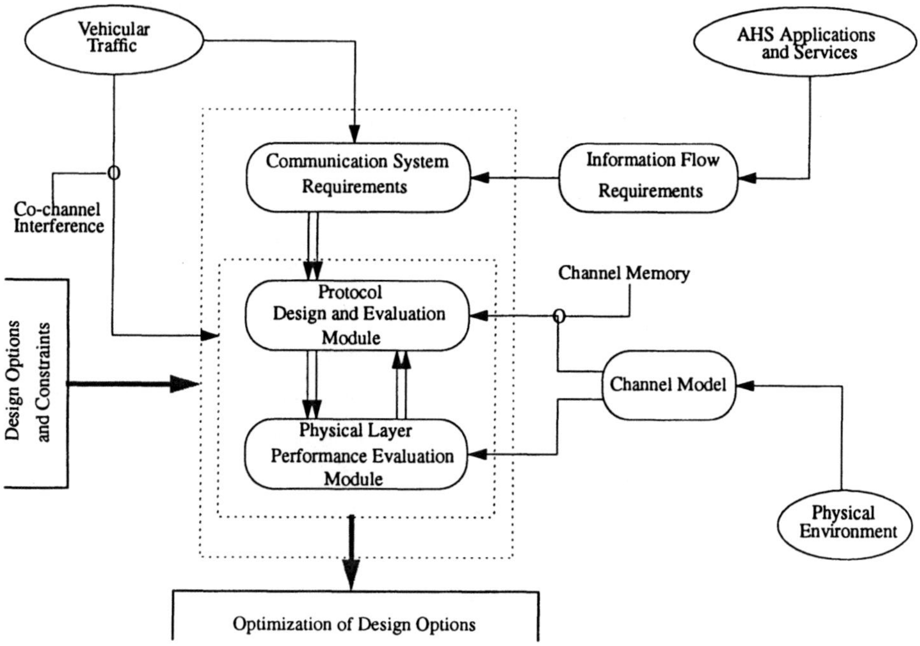

FIGURE 1. Integrated-layer model.

effects of design options, vehicular traffic, physical environment, information flow, and communication system requirements on the evaluation procedure.

In the Protocol Layer Design and Evaluation Module (Figure 2), the suitability of existing protocols for AHS communications is evaluated. The inputs to this module include the communication system requirements, the memory of the radio channel, and the physical layer parameters such as the bit error rate and the packet error rate. The measures for evaluation of a protocol's performance are the throughput-delay for the vehicle-to-base station link (nonperiodic data traffic), and the deadline failure probability (i.e., the probability that the delay of a real-time packet is longer than the deadline period) for the vehicle-to-vehicle link (periodic data traffic).

The Physical Layer Performance Evaluation Module (Figure 3) is a general analytical tool that handles a wide range of options. The inputs to this module are the channel characteristics (propagation path loss, shadowing, multipath fading, duration of fades, delay spread), the users' spatial distribution (vehicular traffic), and the performance measures generated in the Protocol Layer Design and Evaluation Module. The performance measures of interest are the bit-error probability, the packet-error probability, the capture probability, and the outage probability. This module can be used for the analysis of both spread spectrum and narrowband systems in combination with various coherent and noncoherent modulation schemes. The coding, synchronization, diversity (micro, macro), and power control issues are also taken into account for the evaluation of the performance measures.

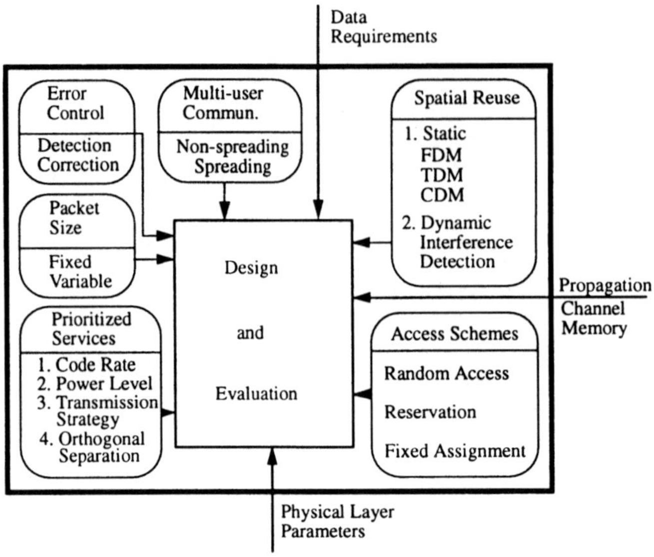

FIGURE 2. Protocol Layer Design and Evaluation Module.

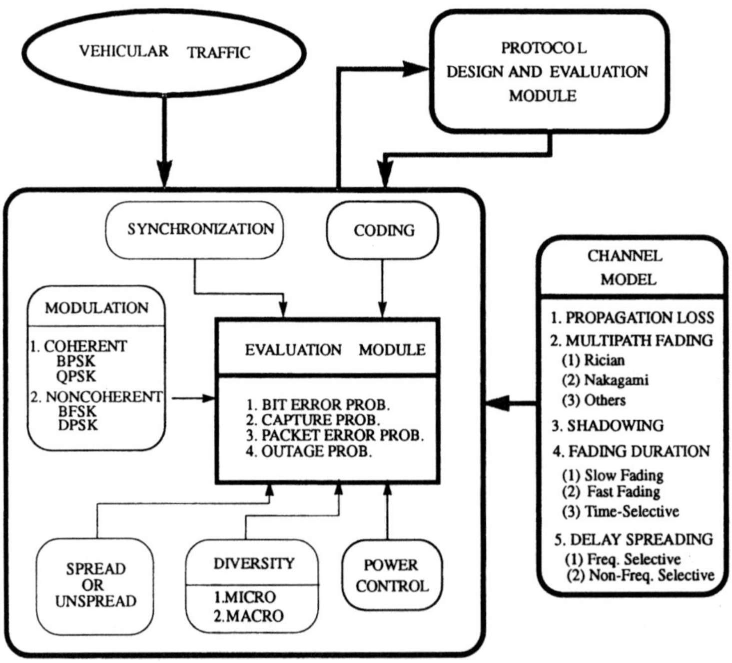

FIGURE 3. Physical Layer Performance Evaluation Module.

This model reveals the strong interaction between the Physical and the Link-Access layers of a packet radio system. Furthermore, the external inputs affect both layers, thus making the separate evaluation of some performance measures in each layer a difficult task. Even if one narrows down the question to one of understanding a particular signal design option, the answers are far from clear. Therefore, in order to optimize the overall system, we must jointly optimize the two layers, rather than design, analyze, and optimize each layer independently from the other. This strong interaction between the layers of a wireless network is distinct from the experience of the fixed, wired-network communication community with its fairly neat separation of layers. This is due to the open-air, wireless nature of the transmission which tends to distort the signals as they propagate and potentially mixes them with everybody else's on the air at that time.

The above conclusion indicates that the analysis and performance evaluation of multiple access schemes that can accommodate the heterogeneous (periodic and nonperiodic) data traffic of integrated systems (e.g., ATMIS/AVCS) can be a difficult and challenging task. Taking into account this interaction between the two layers, the performance of four multiple access schemes (Slotted-ALOHA, Reservation-ALOHA, TDMA, CDMA) has been evaluated in Refs. 27 and 28. In terms of ease in expandability (i.e., integrated data traffic scenarios), Reservation-ALOHA and CDMA seem to be the two most promising technologies. These two schemes have also the advantage of being amenable to decentralized optimization and expansion (provided that the issues of power control and code distribution in CDMA can be resolved in a decentralized manner). On the other hand, some type of external control is required for Slotted-ALOHA and TDMA. Finally, regarding robustness to channel and traffic variations, attention should be paid to schemes that are inherently adaptive to the access protocol (Reservation-ALOHA) and to channel equalization such as advanced multipath mitigation techniques (RAKE/CDMA),[29] or pursue hybrid versions.

6. CONCLUSIONS

Communication technologies are an integral part of the evolution and successful deployment of AHS. In this chapter, we have presented candidate communication technologies that can accommodate the communication requirements of AHS applications. A large part of the chapter was devoted to packet radio, an emerging technology that has the potential to be a feasible, viable, and (most likely) cost-effective solution for the communication needs of AHS. Furthermore, we have identified the design issues that are unique to such a system.

We believe that future research trends in the area of AHS communications should be towards communication systems that incorporate three important features: interactivity (namely, two-way real-time exchange between any two designated peers, including a possible infrastructure, if it exists), autonomy (namely, its ability to self-organize and manage the dynamically formed network in a distributed, self-sufficient fashion, even in the absence of any external controlling agents such as an infrastructure), and integration of services (namely, the ability to offer a variety of services in a compendious integrated system). In addition, major design drivers for such a system will be the inclusion of desirable features such as resistance to extraneous/unpredictable interference (inherent to multiuser systems), robustness to prevailing natural and man-made conditions, graceful expandability to an ever-increasing network scale, adaptability to new and evolving services and requirements, prioritization feasibility, and openness to specific technologies at the core. Finally, a major challenge for the AHS community is to ensure adequate spectral allocation (and a "clean," interference-free environment at that) within which such systems can be deployed.

REFERENCES

1. R. K. JURGEN, Smart Cars and Highways Go Global, *IEEE Spectrum* **28**(5), 26–36 (1991).
2. A. POLYDOROS, K. DESSOUKY, J. M. PEREIRA, C. M. SUN, K. C. LEE, AND T. D. PAPAVASSILIOU, Vehicle to Roadside Communications Study, Communication Sciences Institute, University of Southern California, California PATH Research Report, UCB-ITS-PRR-93-4 (1993).
3. L. A. KLEIN *et al.*, IVHS Architecture Development and Evaluation Process, *IVHS J.* **1**(1), 13–34 (1993).
4. Transportation Research Board, Communicating With Vehicles, *Circular,* No. 411 (July 1993).
5. P. VARAIYA AND S. E. SHLADOVER, Sketch of an IVHS System Architecture, University of California, Berkeley, Institute of Transportation Studies, California PATH Research Report, UCB-ITS-PRR-91-3 (1991).
6. S. E. SHLADOVER, *Draft Glossary of AVCS Terminology,* University of California, Berkeley, Institute of Transportation Studies, California PATH Technical Memorandum, 1993.
7. A. POLYDOROS, A. ANASTASOPOULOS, T. K. LIU, P. PANAGIOTOU, AND C. M. SUN, An Integrated Physical/Link-Access Layer Model of Packet Radio Architectures, Annual Report, Communication Sciences Institute, University of Southern California, California PATH Research Report, UCB-ITS-PRR-94-20 (1994).
8. S. L. STREISAND AND J. WALRAND, A Communication Architecture for IVHS, University of California, Berkeley, Institute of Transportation Studies, California PATH Research Report, UCB-ITS-PRR-92-10 (1992).
9. S. SACHS AND P. VARAIYA, A Communication System for the Control of Automated Vehicles, University of California, Berkeley, Institute of Transportation Studies, California PATH Research Report, UCB-ITS-PRR-93-5 (1993).
10. Transportation Research Board, Intelligent Vehicle Highway Systems (IVHS) Communications Standards, *Circular,* No. 383 (Dec. 1991).
11. D. J. CHADWICK *et al.*, A Communications Architecture for Early Implementation of Intelligent Vehicle Highway Systems (IVHS), Transportation Research Board, 72nd Annual Meeting, Jan. 1993, Washington, DC.
12. O. YAMADA *et al.*, Traffic Information Services Using FM Multiplex Broadcasting, *IVHS J.* **1**(1), 35–43 (1993).
13. D. J. CHADWICK *et al.*, Communications Concepts to Support Early Implementation of IVHS in North America, *IVHS J.* **1**(1), 45–61 (1993).
14. R. AIKAWA AND H. HABAL, IVHS Architecture—Physical Architecture, Rockwell International. Technical report submitted to Federal Highway Administration.
15. N. MCKEOWN AND J. WALRAND, An Infrared Communication Link for the PATH Project, Research Report, EECS Department, University of California at Berkeley (1990).
16. R. L. PETERSON, R. E. ZIEMER, AND D. E. BORTH, *Introduction to Spread Spectrum Communications* (Prentice-Hall, Englewood Cliffs, NJ, 1995).

17. R. STEELE, *Mobile Radio Communications* (Pentech Press Ltd., London, 1992).
18. K. PAHLAVAN AND A. H. LEVESQUE, Wireless Data Communications, *Proc. IEEE* **82**(9), 1398–1430 (1994).
19. B. Z. KOBB, Personal Communications Services, *IEEE Spectrum* **30**(6), 20–25 (1993).
20. N. J. COLMENARES, The FCC on Personal Wireless, *IEEE Spectrum* **31**(5), 39–46 (1994).
21. I. A. GETTING, The Global Positioning System, *IEEE Spectrum* **30**(12), 36–47 (1993).
22. W. C. COLLIER AND R. J. WEILAND, Smart Cars, Smart Highways, *IEEE Spectrum* **31**(4), 27–33 (1994).
23. K. S. CHANG *et al.*, Automated Highway System Experiments in the PATH Program, *IVHS J.* **1**(1), 63–87 (1993).
24. S. E. SHLADOVER, On the Path to Automated Highways and AVCS, *IVHS Rev.* 95–110 (1993).
25. R. D. ERVIN, An American Observation of IVHS in Japan, Technical Report, University of Michigan (1991).
26. Transportation Research Board, Spectrum Needs for IVHS Applications *TRB Workshop,* July 1993.
27. A. POLYDOROS, A. ANASTASOPOULOS, T. K. LIU, P. PANAGIOTOU, AND C. M. SUN, An Integrated Physical/Link-Access Layer Model of Packet Radio Architectures, Final Report, Communication Sciences Institute, University of Southern California, California PATH Research Report, UCB-ITS-PRR-95-33 (1995).
28. A. POLYDOROS, P. PANAGIOTOU, A. ANASTASOPOULOS, T. K. LIU, C. M. SUN, AND R. GERGES, Integrated-Layer Packet Radio Study for AHS, *Proceedings of the Sixth International Symposium on Personal, Indoor and Mobile Radio Communications (PIMRC '95),* pp. 870-875.
29. J. G. PROAKIS, *Digital Communications,* 2nd ed. (McGraw-Hill, New York, 1989).

CHAPTER 10

Control and Sensor Requirements and Issues in AHS

Petros Ioannou

1. INTRODUCTION

The current highway traffic system consists of the following basic elements and decision makers: the driver who controls the motion of the vehicle and makes decisions regarding the path of the vehicle, the route, destination, and so forth; the vehicle that obeys the commands of the driver whose on-board diagnostics often provide information to the driver about the health of the vehicle components; and the roadway or infrastructure that controls the flow of traffic using ramp metering, traffic lights, and roadway signs. These three basic elements interact and may influence each other considerably. For example, an inappropriate action by a driver of a single vehicle could have a dramatic effect on the performance of the overall traffic system. Similarly, a broken or low-performance vehicle could cause a large disturbance in the flow of traffic that may lead to heavy congestion. The roadway commands via road signs, ramp metering, and the like may equally have a negative effect on traffic flow if they are not the appropriate ones for the state of the traffic system the moment they are applied due to inaccurate sensing and/or incorrect control decision making. Inappropriate actions by any one of the three basic elements

Petros Ioannou • Center for Advanced Transportation Technologies, University of Southern California, Los Angeles, California 90089-2562.

Automated Highway Systems, edited by Ioannou. Plenum Press, New York, 1997.

may not only lead to deterioration of performance and congestion but also to accidents and loss of human life.

On the macroscopic level the current highway traffic system operates most of the time as an open-loop control system. As a result, it does not have the capability to correct itself in order to accommodate disturbances such as those caused by driver behavior or inaccurate sensing. Ramp metering and occasional road signs exert a limited form of control that has a reduced and often ineffective level of feedback. On the vehicle level, the randomness of driver behavior causes disturbances that influence the macroscopic behavior of traffic flow. It is well known that an open-loop controlled system is lacking robustness in the sense that small disturbances may get amplified and destroy the expected performance of the system. Feedback is invented to correct this problem. Feedback control systems are widely used in all areas of science and engineering for changing or correcting the behavior of dynamic systems in order to meet desired performance criteria. The lack of feedback in the current traffic system is responsible most of the time for the instability phenomena that manifest themselves as congestion, and observed in today's highway traffic. The use of feedback, however, in the current traffic system may not always be effective because of the random behavior of drivers.

The main objective of AHS is to change the sluggish, unpredictable, and inefficient behavior of the current traffic system through the use of automation. Even though the precise configuration of AHS has not yet been decided, there is a common understanding that AHS will involve fully automated vehicles equipped with sophisticated sensors and computer-controlled systems, and the infrastructure will be actively involved in managing and controlling the flow of traffic by using sensor technologies and control techniques. Automation will enable the use of feedback on the macroscopic level that could improve the traffic flow characteristics considerably. On the vehicle level, the use of sensors and controls could improve safety and allow vehicles to travel at higher speeds with small intervehicle gaps. This in turn could increase the capacity and efficiency of the highway system considerably.

The purpose of this chapter is to examine the control and sensor issues and requirements that need to be addressed in the design of AHS. Section 2 discusses an overall control system architecture[1] that is flexible enough to be applicable to a wide class of AHS configurations. The structure indicates the interaction of the various control layers from the vehicle to the infrastructure level. In Sections 3 and 4 the sensor and control issues associated with automated lateral and longitudinal control of AHS vehicles are presented. The control and sensor issues associated with the infrastructure and the macroscopic control of traffic flow are presented in Section 5.

2. AHS CONTROL STRUCTURE

The basic functions and control tasks that are likely to be present in an AHS configuration are the following:

- *Check-in*. The capability of the vehicle to operate in AHS is checked using on-board diagnostics and/or roadside check-in stations before entering the AHS lanes. At this stage, the identity and destination of the vehicle may be communicated to the infrastructure.
- *Entry to AHS*. Once the vehicle passes the check-in test, the infrastructure and the on-board computer control system take over control of the vehicle and guide it to the appropriate automated highway lane.
- *Lateral and longitudinal control*. Once the computer control system takes over, the motion of the vehicle is dictated by the commands generated by the computer and sent to the steering, brake, and throttle actuators. These control commands or inputs are determined by processing the sensor outputs that provide information about the status of the motion and position of the vehicle relative to the other vehicles and the environment.
- *Maneuver coordination*. The automated vehicles and infrastructure may use communications to negotiate lane changes and other maneuvers that may be essential for safety and performance.
- *Roadway control*. The roadway may use roadside beacons and local control stations to issue commands to vehicles in the lanes in order to control traffic flow density and speed by interacting with the higher level network controller.
- *Network control*. The infrastructure may view the overall traffic system as a network with inlets, outlets, nodes, and so forth and use optimization techniques to come up with desired densities and speeds to be followed by vehicles at the various branches of the network. The desired densities and speeds may be communicated to the local roadway controllers whose task is to make sure that the vehicular traffic flow in their area can meet them.
- *Emergency control*. Under normal operation the throttle, brake, and steering subsystems will be used to control the motion and position of the vehicle relative to the environment in a smooth fashion in order to maintain comfortable ride. In an emergency the primary objective is safety and therefore higher-bandwidth throttle, brake, and steering control inputs may have to be applied. The design of controls systems for emergency may be considerably different from that for normal operations.
- *Check-out and exit from AHS*. At the end of the trip the automated vehicle functions will be transitioned to manual ones by following a check-out procedure where the driver takes over control of the vehicle. The vehicle may exit AHS before or after the check-out procedure depending on the AHS configuration.

An overall AHS control structure that can perform the overall functions and tasks described above is shown in Figure 1.

The general structure shown in Figure 1 was originally proposed by Varaiya's group[1] and has undergone very minor modifications since then. It shows the vari-

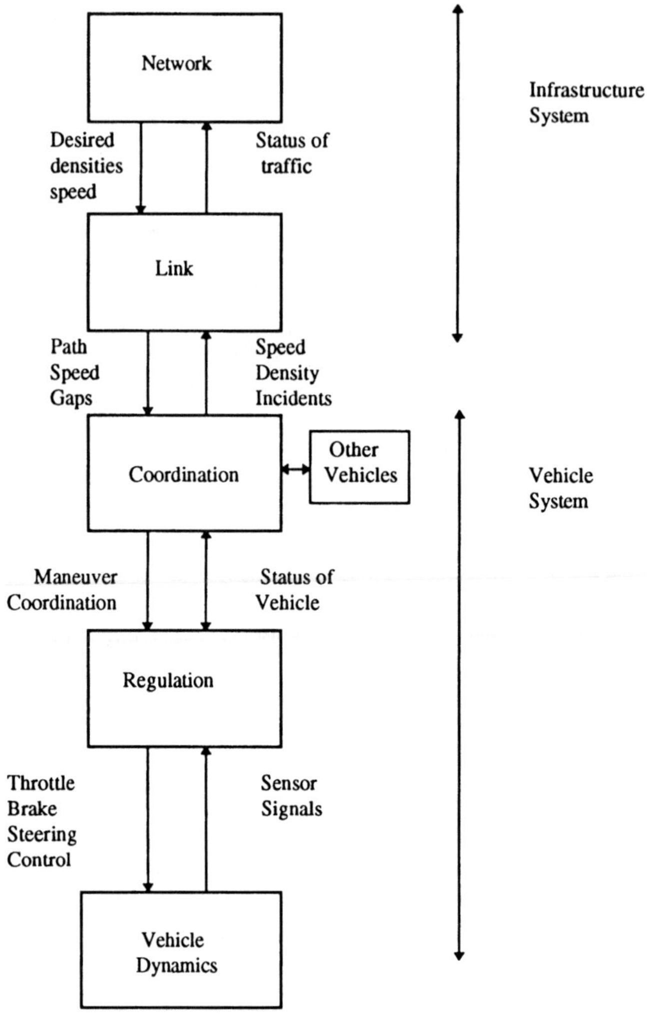

FIGURE 1. AHS control structure.

ous layers of control and decision-making tasks. At the bottom are the vehicle dynamics and the vehicle sensors and actuators. The sensor signals are fed into the regulation layer which contains the computer control system and control algorithms. The regulation layer computes the necessary commands that are sent to the throttle, brake, and steering actuators.

The coordination layer coordinates maneuvers with neighboring vehicles for lane changing, entry, exit (depending on the AHS configuration). It accepts as inputs the status and intentions of the vehicles in the neighborhood and generates the input commands for maneuvering.

The link layer sends speed, gap, and path commands to vehicles in each section of the lane in an effort to achieve the desired density distributions communicated to the layer by the network control layer. These control commands are calculated using sensor information about the status of the traffic in each section and the intentions of vehicles to (say) change lanes.

The network layer treats the overall traffic system as a network. It evaluates the status of traffic at the various branches and uses optimization to calculate desired densities and speeds and sends the appropriate commands to the link layer and the traffic lights at the ramps and intersections.

3. VEHICLE CONTROL SYSTEM

The vehicle control system (VCS) consists of three layers as shown in Figure 1. We assume that VCS is on board the vehicle even though in some AHS configurations parts of it could be with the roadway infrastructure.(2)

An overall block diagram that shows the main inputs, outputs, and components of VCS is shown in Figure 2.

The executive controller is the "brain" of VCS and contains the coordination layer and parts of the regulation layer. It receives inputs from the lateral and longitudinal sensors that provide information about the position and motion of the vehicle relative to the other vehicles and the lanes. It communicates with the link layer and other vehicles in the vicinity and interfaces with the driver during transitions

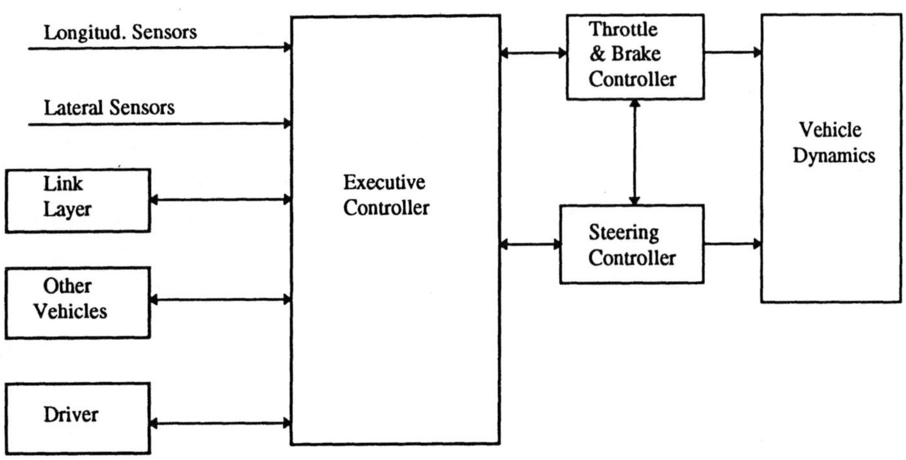

FIGURE 2. Overall vehicle control system.

from automatic to manual and vice versa. It evaluates all inputs received, does self-diagnostics, and generates the appropriate commands to the throttle/brake/steering controllers. The throttle controller uses engine torque to control the speed and relative spacing of the vehicle during normal operation. When engine torque is not sufficient, the brake controller gets activated in order to reduce vehicle speed and control intervehicle spacing. While the throttle/brake controllers are mainly responsible for the longitudinal motion of the vehicle, the steering controller is responsible for the lateral motion. The lateral control task includes lane keeping and lane changing. The throttle/brake and steering controllers could interact during lane changing or when the vehicle goes around curves. The interaction of the three controllers is much stronger and more crucial during collision avoidance and emergency situations.

The main tasks of VCS that affect safety and performance are the following:

- Spacing and speed control
- Lane keeping and lane changing
- Emergency control
- Maneuver coordination

We discuss the control and sensor issues associated with these tasks in the following subsections.

3.1. Spacing and Speed Control

In AHS the vehicles will be required to maintain certain speed and intervehicle spacing. Both vehicle speed and spacing affect safety and performance. If V is the speed of vehicles, L the vehicle length, and S the intervehicle spacing measured from the front of the vehicle to the back of the leading vehicle, the maximum vehicular flow that can be achieved in the lane referred to as capacity C is given by the formula

$$C = \frac{V}{S + L}$$

Capacity is measured in number of vehicles per unit time. It is clear that for a given S and L, the higher the speed V the higher the capacity will be. Similarly, for a given speed V, the smaller the intervehicle spacing, the higher the capacity will be. Therefore, from the performance point of view we require the capacity to be as high as possible, which implies high V and small S. High V and small S, however, may have an adverse effect on safety because high V implies shorter time for corrective actions to take place and small S implies smaller space for the corrective action to ensure our effect before the vehicle collides with the preceding one. Safety puts an upper bound on V and a lower bound on S. These bounds depend on a number of factors that include the braking capabilities of vehicles involved, their sensor/actuator characteristics, and the friction coefficient between tires and road surface, etc.[3] Despite the numerous research efforts in the area, there is still some controversy

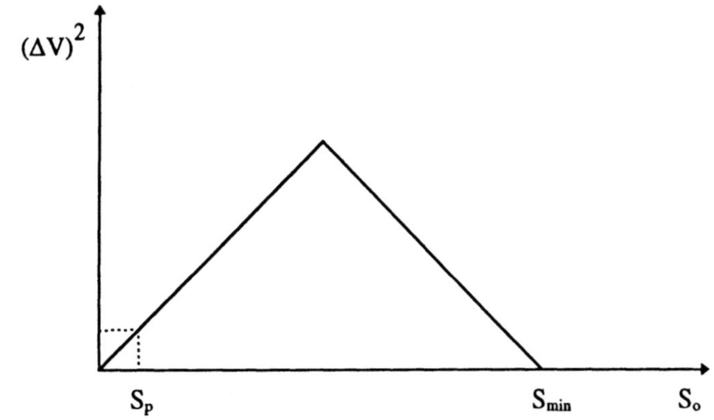

FIGURE 3. Collision curve.

about acceptable bounds for the minimum intervehicle spacing. The controversy is between two approaches for selecting intervehicle spacing that are explained as follows:

Let us consider two vehicles following each other in a single lane with an initial intervehicle spacing S_0. At time $t = 0$ the leading vehicle performs emergency braking and the following vehicle does the same with a small delay. If S_0 is small, it is likely that the two vehicles will collide with a relative speed (say) ΔV. If S_0 is large enough, no collision will take place. The severity of collision is proportional to the square of the relative speed at the time of collision $(\Delta V)^2$. A typical plot of $(\Delta V)^2$ versus the initial intervehicle spacing S_0 is shown in Figure 3.

It is clear from Figure 3 that for $S_0 > S_{min}$, no collision will take place, whereas for S_0 sufficiently small, say $S_0 < S_p$, the impact of the collision could be so small that it can be argued that the collision can be sustained by the vehicles involved without any damage. The value of S_{min} and the peak of the collision curve in Figure 3 depend on the braking capabilities of the two vehicles, the initial relative velocity, the friction coefficient between the tires and road surface, the time delay between the onset of braking by the following vehicle and that of the leading, and other factors.[(3)] Furthermore, S_{min} is approximately proportional to the speed of the following vehicle V, i.e., $S_{min} = hV$. The proportionality constant h is referred to as the time headway.[(3)] The value of S_p, however, that corresponds to a $|\Delta V| < 5$ mph (say) is insensitive to the variations of the factors that affect S_{min}. This information gives rise to the following two approaches:

In the first approach, $S_0 = h_0 V$ and the time headway h_0 is chosen so that $h_0 > h_{min}$ where $S_{min} = h_{min}V$. In the second approach, $S_0 \leq S_p$ is very small, e.g., about 1 to 2 m and is independent of the vehicle speed. The first approach guarantees that no collision will take place during emergency stopping that may arise by certain failures within AHS. The second approach guarantees that during emergency stopping, rear-end collisions are possible but such collisions are mild enough to be accommodated. As an additional safeguard, the second approach requires the vehicles to be organized in platoons of (say) 10 to 25 with the interplatoon spacing to be greater than S_{min} so that minor collisions within a platoon do not lead to collisions between platoons. It is clear that the second approach will lead to a much higher capacity[1,4] for platoons of 10 to 25 vehicles because of the much smaller intraplatoon spacing S_0 as discussed and simulated in Chapter 8. From the control design and sensor point of view the choice of the spacing approach gives rise to different requirements and issues that need to be addressed. We discuss these issues and requirements in the following subsections.

3.2. Longitudinal Control Based on Time Headway

The desired intervehicle spacing S_{d} in meters, measured from the front of the following to the rear of the leading vehicle, is chosen as

$$S_{\text{d}} = hV_{\text{f}} + S_{\text{c}} \tag{1}$$

where V_{f} is the speed of the following vehicle in meters per second, $S_{\text{c}} \geq 0$ is a constant in meters, and h is the time headway in seconds. The control objective is to use the throttle and brakes so that S_{d} is maintained between consecutive vehicles at any speed of flow V_{f}. During normal driving (no emergencies) the tracking of S_{d} has to be maintained under the following constraints:

(i) $a_{\text{min}} \leq \dot{V}_{\text{f}} \leq a_{\text{max}}$

(ii) $\left|\ddot{V}_{\text{f}}\right|$ as small as possible.

The first constraint limits the level of acceleration/deceleration and the second limits the amount of "jerk" to be as small as possible. These constraints are necessary in order to guarantee comfortable ride during normal vehicle following.[5]

When vehicles follow each other automatically in a lane, they behave as a coupled system. The behavior of the system depends on the control actions of the individual vehicles within the system. A phenomenon known as the "slinky effect" where a small tracking error or disturbance in the response of a lead vehicle gets amplified as it propagates along the platoon or string of vehicles is commonly observed in today's driving.[6,7] The longitudinal control system of each vehicle has to be designed to guarantee platoon or string stability, which in turn implies the absence of slinky effects. A block diagram of the longitudinal control system that can meet the control objective and the constraints described above is shown in Figure 4.

The ranging sensor provides measurements of the relative speed and spacing between the preceding and its own vehicle. The vehicle measures its own speed and acceleration and may receive the acceleration of the preceding vehicle via communications or by processing the measurements of the ranging sensor. The time headway h and S_{c} are calculated based on the braking capabilities of the vehicles involved, the accuracy and speed of response of the sensors and actuators used, and the other factors discussed earlier.

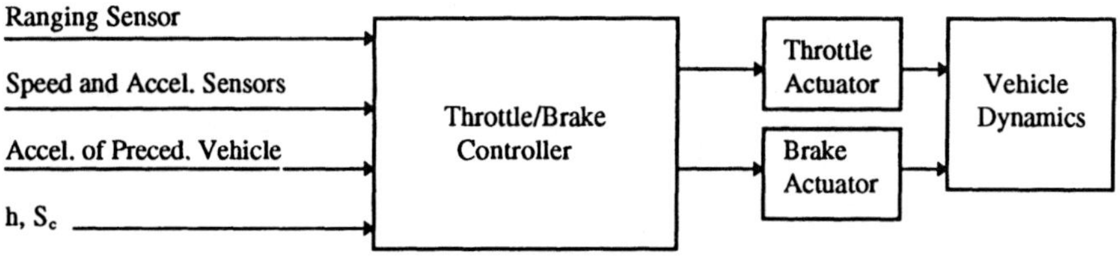

FIGURE 4. Longitudinal controller.

The throttle/brake controller calculates the appropriate commands that go to the throttle or brake actuator in order to meet the control objective. The throttle and brake controllers have to be designed separately since the throttle and brake actuators should not be activated simultaneously at any instant of time.[5] A switching logic has to be used to dictate the sequence of switching from brake to throttle actuators. This logic may include a hysteresis in order to eliminate possible limit cycles and jerks because of frequent switching between the two actuators.[5] In Ioannou and Xu,[5] three different throttle controllers and a brake controller are designed and tested with two different throttle and brake actuators and four different radar-based ranging sensors at the Ford Motor Company test track. In these experiments no acceleration measurements or vehicle-to-vehicle communication were used. String stability and tight vehicle following has been demonstrated for time headways as low as 0.3 s, which at 60 mph correspond to about an 8-m intervehicle spacing. The error in the tracking of the desired spacing was less than 12%. This error could be reduced considerably if the delay of the throttle actuator was reduced from 0.2 s to a smaller value. The same experiments were repeated at PATH and on interstate I-15 with different sensors and actuators. In this case the acceleration of the preceding vehicle obtained using vehicle-to-vehicle communications was also used by the throttle controller. The experiments demonstrated tight vehicle following with very small position error that satisfied all of the constraints discussed earlier.

Despite the success of these experiments, the deployment of vehicles that can follow each other automatically in an AHS environment raises several issues and creates additional requirements that need to be addressed. The current ranging sensors use a narrow-angle beam to measure relative speed and spacing. As a result the beam may miss the preceding vehicle around curves and may fail to detect a vehicle cutting in front of it. Sensors with a wide-angle beam, on the other hand, may lock-in to the wrong target such as vehicles in neighboring lanes. The ranging sensor has to emulate the human eye and be able to provide information, in addition to relative speed and spacing, that could be used to evaluate on line the vehicle following operation. Sensors based on vision together with narrow-beam radar may meet this requirement. With vision sensors, however, a considerable computing power is required to process the images generated. In addition, vision sensors may not be accurate during fog, rain and at night. The development of sensors that emulate the human eye and provide measurements much faster and more accurate than a human driver is a requirement that needs to be satisfied for the deployment of full-authority longitudinal control to be possible.

The structure of control algorithms for vehicle following is simple despite the complexity of the vehicle dynamics. The controllers are proportional-integral-derivative (PID) type with nonlinear elements.[5] This is not surprising for normal smooth vehicle following because the vehicles are designed to be easily controlled by a human driver. After all, most drivers treat the vehicle as a constant gain system. The complexity of the control algorithms, however, increases as the performance requirements for accurate vehicle following become more strict.

Additional important requirements and issues that are not obvious from the success of vehicle following experiments are reliability and human factor considerations. Reliability is one of the most crucial factors that will affect cost and safety. How reliable the system should be is a trade-off between cost and safety. Reliability may imply redundancies in sensors, actuators, and other hardware and software elements that will increase cost to the point that cost may outweigh the expected benefits. On the other hand, the lack of redundancies may reduce the cost of the vehicle but affect safety and raise liability issues. Whom does one sue in an AHS accident? Based on today's legal system, the answer is "all parties involved." Liability lawsuits may prove to be costly and a serious obstacle to the deployment of full-authority longitudinal control systems. Reliability considerations will require the design of different control laws on the executive and on the throttle and brake level. For example, if redundancies are present, the switching from a failed or partially failed path to a healthy one has to be done in a way that does not affect performance and safety. The speed of detection of the failure and the design of controllers to accommodate the failure are important elements of a reliable design.

Human factor considerations are also important in the control design. Some of these considerations led to constraints (i) and (ii) given above and used in the control design and experiments presented in Ioannou and Xu.[5] Some equally important considerations are the interface of the driver with the vehicle during transitions from manual to automatic and vice versa. The interface should appear to the human driver as simple as possible and transitions should be smooth and compatible with the human senses. These requirements will put additional constraints on the executive controller of VCS and on the throttle/brake control design.

3.3. Platooning

The concept of platooning has been associated with tight vehicle following where the intervehicle spacing is small (about 1 to 2 m) and constant at all speeds and vehicles are organized in platoons of 10 to 25 vehicles with large interplatoon spacings. This approach has been pursued by most researchers at PATH under the support of the California Department of Transportation and FHWA.[8,9] As explained earlier, the principal motivation behind platooning is the short intraplatoon spacing that is insensitive to factors such as the braking capabilities of vehicles and friction coefficients, it is independent of the vehicle speed and leads to high capacity. It has been argued[4] that platooning may increase the capacity of the current highway system by as much as 400%. A block diagram of the throttle/brake controllers for platooning is similar to that shown in Figure 4. The difference is that $h = 0$ and S_c is fixed to about 1 to 2 m and the speed and acceleration of the leader of the platoon is an input to the throttle and brake controller.

The control objective in platooning is to maintain tight vehicle following within the platoon and platoon stability under the constraint of comfortable ride. Since the desired intervehicle spacing is very small, the allowable position error is also small, which implies very accurate tracking of the desired spacing and speed trajectories. This accuracy puts constraints on the performance of the sensors and actuators as well as on the controller bandwidth. Because of the small intervehicle spacing, problems associated with cutting in or going around curves are not present even though such problems may still arise during joining or splitting from a platoon. The ranging sensor has to provide accurate measurements without significant delays. For example, for a 1-m spacing with a 10% allowable position error the total time associated with sensor measurement, actuator response, and computations should be less than 4 ms at 60 mph. This requirement puts serious constraints on the sensor and actuator designs and on the computer system and control algorithms.

For platoon stability it has been shown that the speed and acceleration of the leader of the platoon has to be used by the throttle/brake controllers of each vehicle.[(6,7,10)] As a result, vehicle-to-vehicle communication is essential. Because of the high performance requirements the throttle/brake controllers should have a higher bandwidth than those used with the time headway approach. A throttle controller based on feedback linearization designed using a sliding mode approach is designed and tested with a platoon of four vehicles.[(11)] Spacings as low as 4 m were used. Despite the success of these experiments, going from a spacing of 4 m to 1 or 2 m will require a considerable upgrade of the performance of the sensors and actuators.

There is no doubt that platooning has the potential of increasing capacity considerably. It is, however, a high-performance system and requires very precise control actions. Most of the research studies at PATH that include architectural designs,[(12,13)] safety,[(14)] longitudinal control laws,[(6,9)] link layer control laws,[(15,16)] consider platooning as the potential concept for increasing capacity considerably.

The opponents of platooning argue that the underlying principle that motivates platooning is based on the "faulty" assumption that low-ΔV collisions between vehicles could be accommodated within the platoon. As an example, they use train cars whose ΔV is always zero yet derailing is still possible. Another argument is that a low-ΔV collision between two vehicles in the platoon may lead to a high ΔV between two other vehicles in the same platoon especially in the case where such collisions are more elastic than plastic. In addition, during joining of the platoon the spacing of the vehicle that is about to become a member of the platoon goes from S_{min} to S_p, i.e., it goes through the region that corresponds to high ΔV (see Figure 3). A similar situation occurs during splitting of the platoon. In Chapter 2, Shladover, an expert and proponent of platooning, addresses these concerns.

Despite opposition, the concept of platooning can be made to work if the controllers and sensors, actuators are designed to meet the strict performance requirements in a reliable manner. Both sensors and actuators need to have a fast response with very small time delays. The communications must have high bandwidth in order to guarantee robustness and the communication delay has to be small. The control algorithms have to be of high bandwidth for fast response and simple enough in order to reduce the computational delay. The sum of all time delays should be less than the allowable position error expressed in units of time (i.e., the time required by the vehicle to cover the space that corresponds to the position error). A 1-m intraplatoon spacing puts a tough-to-meet constraint on the allowable total time delay especially at high speeds.

The reliability and human factors issues could be more serious in the case of platooning. Because of the small spacing, there is no room for errors and therefore all hardware and software components have to be highly reliable and capable of accommodating failures without degradation of performance that jeopardizes safety. How many of the failures could be accommodated because of the low-ΔV scenario and how many would lead to severe multiple-vehicle collisions is an issue that needs to be resolved. An associated and equally important issue that needs to be studied is the probability of occurrence of catastrophic failures.

3.4. Lane Keeping and Lane Changing

The lateral control system (LCS) of the vehicle during normal operation is responsible for lane keeping and smooth lane changing. In lane keeping the control objective is to steer the wheels so that the vehicle tracks the center of the lane under the constraint of good ride quality at all possible longitudinal speeds, loads, road surface conditions, and other disturbances. In lane changing the LCS is responsible for steering the vehicle from one lane to another in a safe and comfortable manner by interacting with the longitudinal control system for adjusting the longitudinal speed and acceleration as needed. A block diagram of LCS is shown in Figure 5.

The LCS accepts inputs from the following sensors: a road reference/sensing system that measures the position and orientation of the vehicle relative to the road, a lateral accelerometer for measuring lateral acceleration, and a yaw-rate sensor for measuring yaw-rate. In addition, it accepts and responds to lane change commands. The executive controller is the "brain" of the LCS that makes the high-level decisions and self-diagnostics. It activates the steering controller for lane keeping and

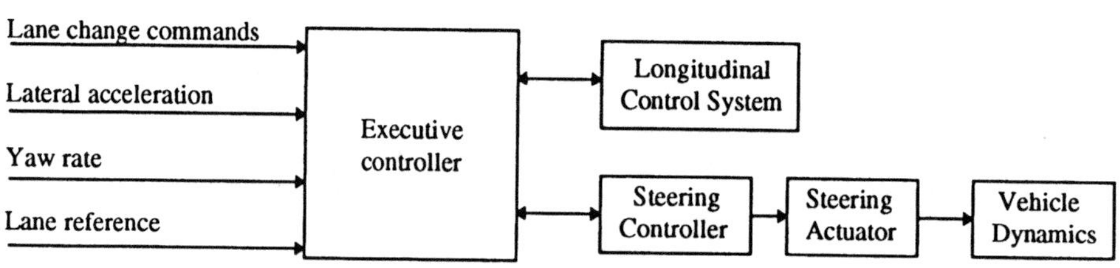

FIGURE 5. Lateral control system.

lane changing. The steering controller contains the control algorithms that generate the appropriate commands to the steering actuator that moves the wheels of the vehicle.

One of the most important parts of the LCS is the roadway reference system. Several reference sensing systems have been proposed and tested in the past. These include wire reference schemes, an optical marker system, an optical line-following system, vision-based systems, a side-looking radar system that uses a reference side wall, a and permanent magnetic marker system.[17] A review of the advantages and drawbacks of these technologies concluded that the permanent magnetic marker system is the most feasible one.[17] This system has been implemented and tested at Richmond Field Station of the University of California by PATH researchers.[8,9,18] In this study magnetic markers are installed along the center of the test track at about 1-m intervals. Two magnetometers mounted beneath the middle of the front bumper measure the vertical and horizontal components of the magnetic field generated by each marker. These measurements are processed and converted to a lateral error relative to the center of the lane. In this setup the road curvature information can be encoded in the markers and provided to the steering controller as preview information.

Several lane keeping controllers are designed and tested at PATH using an actual vehicle. These controllers are based on a fourth-order linear model of a front-wheel-steered vehicle and they employ preview information of the road curvature. In these experiments it was demonstrated that the lane tracking error was within ±1 m and the controllers were robust with respect to increased vehicle loads, reduced tire pressure, missing magnetic markers, and road conditions.

Vision-based sensing systems are used by several organizations to provide measurements for the lateral controller.[19,20] Such systems consist of a forward-looking camera that senses the vehicles lateral deviation and lane geometry. The image signals are sent to an on-board real-time image processing unit that processes the image data. The image processor extracts lane parameters by finding the edges of the lane and fitting these edges to a lane model. The camera needs to cover wide angles and large distances in order to provide the required preview information. Vision-based sensing requires few infrastructure changes. However, it requires more instrumentation on the vehicle. The lane reference aid can be as simple as standard lane markers (stripes) on the roadway to indicate the boundaries of the lane. The lane markers have to provide a clear contrast for the video camera. Experiments[19] indicate that standard wide markers 9 feet long, 4 inches wide, with 15-foot spacing on two edges of the 12-foot-wide freshly sealed asphalt lane can provide excellent contrast. While lane lines can be easily furnished and maintained, vision systems may not be effective under some weather conditions such as snow, heavy rain, or fog, or at night. Several lateral control systems are designed and tested using vision-based reference systems.[19-21] The steering controller in these experiments is a rather simple one, which for normal operation is good enough to guarantee good lane tracking under the constraints of comfortable ride. This is not surprising because the vehicle is designed to be easily controlled by a human driver. The control design will become more complex and challenging when collision avoidance and reliability requirements are introduced. In these situations, a higher-bandwidth steering controller will be needed and consideration of the higher-order lateral vehicle dynamics is essential, especially when they become strongly coupled with the longitudinal ones during rapid maneuvers.

Radar-based sensing systems are used to track the vehicle's lateral position with respect to reflector(s) mounted along the road lane. The reflector can be as simple as aluminum foil glued to a plywood wall surface.[22] Experiments[22] showed that the vehicle can stay within ±2.5 cm of the desired lateral position when the vehicle is traveling at about 31 mph or 14 m/s. One of the disadvantages of this method is that the lane needs to be next to a wall, which is not possible for highways with multiple lanes or without a wall. Corner cubes can also be used as reflectors.[23] In this case the vehicle is equipped with a transceiver that emits and receives laser beam signals. When the vehicle passes by the corner cubes, it recognizes them by detecting the reflector laser beam. This knowledge is used to measure the moved distances and estimate the lateral position and heading relative to the corner cubes based on the triangulation principle. Experimental results showed that tracking errors within millimeters can be achieved at speeds of 20 km/h.[23]

The main disadvantage of the radar-based sensing systems tested is that they provide little preview information. As a result the tracking performance deteriorates considerably with increasing speed.

The up-to-date lane keeping or lane tracking experiments performed demonstrate the concept. They do not imply that an automatic lane keeping system is ready to be deployed. The lane reference sensing system is one of the most important components of the lateral controller. Faults in the sensing system must be detected and accommodated before the vehicle deviates from the center of the lane considerably. For robustness a redundant lane sensing system based on a different principle may be necessary in order to obtain independent measurements. The steering subsystem that involves the steering actuator, computer system, and other parts also needs to be robust and free of catastrophic failures. The time required for the vehicle to go off the lane in case of some failures is so short that the detection and accommodation of these failures becomes difficult if not impossible. Redundant paths in the steering subsystem and fast control algorithms will be necessary to ensure that the vehicle can stay in the center of the lane at all speeds and under different road conditions and despite reasonable vehicle failures such as tire blowout or failed actuator.[24]

In contrast to lane keeping, not much work or experiments have been performed in lane changing maneuvers. The automatic lane changing function moves the vehicle from the center of one lane to the center of an adjacent lane provided the maneuver is safe. The sensor requirements in this case are enormous. The vehicle should be able to detect the position, speed, and possibly acceleration of the vehicles in the adjacent lane as well as of the vehicles in its own lane. This information will be used to calculate the trajectory of the lane changing maneuver. The range of the vehicle sensors should be large enough relative to the possible vehicular traffic flow speeds and accelerations. Some of these sensor requirements may be relaxed in a cooperative AHS environment where vehicle-to-vehicle communications may be used to negotiate maneuvers, synchronize speeds, and create the necessary gaps for the lane change to be possible. Communications, however, will require a large bandwidth in order to guarantee robustness. Lane changing requires a combined lateral and longitudinal control design that is also needed for going around curves. For smooth maneuvers the lateral/longitudinal vehicle dynamics are relatively simple and easy to control.

Research on combined lateral/longitudinal control for fully automated vehicles has been performed in industry and academia.[25,26] Most of the experiments performed employ image processing techniques for lane following and obstacle detection. Some of the most significant experiments include those performed by Toyota[27,28] and Daimler Benz.[29] In Toyota's experiments an inductive cable combined with stereo vision is used for lane detection and a laser radar is used for obstacle detection and vehicle following. The Daimler Benz test vehicle is equipped with about 12 video cameras and advanced image processing computers to handle all of the tasks of lane tracking, vehicle following, obstacle avoidance, and lane changing.[29]

Despite these experiments the development of a combined lateral/longitudinal control system that is affordable and reliable in an AHS environment requires considerable research efforts and experiments. The sensor requirements in an environment as dynamic as AHS will be considerable. The design of control algorithms for normal operation where maneuvers are smooth is expected to be relatively simple. This design, however, will become very challenging if reliability and collision avoidance considerations are introduced.

3.5. *Emergency Control*

Emergency control is a crucial part of VCS and its design is based on different considerations than those of lateral and longitudinal control. Here safety is the primary objective and ride quality and performance are secondary considerations.

An emergency controller should have the proper diagnostics and sufficient sensor measurements. It should continuously monitor the vehicle motion and its surroundings, identify possible emergency situations, and take the proper actions in order to prevent collisions.

The design of such a controller depends on the specific AHS configuration and it is not straightforward. The design of emergency controllers can be considered provided all of the control and sensor issues associated with the automated vehicle during normal operation are resolved. Some of the most challenging tasks of the emergency controller are to identify the threatening situations. In an AHS environment where all vehicles are automated, moving at high speeds with small intervehicle spacing and changing lanes automatically, false alarms may be difficult to avoid. In the up-to-date lateral and longitudinal control experiments, emergency situations are handled by restricting the experiments to be in a test track where the appropriate precautions can be taken and/or have the driver closely monitoring the vehicle operation so he can take over in case of failures or emergencies.

3.6. Maneuver Coordination

The vehicle maneuvers during entry to AHS and during lane changing must be such that the vehicle or other vehicles are not put in a situation that their on-board controller cannot handle in case of emergency. For such maneuvers to be possible, synchronization of the speed of vehicles in the lanes affected and the creation of sufficient intervehicle spacing will be necessary. One way to achieve this coordination is to use vehicle-to-vehicle communications and protocols that assign priorities and govern the logic of the maneuvers. Researchers at PATH addressed this problem for an AHS configuration based on the concept of platooning[1] and proposed a set of protocols to be used for lane change, join a platoon, or split from a platoon.[1,30] The case where there is no vehicle-to-vehicle communication has also been addressed by some researchers.[24] In this case the vehicle relies on its own sensors and its interpretations of the actions of other vehicles in its vicinity in order to make decisions about a maneuver. The work performed up to date is mainly theoretical and is demonstrated using simulations. No experiments have been performed to demonstrate the effectiveness of the proposed communication protocols or other techniques.

In the case of vehicle-to-vehicle communication, a sufficient bandwidth is necessary in order to guarantee robustness. It is unlikely that the vehicle will rely solely on communications for maneuvers. Instead the vehicle will also use its on board sensors to verify the safety of the situation before proceeding with the maneuver. In the absence of communications, vehicle maneuvers such as lane change will be difficult to perform in a dense traffic situation or in a situation where the speed differential between the lanes is high.

4. LINK AND NETWORK CONTROL

On the vehicle level, high speeds and small intervehicle spacing imply high theoretical capacities. Whether such capacities can be achieved and be sustained for a reasonable period of time will very much depend on the behavior of the traffic flow on the macroscopic level. As discussed in the Introduction, the current highway traffic system operates as an open-loop system. As a result, small disturbances on the vehicle level get amplified and the traffic flow rate remains well below the maximum possible capacity. In AHS most of the disturbances on the vehicle level that are associated with the driver will be eliminated. Disturbances, however, associated with the vehicle response, accidents, and so forth will be present. Such disturbances may have a significant effect on the open-loop traffic system despite automation on the vehicle level. One way to reduce the effect of disturbances and maintain traffic flow rates that are close to the maximum possible capacity is to use feedback control techniques that could force the traffic flow to achieve the desired characteristics.

The purpose of the link layer is to provide the appropriate speed, gap, and path planning (depending on the AHS configuration) commands to the vehicles at each section of the lane so that their response leads to certain desired traffic density and speed distribution. The desired density and speed profile may be calculated locally or may be provided by the network layer. The profile depends on the status of traffic system or traffic network and could be calculated using optimization techniques. Attempts have been made to design link layer controllers using macroscopic traffic flow models.[15,16,31] The results of these studies and simulations demonstrate dramatic improvements in the stability properties of traffic flow. In particular by sending the appropriate speed commands to the vehicles at each section of the automated lanes, a desired traffic density can be tracked and congestion can be managed and prevented from getting amplified. Additional work is necessary both in the area of modeling the AHS traffic on the macroscopic level and on link layer control techniques.

The highest layer of the automated control system of AHS is the network layer. At this level the overall traffic network is to be considered and its operation to be optimized. This optimization may lead to desired density and speed distributions along different branches, to ramp metering, traffic light control, and so forth. The network layer may be a collection of decentralized controllers that are closely interacting with the local link layers or an overall centralized controller. Despite some earlier work in this area, no particular approach to network control has been adopted. The rapid developments in the area of computers, in particular on processing speed and memory and software in recent years, make it possible to implement network controllers dealing with relatively large traffic networks. The development of such controllers will very much depend on the AHS configuration and it is a challenging problem.

REFERENCES

1. P. VARAIYA, Smart Carts on Smart Roads: Problems of Control, *IEEE Trans. Autom. Control* **38**(2) (1993).
2. J. R. BISHOP AND W. STEVENS, Results of Precursor Systems Analyses of Automated Highway Systems, *Proc. First World Congress on Applications of Transport Telematics and Intelligent Vehicle–Highway Systems,* Paris, 1994, pp. 1959–1966.
3. Y. SUN AND P. IOANNOU, A Handbook for Inter-Vehicle Spacing in Vehicle Following, California PATH Research Report UCB-ITS-PRR-95-1, University of Southern California (1995).
4. S. E. SHLADOVER, Potential Freeway Capacity Effects of Advanced Vehicle Control Systems, *Second International Conference on Applications of Advanced Technologies in Transportation Engineering,* Minneapolis, 1991.
5. P. IOANNOU AND Z. XU, Throttle and Brake Control Systems for Automatic Vehicle Following, *IVHS J.,* **1**(4), 345–377 (1994).
6. S. SHEIKHOLESLAM AND C. A. DESOER, Longitudinal Control of a Platoon of Vehicles, *Proceedings of 1990 ACC,* May 1990.
7. C. C. CHIEN AND P. IOANNOU, Automatic Vehicle-Following, *Proc. 1992 American Control Conference,* Chicago, pp. 1748–1752.
8. S. E. Shladover *et al.*, Automatic Vehicle Control Developments in the PATH Program, *IEEE Trans. Veh. Technol.* **40**(1), 114–130 (1993).
9. J. K. HEDRICK, M. TOMIZUKA, AND P. VARAIYA, Control Issues in Automated Highway Systems, *IEEE Control Sys.* **14**(6), 21–33 (1994).

10. P. IOANNOU AND C. C. CHIEN, Intelligent Cruise Control, *IEEE Trans. Veh. Technol.* **42**(4), 657–672 (1993).
11. K. S. CHANG, J. K. HEDRICK, W.-B., ZHANG, P. VARAIYA, M. TOMIZUKA, AND S. E. SHLADOVER, Automated Highway System Experiments in the PATH Program, *IVHS J.*, **1**(1), 63–87 (1993).
12. P. VARAIYA AND S. E. SHLADOVER, Sketch of an IVHS Systems Architecture, *Proc. Vehicle Navigation and Information Systems Conference,* Dearborn, MI, Oct. 1991, pp. 909–928.
13. W.-B. ZHANG, S. E. SHLADOVER, R. W. HALL, AND T. PLOCHER, Functional Definition of Automated Highway Systems, TRB Paper 940988, 73rd Annual Meeting, Jan. 1994, Washington, DC.
14. A. HITCHCOCK, Intelligent Vehicle/Highway System Safety: Multiple Collisions in AHS Systems, PATH Technical Report, Institute of Transportation Studies, University of California, Berkeley (1994).
15. U. KARAASLAN, P. VARAIYA, AND J. WALRAND, Two Proposals to Improve Freeway Traffic Flow, PATH Report 90-6 (1990).
16. C. C. CHIEN, Y. ZHANG, A. STOTSKY, AND P. A. IOANNOU, Roadway Traffic Controller Design for Automated Highway Systems, *Proc. IEEE Conference on Decision and Control,* Orlando, FL, Dec. 1994, pp. 2425–2430.
17. W. B. ZHANG AND R. E. PARSONS, An Intelligent Roadway Reference System for Vehicle Laser Guidance/Control, *Proc. 1990 American Control Conference,* San Diego, pp. 281–286.
18. H. PENG AND M. TOMIZUKA, Preview Control for Vehicle Lateral Guidance in Highway Automation, *ASME J. Dyn. Sys. Meas. Control* **115**, 679–686 (1993).
19. O. D. ALTAN, D. B. CRAIG, B. B. LITKOUHI, AND L. M. OBERDIER, LANETRAK: A Vision-Based Automatic Vehicle Steering System, GMR, Research Publication GMR-7835 (1992).
20. D. POMERLEAU *et al.*, AVCS Research at Carnegie Mellon University, *Proc. IVHS America 1994 Annual Meeting,* Atlanta, Paper #94100.
21. B. ULMER, Autonomous Automated Driving in Real Traffic, *Proc. First World Congress on Applications of Transport Telematics and Intelligent Vehicle–Highway Systems,* Paris, 1994, pp. 2118–2125.
22. C. CUGIANI *et al.*, Millimetre-Wave Guard Rail Tracking System for Vehicle Lateral Control, *Proc. IEEE Vehicle Navigation and Information Systems Conference,* Ottawa, Canada, 1993, pp. 521–524.
23. T. TERAMOTO *et al.*, Study of Laser Radar, *Proc. 12th International Technical Conference on Experimental Safety Vehicles,* 1989, pp. 888–895.
24. J. DICKERSON *et al.*, Lateral and Longitudinal Control Analysis, Activity Area D, Precursor Systems Analyses, Final Report, submitted to FHWA (1994).
25. H. A. PHAM, J. K. HEDRICK, AND M. TOMIZUKA, Combined Lateral and Longitudinal Control of Vehicles for Automated Highway Systems, *Proc. 1994 American Control Conference,* Baltimore.
26. S. SHEIKHOLESLAM AND C. A. DESOER, Combined Longitudinal and Lateral Control of a Platoon of Vehicles, *Proc. 1992 American Control Conference,* Chicago, pp. 1763–1767.
27. K. OHNISHI, J., KOMURA, AND T. ISHIBASHI, Development of Automatic Driving System on Rough Road—Realization of High Reliable Automatic Driving System, *Proc. Intelligent Vehicles 92 Symposium,* IEEE, Detroit.
28. T. SUZUKI AND A. TACHIBANA, A Fully Automated Intelligent Vehicle System Based on Computer Vision for Automated Highway System, *Proc. First World Congress on Applications of Transport Telematics and Intelligent Vehicle–Highway Systems,* Paris, 1994, pp. 1925–1932.
29. B. ULMER, Autonomous Automated Driving in Real Traffic, *Proc. First World Congress on Applications of Transport Telematics and Intelligent Vehicle–Highway Systems,* Paris, 1994, pp. 2118–2125.
30. A. HSU, S. SACHS, F. ESKAFI, AND P. VARAIYA, The Design of Platoon Maneuvers for IVHS, *Proc. 1991 American Control Conference,* Boston, pp. 2545–2550.
31. P. LI *et al.*, An AVHS Link Layer Controller for Traffic Flow Stabilization, PATH Technical Note 95-7 (Nov. 1995).
32. R. K. BOYD AND M. P. LUKAS, How to Run an Automated Transportation System, *IEEE Trans. Syst. Man Cybern.* **2**(3), 131–141 (1972).
33. R. G. RULE, The Dynamic Scheduling Approach to Automated Vehicle Macroscopic Control, Transportation Control Lab, Ohio State University, Columbus, Report EES 276A-18 (Sept. 1974).
34. J. G. BENDER, An Overview of Systems Studies of Automated Highway Systems, *IEEE Trans. Veh. Technol.* **40**(1), 82–99 (1991).

CHAPTER 11

COMMERCIAL TRUCKS AND BUSES IN AUTOMATED HIGHWAY SYSTEMS

IOANNIS KANELLAKOPOULOS AND MASAYOSHI TOMIZUKA

1. INTRODUCTION AND MOTIVATION

As indicated throughout this volume, research and development on Advanced Vehicle Control Systems (AVCS) for Intelligent Transportation Systems (ITS) and Automated Highway Systems (AHS) to date has been primarily focused on passenger vehicles,[1–3] while commercial heavy vehicles (CHVs) such as heavy-duty freight trucks (including tractor–trailer combinations) and commuter buses have been largely ignored. The obvious justification is that there are many more passenger vehicles on the road, and thus their automation will have the largest possible impact on the desired increase of highway traffic flow.

However, this argument does not take into account the many differences in the operational modes between commercial and passenger vehicles, which render the relative impact of CHVs on traffic congestion and economic growth much larger than their numbers suggest. In fact, the motivation for automating CHVs goes far beyond the need to include these vehicles in the automated highways of the future;

IOANNIS KANELLAKOPOULOS • Department of Electrical Engineering, University of California, Los Angeles, California. MASAYOSHI TOMIZUKA • Department of Mechanical Engineering, University of California, Berkeley, California.

Automated Highway Systems, edited by Ioannou. Plenum Press, New York, 1997.

it is often argued that CHVs have the potential of becoming the flagships of AHS efforts, due to several economic and policy issues.

Heavy-duty commercial freight trucks, for example, are big business in the United States. The following data, taken from Ref. 4, are quite illustrative. More than 10 heavy truck/tractor manufacturers and more than 100 trailer/semitrailer manufacturers produce the equipment used by more than 50,000 commercially operated fleets in this country, and generate around \$3.2 billion in truck exports (and \$170 million in bus exports) annually. Based on vehicle registration records, in 1992 there were 1.28 million truck tractors on U. S. roads (113,000 of them in California alone) and 3.8 million trailers (460,000 in California), with 300,000 new trucks being added every year. When lighter trucks and buses are taken into account, the numbers rise to 660 truck/bus body manufacturers with 37,000 employees and a total payroll of \$1 billion (100 manufacturers with 3000 employees and a \$70 million payroll in California), and 327 trailer manufacturers with 22,000 employees and \$500 million in payroll (34 companies with 1500 employees and a \$38 million payroll in California).

While these numbers are impressive, they are much lower than the corresponding numbers for passenger vehicles, and hence they don't convey the true impact of CHVs on the economy. To see that, one must look at these vehicles from the point of view of transportation and traffic. In 1992, for instance, an 80,000-lb diesel-powered five-axle tractor–semitrailer truck using 14,000 gallons of fuel and traveling 80,000 miles annually, paid an average of \$1200 in registration and weight fees, \$2700 in fuel taxes, and \$4500 in highway-user fees. In fact, trucks pay around 28% of state and 51% of federal highway-user taxes, \$1.2 billion annually in California alone. Even more importantly, trucks absorb approximately 75% of the funds spent on all modes of freight transportation, even though they account for only 29% of total intercity freight-ton miles, with railways accounting for 37%, pipelines for 19%, and inland waterways for 15%. In 1993, trucks had a total of 880 billion ton-miles, up from 735 billion in 1990 and 555 billion in 1980. As far as traffic and fuel consumption are concerned, even though truck/trailer combinations account for only 1% of the vehicles on the road, they account for 5% of the total miles traveled (100 billion annually, with cars traveling 1.6 trillion and buses 6 billion miles), and for 13.3% of the total fuel consumption (17.7 billion gallons annually, with cars consuming 73.8 billion and buses 877 million gallons). But the most important numbers from the ITS/AHS point of view are the average mileage and fuel consumption per vehicle: In 1992, the average passenger car traveled 11,063 miles and consumed 512 gallons of fuel (21.61 miles per gallon), while the average bus traveled 8901 miles on 1360 gallons of fuel (6.54 mpg) and the average truck traveled 59,846 miles on 10,695 gallons of fuel (5.60 mpg). So, the fuel consumption improvements that are likely to result from automation would noticeably increase operating profits and would significantly shorten the time to recover the initial investment.

Before proceeding to additional factors contributing to the importance of CHVs for AHS, let us note that AHS deployment is likely to proceed in several stages. In the first stage, for example, vehicles have only longitudinal control capabilities for vehicle following without intervehicle communication, while the driver is responsible for steering and emergency situations. This stage represents the Autonomous Intelligent Cruise Control (AICC) concept, which is already being deployed in Japan with the commercial introduction of the AICC option by a luxury car manufacturer. The second stage adds communication devices that allow vehicles to exchange information about their current operating state, thereby enhancing their longitudinal control capabilities. In the third stage, vehicles are also equipped with lateral control capabilities, which can be autonomous (if a vision-based system is used) or dependent on the roadway infrastructure (magnetic markers, reflective siderails). The fourth and subsequent stages add communication with the roadway and correspond to fully automated highways.

Commercial vehicles will benefit from automation in all of these stages, with new advantages added from stage to stage. The above-mentioned fuel savings are going to appear in the first stage, when two or more trucks, each with its own driver, form a platoon. In this scenario, the leading truck is controlled completely by its driver, while following trucks are only steered by their drivers. The resulting smoother operation and reduced aerodynamic drag will decrease the fuel consumption of the following trucks. These savings will be even more significant in the second stage, because intervehicle communication allows stable vehicle following with smoother throttle/brake effort and smaller intervehicle spacings (hence, reduced aerodynamic drag) than autonomous operation. The most significant savings, however, will be realized in the third automation stage, in which vehicles have integrated longitudinal and lateral control systems, and hence are capable of driverless (unmanned) operation. While this is insignificant for passenger cars, it is of crucial importance for heavy-duty commercial trucks, since it allows the elimination of drivers from the following vehicles, and requires their presence only in the lead vehicle. In truck fleets drivers account for almost *half* of the total cost of owning and operating a vehicle.[5] Hence, if a fleet operator can platoon four trucks using only two drivers in the lead truck, who can take turns driving the platoon in order to decrease the total travel time, the cost of the transport is cut by almost 25% just by the reduction of the number of drivers from four to two, not to mention the benefit from the substantially reduced travel time.

The above discussion gives the most important reason why CHVs, and in particular heavy-duty freight trucks, are likely to be the first semi- or fully automated vehicles:

- *Profit-driven operation.* The average truck travels *6 times the miles* and consumes *27 times the fuel* of the average passenger car, and these numbers are even higher if restricted to heavy commercial trucks, rated at 40,000 lb or more. Thus, even a small improvement in fuel efficiency can be incentive enough for truck fleet operators to invest in AVCS equipment. Furthermore, the possible reduction in the number of drivers will yield significant savings (25% or more) in operating costs, thereby reducing the time to recovery of the initial investment and increasing the profit margin.

The effect of these savings through platooning is further compounded by the following two facts.

1. Truck traffic follows much more regular patterns than passenger car traffic; trucks usually travel on well-established commercial routes, mostly between major cities. Thus, fleet operators can easily compose platoons of trucks with the same origination and destination points which will travel together for the entire trip. Under this scenario, in the first stages of AHS deployment, freight transport companies will operate departure/arrival stations in major cities. Individual trucks will be driven manually from all over the city to the station, where they will join the platoon departing for their destination city. In the meantime, the drivers who are not on the departing platoons will manually drive the trucks from arriving platoons to their individual destination points within the city. The absence of the requirement for split/merge maneuvers will significantly simplify the control problem for these platoons.

2. The ratio of intercity to intracity miles traveled by freight trucks is much higher than for passenger cars. In 1993, truck traffic accounted for 5% of the total miles traveled in the United States, but that percentage jumps to 28% if only highway miles are considered,(6) and it is even higher if we exclude intracity highway traffic. Hence, CHVs will utilize their automation capabilities much more than passenger cars, since they spend most of their travel time on interstate highways, where they will be able to travel in platoons.

Another reason is easily obtained by considering the difference in initial cost between passenger cars and CHVs:

- *Relative equipment cost.* While the average cost of a new passenger car is around $20,000, the average cost of a new commercial truck or bus approaches $100,000. Consequently, the ratio of AVCS equipment cost to vehicle cost is much lower (and thus much more favorable) for CHVs. Alternatively, if this ratio is increased to levels comparable to those of passenger vehicles, then CHVs can be equipped with much more accurate, reliable, and sophisticated sensors and actuators, and a lot of redundancy can be introduced to guarantee fault-tolerant operation.

Finally, from the safety and human factors points of view, which are crucial to the success of any novelty in highway transportation, CHVs stand to gain a great deal from automation.(5) First, driver acceptance of and familiarity with the automated systems will be much easier to achieve. Truck and bus drivers are professionals who are far better trained than car drivers, and will thus be (1) more likely to accept innovations that make their driving task easier and (2) better prepared to assist the automated system and handle any emergency situations caused by system malfunctions. Second, automated longitudinal and lateral control will make the operation of long truck–trailer combinations much safer. This is especially true for double and triple combinations, which can be up to 48 or even 53 feet long and require turning radii and passing and sight distances that most roads, including many interstate highways, do not meet. Nevertheless, the trend toward lighter-weight and higher-value freight to be hauled by trucks pushes fleet operators to use these long combinations for increased productivity. While automation will undoubtedly make more highways accessible to these trucks, it may also make them unnecessary. After all, if fleet operators can increase productivity by platooning several regular-size trucks, which are also drivable in the city, they will have no incentive to use these long trucks whose cargo will have to be reloaded onto smaller trucks (a very expensive operation) for final distribution. These safety benefits will also help public acceptance of truck automation. Public opinion about commercial truck operators has been steadily declining in recent years, and this trend is reinforced by several media reports that show truck drivers regularly violating sleep/driving regulations and exceeding speed limits in the name of profitability. Even more alarming is the fact that since deregulation the number of traffic accidents involving trucks has increased at a much more rapid pace than the total number of traffic accidents, and that almost 25% of all combination vehicles are involved in a reported accident each year. Hence, the third major reason for truck automation is:

- *Safety and public acceptance.* Operational safety enhancements resulting from the automation of commercial trucks and buses will be much more significant than for passenger cars. Moreover, CHV drivers will be much more capable of supervising their automated vehicles, which will also be perceived by the public as much less threatening: They will not only be operated in a safer manner, but will also appear that way, because of the absence of frequent noticeable accelerations/decelerations and abrupt steering maneuvers.

2. CONTROL ISSUES

Having established the reasons why commercial trucks and buses should be automated, and what the trucking industry and the public at large would gain from their automation, we now proceed to discuss the difficulties in designing suitable lateral and longitudinal control systems for CHVs. After all, one could argue that the same algorithms used for passenger cars may be good enough for these vehicles. However, this is not true for the following reasons.

2.1. Actuation-to-Weight Ratio

Because of their increased weight, CHVs cannot achieve the same levels of acceleration and deceleration as passenger cars. This has to be accounted for in the control design, since it significantly affects performance. In fact, as we will see in Section 4, the low actuation-to-weight ratio makes string stability much more difficult to achieve for CHVs than for passenger cars, and necessitates the design of new longitudinal control algorithms that endow the vehicles in the platoon with a "group conscience" and sacrifice the individual performance of some vehicles in order to improve the performance of the whole platoon.

2.2. *Lateral Dynamics*

The directional response of CHVs is different from that of passenger cars in several ways. Because of the weight difference, heavy-duty vehicles are slower to respond to steering inputs than passenger vehicles, and the nonlinear dynamics resulting from road–tire interaction and suspension systems are more pronounced in them. In particular, buses and trucks respond to steering inputs in a much more nonlinear fashion than cars, mainly because of changes in tire cornering stiffness caused by load changes and cargo shift resulting from lateral accelerations. Articulation joints of heavy-duty vehicles introduce another level of complexity. The center of mass is normally high, which mandates a careful consideration of roll dynamics in the analysis and design of control systems.

Heavy-duty vehicles exhibit roll and yaw instability modes that in passenger cars are either insignificant or nonexistent. When following curved roads, CHVs may roll over as a result of centrifugal force. There are two types of yaw instability: divergent yaw response (jackknifing or trailer swing) and nondivergent rearward amplification (fishtailing or snaking). These modes are illustrated in Figure 1.

Several researchers[7] have addressed the yaw instability problem of articulated vehicles. Yaw instability has been studied in the past for improvements in design (involving parameters such as wheel base length, suspension stiffness, number of tires and axles, and damping at articulation joints) and establishment of safer operational criteria (e.g., loading distribution, operational speed).

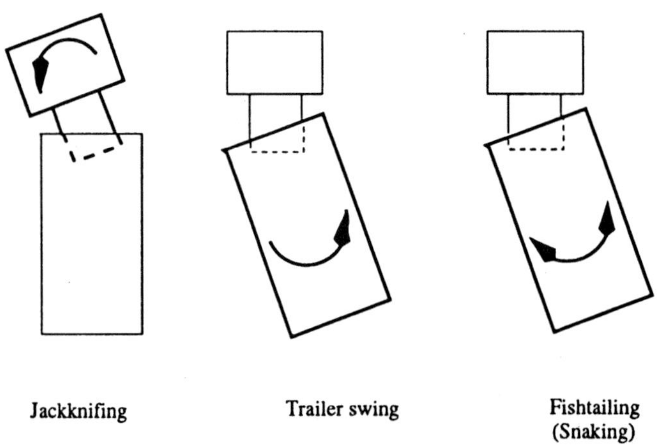

FIGURE 1. Basic instability modes of articulated vehicles.

2.3. Strong Coupling of Lateral and Longitudinal Dynamics

While roll and yaw instabilities are associated with the lateral dynamics of the vehicle, they are clearly affected by vehicle speed, and hence they are coupled to longitudinal dynamics.[8–10] Their high center of mass and high wheelbase-length-to-width ratio make CHVs much more susceptible to this coupling than passenger cars. Another significant factor affecting the nonlinear response of CHVs to steering inputs is tire saturation, which can be caused by braking.[7,12] Hence, when the longitudinal controller commands the use of brakes, especially during downhill driving, it is imperative that it take into account the effect of braking on the lateral dynamics in order to avoid the onset of roll and yaw instabilities.

2.4. Actuator Delays and Nonlinearities

Some of the most undesirable phenomena from the point of view of conventional control system performance are actuator delay and nonlinearities such as saturation, backlash, and hysteresis. These are much more pronounced in CHVs, because of the fact that the primary consideration in the design of their engines, transmissions, and brakes is not performance or smoothness of response, but rather reliability and the ability to cope with the stress caused by heavy loads without excessive wear and tear.

2.5. Disturbance Effects

Compared to cars, CHVs are much less aerodynamically designed, and are thus much more sensitive to head winds (for longitudinal control) and even more sensitive to side winds (for lateral control). The effects of side winds are particularly harmful in the case of long combination trucks, which are perceived by most drivers as extremely dangerous when operated in the presence of strong winds. Furthermore, because of their decreased actuation-to-weight ratio mentioned above, heavy-duty vehicles are severely affected by sloped terrain, both uphill and downhill.

It is therefore clear that the automation of CHVs requires research and development focused on their specific needs and characteristics. In recent years, the study of heavy vehicles for AHS applications has gained considerable interest. Daimler-Benz Research, in cooperation with Freightliner Corporation, is developing a research vehicle called VECTOR (Vision Enhanced/Controlled Truck for Operational Research).[13] Renault and INRETS in France are involved in the study of AHS as applied to freight transport in France and associated issues at various levels (including socioeconomic),[14,15] while Ervin[16] tried to identify the domain of research on IVHS that might be of interest to CHV manufacturers in the United States. Paccar Inc., the manufacturer of Kenworth and Peterbilt trucks, is experimenting with an automated lateral control system to be used for rough-pavement endurance testing of their vehicles.[17] The California PATH program, funded primarily by the California Department of Transportation (Caltrans), initiated in 1993 research efforts in both longitudinal[18-20] and lateral[21] control of CHVs. The remaining sections describe the results of this effort to date.

3. MODELING OF HEAVY VEHICLES FOR AHS

The first step in designing longitudinal and lateral control algorithms for CHVs is the construction of models suitable for simulation and control design. The simulation models need to be realistic enough to capture all of the important characteristics of the vehicle and power-train dynamics, but they cannot be too complex in order to avoid excessive simulation times. These models then need to be further simplified in order to be useful for control design. The validity of the resulting controllers can then be tested with the full-order simulation model. In the following, the longitudinal and lateral models used in the PATH program are described. While these models were derived independently, they have now been incorporated into one complete vehicle model, shown in Figure 2, which is currently being used for the design of integrated lateral/longitudinal controllers.

3.1. Longitudinal Dynamics Model

3.1.1. Turbocharged (TC) Diesel Engine Model. The vast majority of existing internal-combustion engine models serve purposes such as engine performance improvement or diagnostics. Engine models suitable for vehicle control have been developed only for normally aspirated spark ignition (SI) engines. Unfortunately, they cannot be adapted to describe the compression ignition (CI) of the diesel engine and to capture the effect of the turbocharger.

Using several TC diesel engine modeling techniques available in the literature,[23-28] we have compiled a model consisting of three differential and several algebraic equations, as well as four 2-D maps, which are compiled from experimental data available in the literature. Our model is based on the *mean torque production* model developed by Kao and Moskwa.[29] The modeled engine is turbocharged and intercooled, has six cylinders, and 0.014 m^3 (14 liters) displacement volume. The block diagram in Figure 3 gives an overview of the model structure.

As in the SI engine model developed by Cho and Hedrick,[30,31] two of the states are the *intake manifold (IM) pressure* and the *engine speed*. However, because of the different fueling method, the fueling lag is not considered here. In the diesel engine, the fuel is injected directly into the cylinder immediately before the combustion takes place. This eliminates the need to account for the fueling dynamics. The *TC rotor speed* is another state that needs to be introduced because of the presence of the turbocharger.

Some intermediate computations are necessary to determine some of the variables participating in the state equations. Steady-state empirical characteristics and experimental data in the form of 2-D maps, as well as algebraic relations are used, in addition to the state equations, to obtain a complete mathematical description of

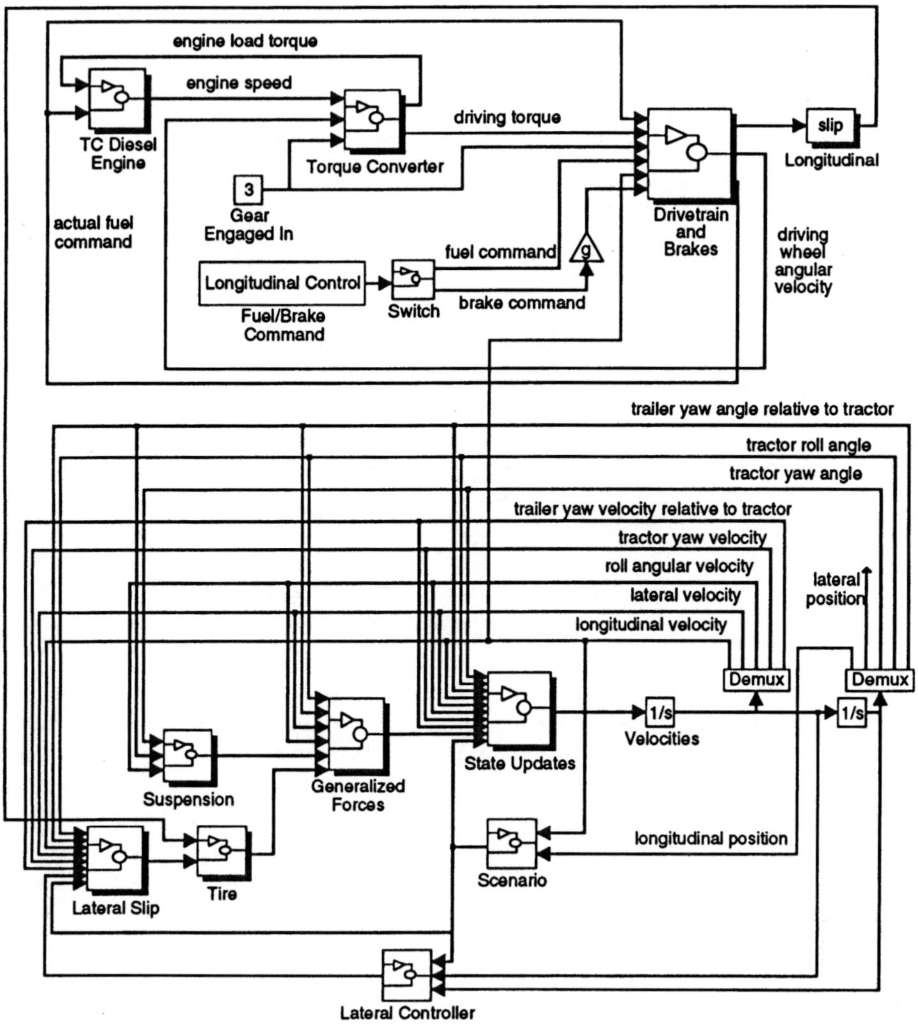

FIGURE 2. Integrated lateral/longitudinal model of truck–semitrailer vehicle.

the system. For a detailed description of our engine and transmission model, the reader is referred to the report.[32]

3.1.2. Torque Converter. The automatic transmission model is based on the assumptions that gear shifting is instantaneous and that there is no torsion in the driveline. The static nonlinear torque converter model derived by Kotwicki[33] has been employed. This model is well suited for vehicle modeling because it provides explicit terminal relations between torques and speeds. Experimentally justified approximation of the exact detailed expressions yields equations that allow us to compute the pump torque and the turbine torque as functions of the pump speed and

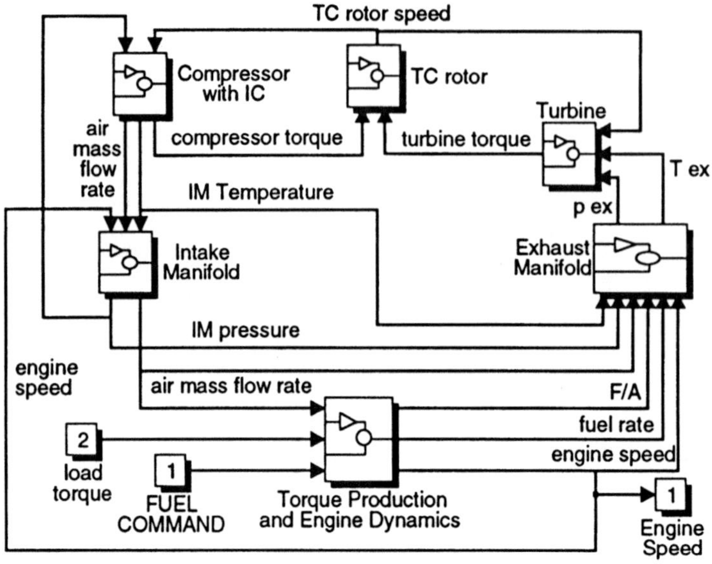

FIGURE 3. TC diesel engine model representation.

the turbine speed. The coefficients of these equations are obtained by regression from experimental data. Although the form of the equations is preserved in both modes of operation, different coefficients have to be determined for the regions when the engine is driving the vehicle and vice versa. The pump and turbine angular velocities are compared to determine the current operating region. The respective coefficients are then used to compute the pump torque, which is the load torque applied to the engine, and the turbine torque, which is the shaft torque transmitted to the drivetrain.

3.1.3. Transmission Mechanicals. The assumption that there is no torsion in the driveline establishes a direct relationship between the angular velocity of the torque converter turbine and that of the vehicle's driving wheels. The model is further simplified by the instantaneous gear shifting assumption, eliminating the need for an additional state which appears during shifting.

3.1.4. Longitudinal Drivetrain. The angular velocity of the driving wheels is determined by the torque converter turbine torque, the tractive tire torque, and the braking torque. Currently, the brake actuating system is represented via a first-order linear system. This is only a crude approximation of the complicated brake dynamics present in heavy-duty vehicles, but it is fairly reasonable for longitudinal control. A more detailed brake model is currently under development.

The tractive force depends linearly on the tire slip up to approximately 15% slip. The aerodynamic drag force (determined from the aerodynamic drag coefficient and the vehicle speed) and the force generated by the rolling resistance of the tires have to be subtracted from the tractive force to yield the force that accelerates or decelerates the vehicle.

Finally, a first-order filter is included in the vehicle model to account for the dynamics of the fuel pump and the actuators that transmit the fuel command to the injectors. The input to this filter is the *index*, defined as the position of the fuel pump rack, which determines the amount of fuel provided for combustion.

3.1.5. Linearization of the Longitudinal Model. The resulting longitudinal vehicle model described so far is highly nonlinear and detailed enough to capture all of the important characteristics of the dynamic behavior of a heavy-duty vehicle. However, it is far too complex to be used as the basis for control design. Because of the presence of several implicit algebraic relations and empirical 2-D maps, it is virtually impossible to obtain a state-space model in a form that is useful for control design. Instead, the model was linearized around several operating points determined by different fuel command/vehicle mass combinations. The results showed that the sixth-order linearized model has the same number of dominant modes throughout the examined range, albeit with significant variations in individual parameter values. The modes associated with the angular velocity of the wheel and the fuel system are always very fast compared to the remaining ones, and can thus be ignored. Of the remaining four modes, those associated with the IM pressure, the engine speed, and the TC rotor speed are much faster than the mode corresponding to the vehicle velocity.

Thus, the longitudinal truck model relating the vehicle speed to the fuel command input can be reduced to a first-order linear model whose coefficients depend on the operating point, i.e., on the steady-state values of the vehicle speed and the load torque. This reduced-order linearized model is the starting point for the design of our control schemes. However, this does not imply that our modeling effort was wasted, since our simulations are always carried out using the full nonlinear model. What this order reduction and linearization do imply is that our controllers do not rely on the particular details of the vehicle model and are thus inherently robust to modeling uncertainties. They are also easier to implement, since they do not contain highly complicated algebraic or dynamic equations. The price paid for this simplicity and robustness is that we can only achieve good performance if the required maneuvers are slower than the slowest neglected modes. But since here we are only interested in maneuvers with time constants of several seconds, this limitation does not play a significant role.

3.2. Lateral Dynamics Model

Two types of dynamic models are required for the design and analysis of the vehicle lateral controller. A complex simulation model represents the vehicle dynamics as realistically as possible, while linear and nonlinear simplified models retain only the significant dynamics (lateral and yaw dynamics). The former is utilized in the simulation study for evaluating the effectiveness of the controller, and the latter in the design of the controller.

3.2.1. Complex Model. In the literature, various mathematical models of heavy-duty vehicles have been proposed and computer simulations presented.[11,34,35] Their goal is to develop a tool for predicting and evaluating the longitudinal and directional response of heavy-duty vehicles. Most of the mathematical models published in the literature adopt the Newtonian mechanics approach to describe the body dynamics of heavy-duty vehicles. In Ref. 21, a complex model of an articulated heavy-duty vehicle is derived by applying Lagrange's equations. This Lagrangian mechanics formulation has an advantage over Newtonian mechanics in that it can eliminate the holonomic constraints at the linking joints of an articulated vehicle by careful choice of the generalized coordinates. The Lagrangian mechanics formulation is particularly suitable for control algorithm synthesis. Furthermore, structural properties, such as the Coriolis and centrifugal forces and the system damping on the lateral dynamics, can be easily analyzed in this approach. It is shown in Ref. 21 that the damping of vehicle lateral dynamics is inversely proportional to the longitudinal speed of the vehicle. A passivity property is also identified in vehicle lateral dynamics and utilized in designing the vehicle lateral velocity estimation algorithm in Ref. 22. Although the Lagrangian modeling approach has been popular in the area of robotics,[36,37] there are two major differences between vehicle and robot applications. First, the generalized force corresponding to each generalized coordinate of a robot manipulator is the actuator torque at the robot joint while the generalized force corresponding to each generalized coordinate of an articulated vehicle needs to be transformed from the tire and suspension forces. This imposes difficulties in designing controllers and analyzing closed loop stability. Second, the sprung mass of a vehicle has roll and pitch motions relative to the unsprung mass, and usually the roll center and pitch center are different from the center of mass of a vehicle, which makes the equations of motion more involved.

3.2.2. Coordinate Systems and Lagrange's equation. In Figure 4, a coordinate system is defined to precisely describe the translational and rotational motion of a tractor–semitrailer vehicle. $X_nY_nZ_n$ in Figure 4 is the inertial reference coordinate frame. $X_uY_uZ_u$ are the tractor unsprung mass coordinates, and $X_sY_sZ_s$ are the tractor sprung mass coordinates. In Figure 5, two generalized coordinates are chosen to satisfy the kinematic constraint at the linking joint (fifth wheel) of the tractor–semitrailer vehicle. The introduction of these coordinates makes it easy to express the kinetic energy, T, and the potential energy, V, of the tractor–semitrailer vehicle.

The equations of motion can then be derived by applying Lagrange's equation,

$$\frac{d}{dt}\frac{\partial L}{\partial \dot{q}} - \frac{\partial L}{\partial q} = F_g$$

where $L = T - V$ is the Lagrangian and F_g is the generalized force. The external forces acting on the vehicle body are from the tire/road interface and the suspension. In applying Lagrange's equation, we need to calculate each generalized force

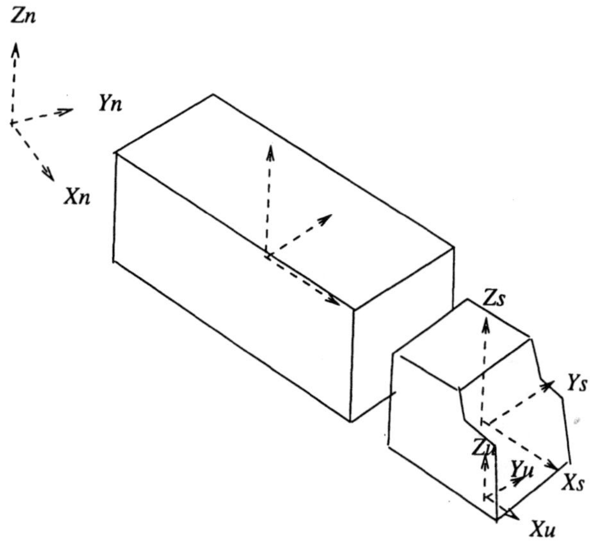

FIGURE 4. A coordinate system to describe the vehicle motion.

corresponding to each generalized coordinate explicitly, that is, we want to transform the external forces from Cartesian coordinates to the generalized force corresponding to each generalized coordinate. Figure 6 shows the definition of the external forces of tractor–semitrailer vehicles. These forces are predicted by the tire and suspension models.

Deriving the position vector in Cartesian coordinates of each wheel in terms of the generalized coordinates, we can calculate the generalized forces in terms of the tire force in Cartesian coordinates by the principle of virtual work.[38]

3.2.3. Tire Model. Modeling the tire/road interaction force is itself an active area of research. Pacejka and Bakker[39] provide a set of mathematical formulas

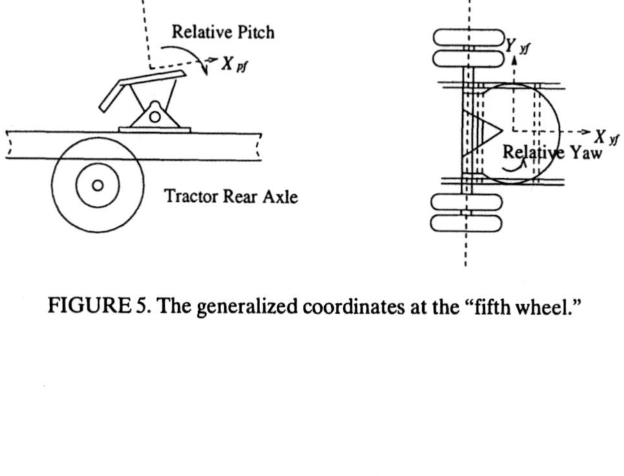

FIGURE 5. The generalized coordinates at the "fifth wheel."

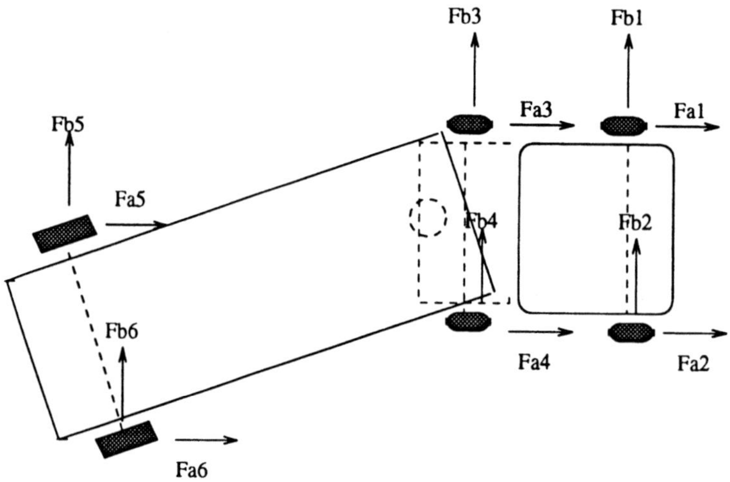

FIGURE 6. Tire force in Cartesian coordinates.

from which the forces and moments acting from road to tire can be calculated at longitudinal, lateral, and camber slip conditions. These formulas are derived from curve fitting of tire data, and a set of "magic formula" tuning parameters constitutes the basis of this model. Fancher and Bareket[40] propose a semiempirical model of truck tires. This model accounts for the influences of tread depth, mean texture depth, and skid number on the sliding friction of truck tires (Figure 7).

3.2.4. Wheel Dynamics. To apply the tire model stated above, we need to specify the operating conditions of the tire, i.e., the longitudinal slip ratio and the lateral slip angle. The tire will generate tractive force when the longitudinal slip ratio is positive and braking force when the longitudinal slip ratio is negative. To calculate the longitudinal slip ratio, we need to include the wheel dynamics in the complex model. Figure 8 shows the free body diagram of the wheel dynamics.

3.2.5. Suspension Model. For the lateral control applications, suspension is modeled analytically as a combination of spring, damper, and a static friction (stiction in the suspension). Since the unsprung mass is negligible for all practical purposes, we will neglect it even for the detailed complex model.

In case of single-vehicle configurations, the model derived with the Lagrangian approach is identical to the model derived with the Euler–Newtonian approach.

3.2.6. Simplified Model. Since the complex model is complicated and unenlightening, a simplification (or reduction) of this model is used for control design purposes. Open-loop simulations reveal that the roll, pitch, and bounce of the vehicle are small in normal highway maneuvers. These state variables are hence eliminated from the complex model to form the simplified model. Thus, the simplified model

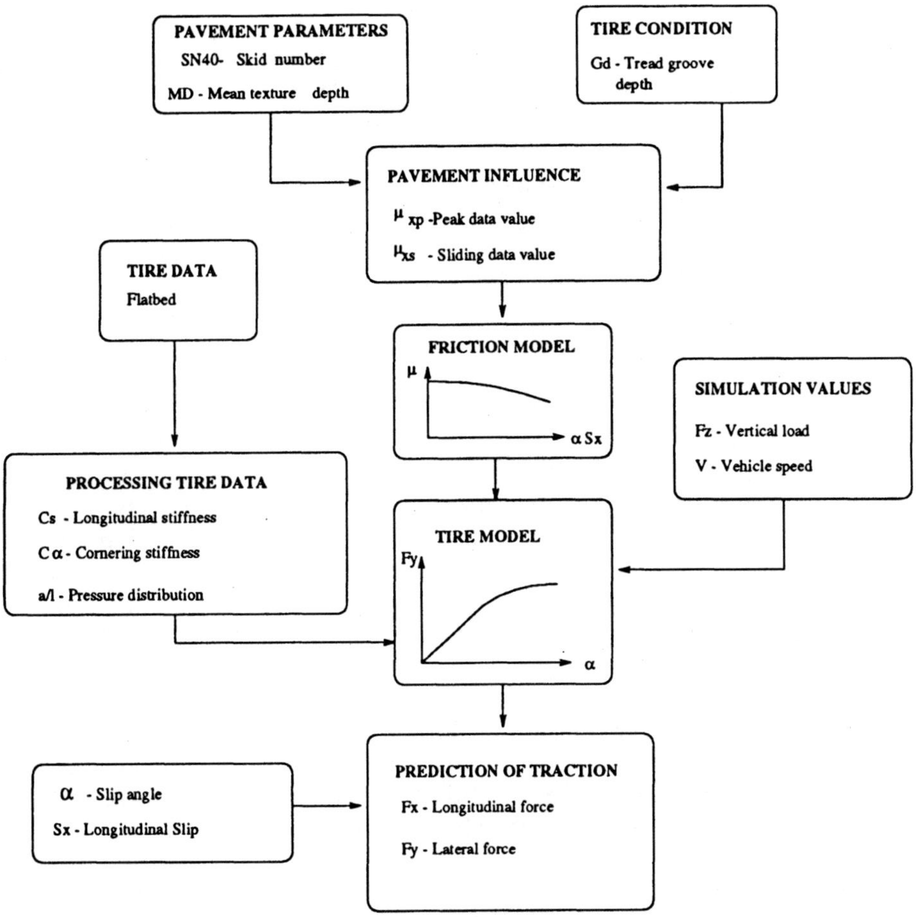

FIGURE 7. Tire model (Fancher and Bareket[40]).

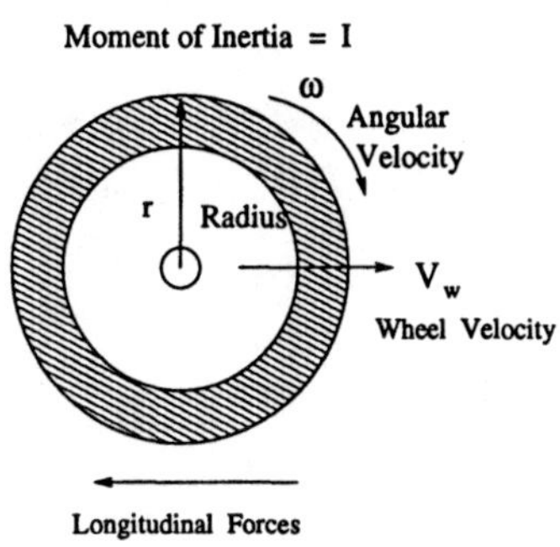

FIGURE 8. Wheel dynamics.

consists of longitudinal, lateral, and yaw dynamics. Later, this simplification is to be justified by looking at the results of closed-loop simulations and observing that the roll and the pitch are small during the operation of the vehicle. A simplified tire model, which represents the cornering force as the cornering stiffness times the slip angle, is used in deriving the simplified vehicle model. Further details of the complex and simplified models can be found in Ref. 41.

4. CONTROL DESIGN FOR AUTOMATED CHVS

4.1. Longitudinal Control Design

The continuous demand for increasing the traffic capacity of highway systems has so far been met by expansion of existing highways and construction of new ones. However, these solutions are quickly becoming unfeasible in many large metropolitan areas. A widely proposed strategy for effectively increasing highway throughput without expansion[42–45] is to group automatically controlled vehicles in tightly spaced formations called *platoons*. Platooning provides significantly higher traffic throughput when combined with small intervehicle spacing. The control design for platoons has to guarantee the desired performance not only for individual vehicles but also for the whole formation. The quantities of interest between any two adjacent platoon members are defined in Figure 9.

One of the key issues here is *string stability*, which ensures that errors decrease in magnitude as they are propagated upstream, thus eliminating the undesirable "slinky effect" associated with heavy traffic patterns. The fact that string stability cannot be achieved for platoons with *constant intervehicle spacing* under autono-

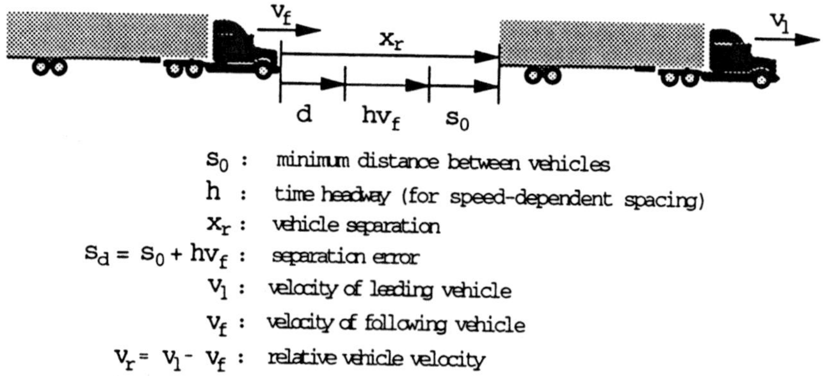

FIGURE 9. Parameters of a truck platoon.

mous operation has been known for more than 20 years.[42,46] String stability can be guaranteed if the lead vehicle is transmitting its velocity[42] or velocity and acceleration[44] to all other vehicles in the platoon. This approach yields stable platoons with small intervehicle spacings at the cost of introducing and maintaining continuous intervehicle communication. String stability can also be recovered in autonomous operation if a *speed-dependent* spacing policy is adopted, which incorporates a *fixed time headway* term in addition to the constant distance.[46,47] In this setup, each vehicle computes its desired distance from the preceding vehicle as the sum of the minimum fixed distance s_0 and the product of the time headway h with its own speed v_f, hence the name speed-dependent spacing. This approach avoids the communication overhead, but results in larger spacings between adjacent vehicles and thus in longer platoons, thereby yielding smaller increases in traffic throughput.

The only previously available results on longitudinal control of CHVs were those reported in Ref. 48. In the past 2 years, we have developed a design framework for longitudinal control of automated CHVs that can be used in all stages of AHS deployment. The resulting controllers command both fuel and brake, thus eliminating the need for separate fuel and brake controllers with ad hoc switching logic. Moreover, they can operate both autonomously as well as with intervehicle communication. This flexibility is desirable for two reasons:

- A vehicle equipped with this controller can take full advantage of future automated lanes, but it can also be operated in AICC mode in nonautomated lanes.
- The AICC capability can act as a fail-safe in automated lanes, in case there is a breakdown in intervehicle communication. In the event of such a failure, the controller can simply increase the time headway and continue to operate with guaranteed string stability, albeit with lower traffic throughput.

This reduction of traffic throughput under autonomous operation is much more pronounced for CHVs than for passenger cars. In fact, if we use a speed-dependent spacing policy with a fixed time headway, that headway has to be much larger to achieve smooth operation of CHV platoons—at least 0.5 s, compared to 0.2 s for cars. This means that the resulting intervehicle spacing will be significantly larger for CHVs. To alleviate this problem, we have developed two new nonlinear spacing policies that yield smoother operation with improved platoon performance in the case of autonomous operation.

To illustrate the various features of our control design framework, we simulated a platoon of ten tractor–semitrailer combination vehicles. Each vehicle weighs 20,000 kg (44,000 lb) and is powered by a 14-liter turbocharged diesel engine capable of producing approximately 1000 Nm (740 ft-lb) in torque and 300 kW (400 hp) in peak power. The platoon starts out at an initial speed of 22 m/s (49.10 mph). At t = 10 s the platoon leader decides to reduce his speed by 10 m/s (22.32 mph) and then, at t = 80 s, to increase it by 5 m/s (11.16 mph). The minimum desired separation between vehicles is s_0 = 3 m. This demanding scenario is representative of the difficulties the system might have maintaining string stability when trying to meet a challenging acceleration/deceleration objective. In all of our simulation plots, different vehicles are represented by lines of different styles: Vehicle 1 is shown by a solid line, while following vehicles cycle through dash-dotted, dashed, dotted, and solid lines (so that Vehicles 5 and 9 are also shown by solid lines).

4.1.1. PI Controller. Since our simplified longitudinal model is a first-order linear system, the proportional-integral (PI) controller is the simplest controller that can achieve regulation of both v_r and δ with some robustness with respect to disturbances and unmodeled effects. Hence, the PI controller is the starting point of our control design. The tasks of regulating the relative vehicle speed v_r and the spacing error δ can be combined into the single control objective $v_r + k\delta = 0$, where k is a positive design constant. This control objective makes sense intuitively: If two vehicles are closer than desired ($\delta < 0$) but the preceding vehicle's speed is larger than the follower's ($v_r > 0$), then the controller in the following vehicle does not need to take drastic action. The same can be said if the vehicles are farther apart than desired ($\delta > 0$) but the preceding vehicle's speed is lower than the follower's ($v_r < 0$). Hence, the error variable used in the PI controller becomes $v_r + k\delta$. The selection of the coefficient k influences the response of the controller, and can be changed depending on the performance requirements.

However, in the demanding scenario of the presented simulations, a simple PI controller is not adequate. Therefore, we introduced several modifications in order to improve platoon performance and smoothness of control effort. These modifications of the control law can be viewed as independent modules that may be used separately or combined. By qualitatively evaluating the contribution of each module, we can add complexity to the control design only as needed to meet the given performance specifications.

4.1.2. Signed Quadratic (Q) Term. The lower actuation-to-weight ratio of heavy-duty vehicles requires a controller that is more aggressive at large errors but does not have the undesirable side effect of overshoot. This is achieved by adding a signed quadratic (Q) term of the form $(v_r + k\delta)|v_r + k\delta|$ to the PI controller, which thus becomes a nonlinear PIQ controller. This Q term was used successfully in the speed control problem of heavy-duty vehicles in Ref. 20, where it proved to be more efficient in avoiding overshoot than an antiwindup term and provided faster attenuation of errors compared to linear controllers. It also performed well in the platoon scenario, where overshoot is even less desirable.

4.1.3. Adaptive Gains. Even if a controller is perfectly tuned for some operating region, it is likely to demonstrate inferior performance in other conditions because of the severe nonlinearities present in the system. A gain scheduling approach could be successful in overcoming the disadvantages of a fixed gain controller, but it would require extensive *a priori* information. Moreover, because of large variations in the mass of heavy-duty vehicles (the change in mass between a lightly loaded and heavily loaded truck can be as much as 300%), such *a priori* information may not even be available to the controller. Hence, adaptation of the control gains is the

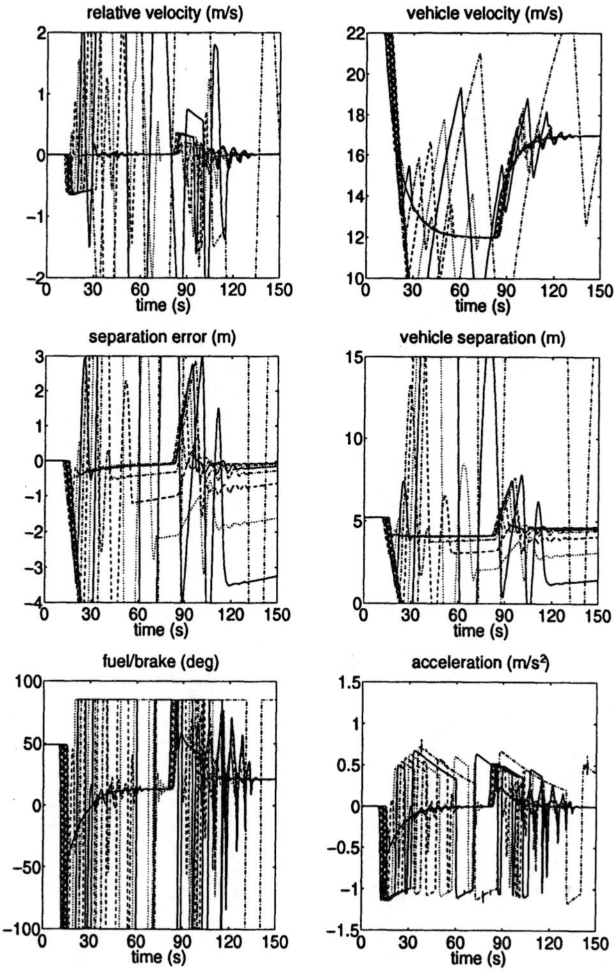

FIGURE 10. Ten autonomous vehicles, constant $h = 0.1$ s.

natural solution, since it makes the closed-loop system response much less dependent on the current operating region and on the specific vehicle characteristics. The latter consideration becomes more significant in the platoon scenario where even if the grouped vehicles are not identical, they are expected to respond uniformly to different commands or disturbances.

The resulting adaptive PIQ controller can operate autonomously using a speed-dependent policy. However, the fixed time headway has to be significantly larger than with passenger cars in order to guarantee good CHV platoon performance. This is illustrated in Figure 10, which shows that with time headway $h = 0.1$ s the errors are significantly amplified as they propagate upstream the platoon (from Vehicle 1 to Vehicle 4). In order to get acceptable performance, one must increase

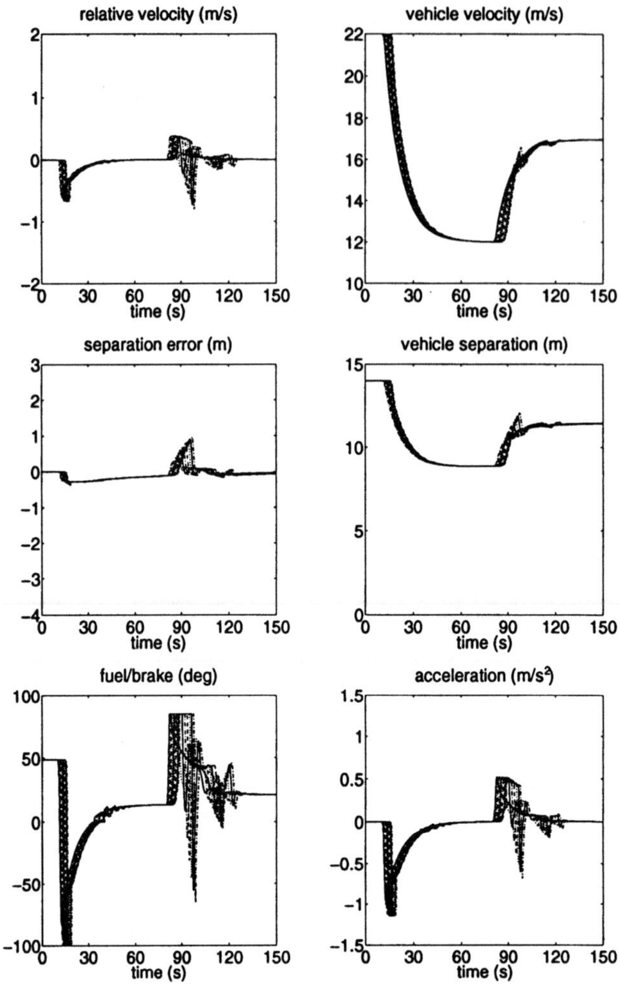

FIGURE 11. Ten autonomous vehicles, $h = 0.5$ s.

the headway to $h = 0.5$ s. Then, as seen in Figure 11, errors become much smaller and the fuel/brake activity much smoother (implying better fuel efficiency), but the intervehicle spacing grows significantly from about 4–5 m to 10–14m.

4.1.4. Intervehicle Communication. While it is expected that the lower actuation-to-weight ratio of heavy-duty vehicles will necessitate larger spacings than those used in car platoons, the spacings resulting from $h > 0.5$ s may be unacceptable. In that case, one good solution for reducing the spacing while guaranteeing small errors is the introduction of intervehicle communication. As demonstrated in Figure 12, the errors are much smaller and string stability is achieved even with

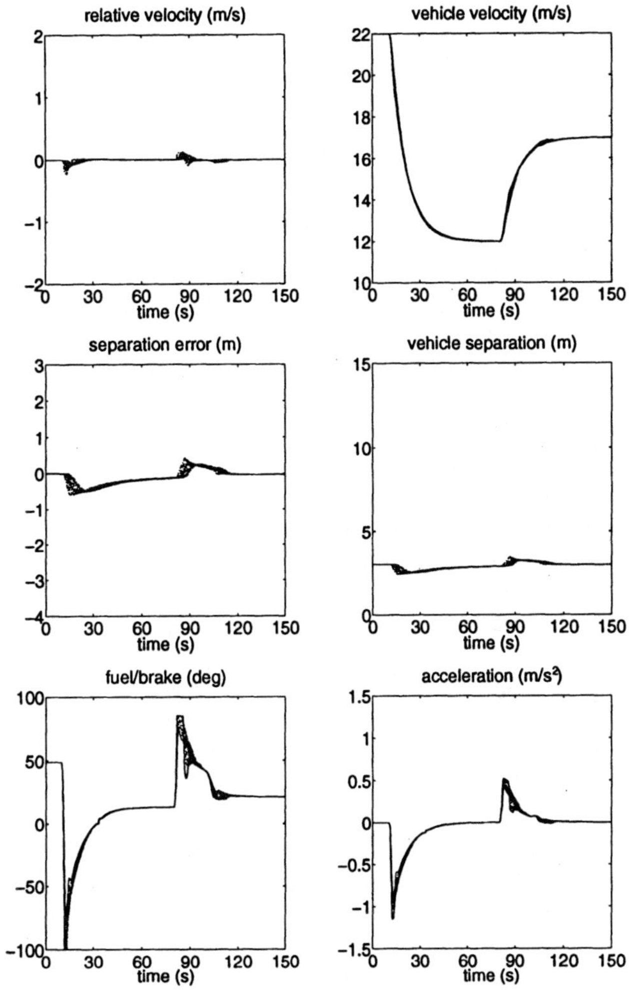

FIGURE 12. Ten vehicles with intervehicle communication, $h = 0$.

$h = 0$, which leads to a significant reduction in intervehicle spacing, albeit at the expense of introducing and maintaining communication between vehicles.

It is worth noting that our notion of intervehicle communication includes only transmission of the *desired* speed of the platoon leader, and can be viewed as a variation of the approach proposed in Ref. 42. In terms of bandwidth and reliability, this is a much less demanding (and hence more robust) scheme than the usually considered ones, which require transmission of the leader's *current speed and acceleration*.

4.1.5. Variable Time Headway. Intervehicle communication may not be implemented in the beginning stages of truck automation, and even if it is implemented it may sometimes fail. Therefore, it would be desirable to develop control algorithms

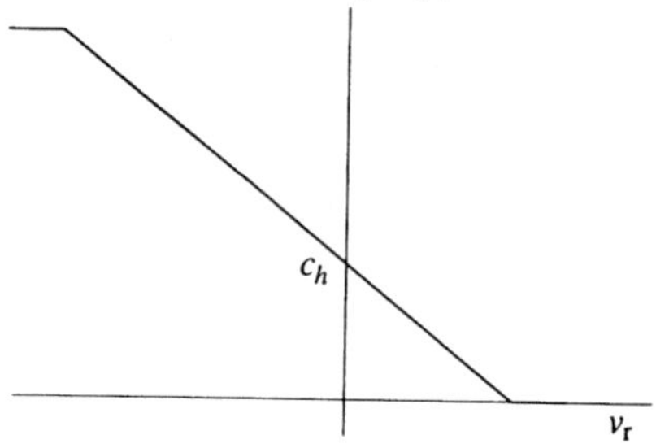

FIGURE 13. Variable time headway $h = \text{sat}(h_0 - c_h v_r)$.

that yield small separation errors without increased intervehicle spacing under autonomous vehicle operation. As it turns out, one can apply some simple modifications to the above adaptive PIQ algorithm to achieve this goal. The first such modification is to use a *variable time headway*, which varies with the relative speed v_r between adjacent vehicles. The intuition behind this modification, which was introduced in Ref. 19, is simple: Suppose that a vehicle wants to maintain a 0.1 s time headway from the preceding vehicle, when both of them are traveling at the same speed. However, if the relative speed between the two vehicles is positive, that is, if the preceding vehicle is moving faster, then it is safe to reduce this headway, while if the preceding vehicle is moving slower, then it would be advisable to increase the headway. For safety reasons, the headway h cannot be allowed to become negative, while very large headways are undesirable as shown in Ref. 20. Thus, we limit the headway in the interval [0,1] and arrive at the form of h shown in Figure 13.

The effect of introducing the variable time headway $h = \text{sat}(0.1 - 0.2v_r)$ in our ten-truck platoon is quite dramatic, as seen in Figure 14. Comparing Figure 14 with Figure 10 reveals an impressive reduction of errors and a considerably smoother control activity without any increase in steady-state intervehicle spacing. In fact, the response is quite similar to that obtained with $h = 0.5$ s in Figure 11, but with much smaller intervehicle spacing.

4.1.6. Variable Separation Error Weighting. Another simple modification is the introduction of a *variable separation error weighting* k. Recall that the intuition behind choosing the control objective as regulation of $v_r + k\delta$ was as follows: If two vehicles are closer than desired ($\delta < 0$) but the preceding vehicle's speed is larger than the follower's ($v_r > 0$), then the controller in the following vehicle does not need to take drastic action, and the same is true if the vehicles are farther apart than desired ($\delta > 0$) but the preceding vehicle's speed is lower than the follower's ($v_r < 0$).

However, when the separation error weighting k is constant, the controller will try to reduce a very large spacing error δ through a very large relative velocity v_r of opposite sign. Hence, if a vehicle falls far behind its preceding vehicle, its controller will react aggressively by accelerating to a very high speed. This behavior is not only

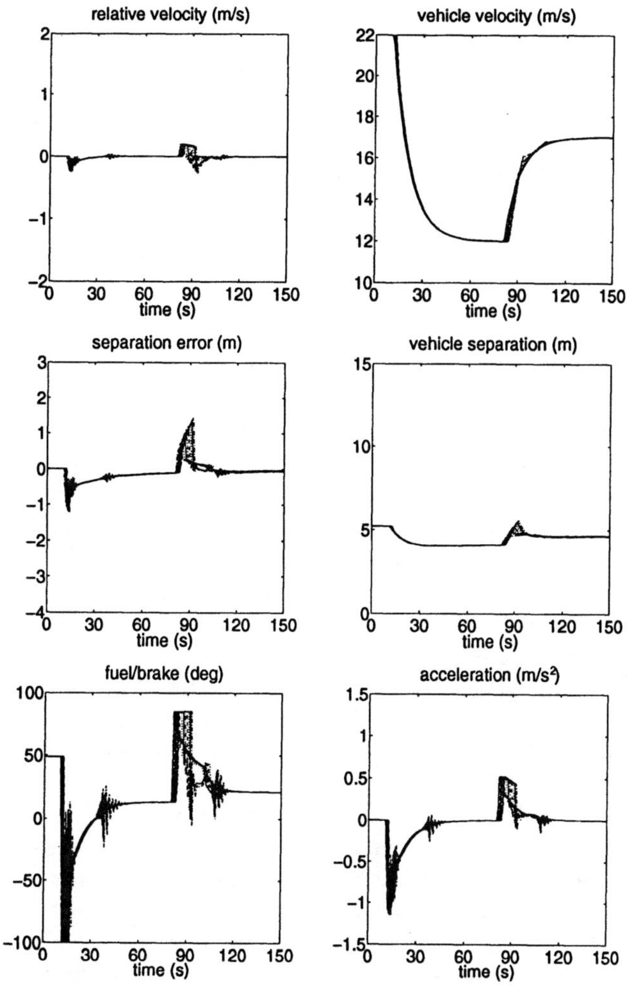

FIGURE 14. Ten autonomous vehicles, $h = \text{sat}(0.1 - 0.2v_r)$.

undesirable, since it increases fuel consumption and can even lead to collisions in extreme situations, but is also counterintuitive. It would be much more natural for the controller to accelerate to a speed somewhat higher than that of the preceding vehicle's, and reduce the spacing error smoothly and progressively. To achieve such a response, we need to decrease the weighting k as δ becomes large and positive, making sure that it remains above some reasonable positive lower bound. The choice of k shown in Figure 15 satisfies this requirement.

However, this choice of k, which makes the control objective $v_r + k\delta = 0$ nonlinear in δ, has another feature that at first glance may seem counterintuitive: The weighting k is reduced even when δ becomes negative. A careful examination of the truck platoon characteristics and of the simulation results yields the following ex-

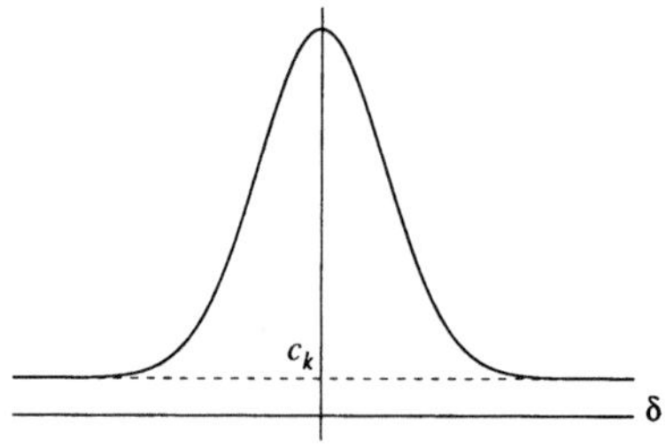

FIGURE 15. Variable k choice.

planation for this choice: In autonomous operation, each vehicle relies only on its own measurements of relative speed and distance from the preceding vehicle. This means that if the preceding vehicle suddenly decelerates, then the following vehicle will have to decelerate even more to maintain the desired spacing. Hence, to maintain string stability, aggressive control actions must be amplified as they propagate upstream the platoon. While this may not be a big problem for passenger cars, it becomes crucial for CHVs because of their low actuation-to-weight ratio, which severely limits the accelerations and decelerations they are capable of achieving. During a sudden braking maneuver, only the first few vehicles in the platoon will be able to achieve the necessary decelerations; the controllers of the next vehicles will quickly saturate, and collisions may occur. In fact, the likelihood of a collision increases with the number of vehicles in the platoon. On the other hand, if the reaction of the first few vehicles is not as aggressive, then the decelerations are not amplified as much, and hence the remaining vehicles can achieve the necessary profiles. In a sense, reducing the weighting k for negative δ endows the controller with a "platoon conscience," which sacrifices the individual performance of the first few vehicles in order to improve the overall behavior of the platoon. This is clearly illustrated in Figure 16, which uses the same fixed headway $h = 0.1$ s as Figure 10. Comparing the two figures we see that the use of variable k yields significantly improved platoon performance: The errors are not amplified as they propagate upstream the platoon, there are no collisions, and the control effort is noticeably smoother. However, the errors of the first five vehicles are larger than in Figure 10, because their controllers did not react as aggressively. The individual performance of these vehicles has been compromised, but the improvement in the performance of the remaining vehicles results in a better overall trade-off.

4.1.7. Qualitative Comparison. According to the AHS Precursor Systems Analysis for Commercial Vehicles and Transit,[49] longitudinal control of truck platoons will be one of the most challenging problems for commercial AHS. The control schemes examined here are promising, but need to be tested in real experiments. Before that, however, their advantages and disadvantages must be clearly understood. Each of the above modifications adds some complexity to the

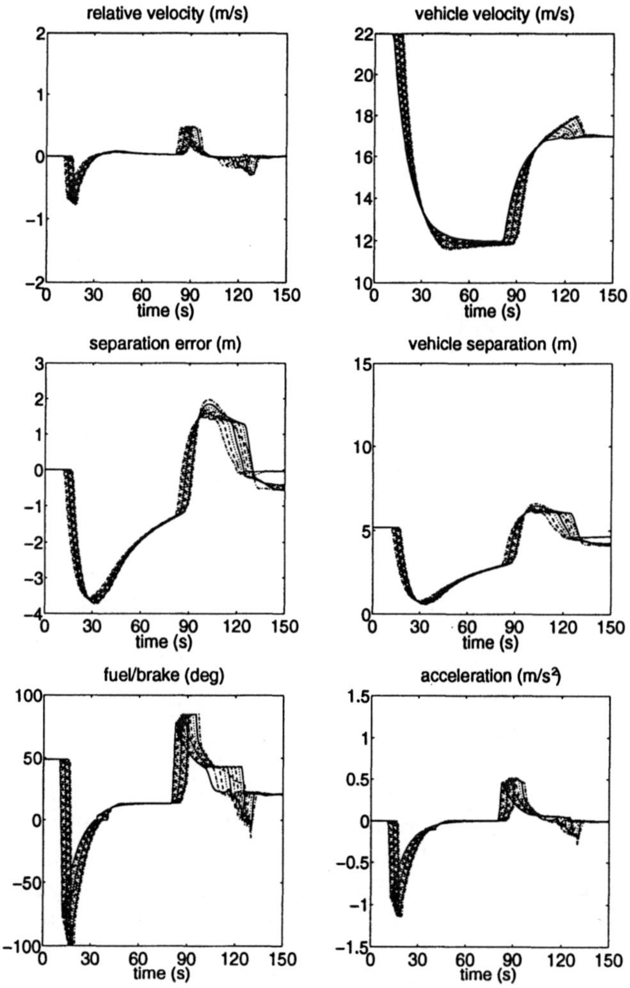

FIGURE 16. Ten autonomous vehicles, $h = 0.1$ s, variable k.

control algorithm, making it more computationally demanding, but at the same time results in improved performance. Comparing all of the presented simulation results is not enough to determine which modifications are worth implementing and which are not, since these simulations are by no means exhaustive. Furthermore, each modification has something different to offer, and no single criterion can be used to compare and rank them. Therefore, we have compiled a qualitative comparison chart, which is based on our analytic results and on our extensive simulations of several AHS scenarios. The two criteria selected for this comparison are the two most important ones from the AHS point of view: *control smoothness*, which is directly related to fuel efficiency and driver comfort, and *platoon performance*, which is related to safety and traffic throughput. In Figure 17, each circle represents

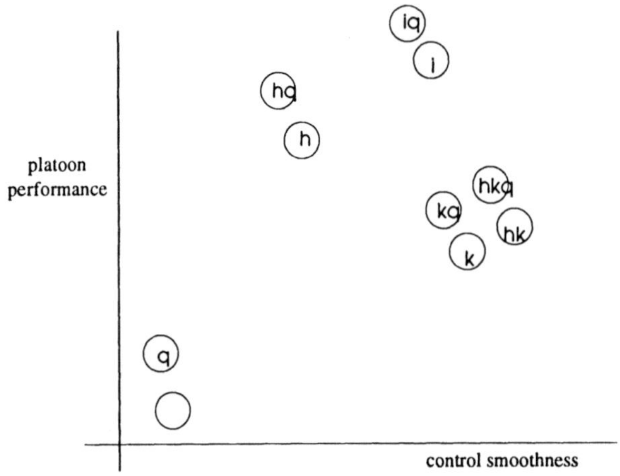

FIGURE 17. Qualitative comparison chart of CHV longitudinal control schemes.

a different controller configuration. All configurations consist of a proportional and an integral term with adaptive gain, and in addition may feature a Q term (denoted by q), variable time headway (denoted by h), variable separation error weighting (denoted by k), or intervehicle communication (denoted by i).

Figure 17 should be used as a "visual aid" in determining the most appropriate longitudinal control scheme for a CHV platoon. Each choice involves a trade-off between control smoothness, platoon performance, and controller complexity. In autonomous operation, for example, it is always worth using either variable time headway or variable separation error weighting. With either of these modifications, both control smoothness and platoon performance are much better than without any of the two, and this benefit justifies the additional control complexity. If platoon performance is the primary consideration, then variable time headway is the modification of choice, while variable separation error weighting should be preferred when control smoothness is more important. It is also clear that if any of these two is used, the additional complexity of a nonlinear Q term may not be justified, since the resulting change is barely noticeable. And if variable k is used, the addition of variable h yields only a marginal improvement in performance and smoothness. Finally, the schemes with intervehicle communication appear to have a distinct advantage over the autonomous schemes. However, this must be weighed against the considerable additional complexity of establishing and maintaining reliable communication between the vehicles in the platoon.

4.2. Lateral Control Design

The history of published research on steering control for lane following for passenger vehicles goes as far back as the late 1950s by General Motors and RCA.[50] An important issue in lane following is the selection of the lateral reference system. Early developments in lateral reference systems are discussed by Fenton and Mayhan.[1] The VECTOR truck of Daimler-Benz Research and Freightliner Corporation is currently undergoing testing of a vision-based control system for lateral guidance.[13] Vision-based lateral guidance system is discussed in Refs. 51 and 52. Bishel[17] used aluminum reflectors and radar sensors as the lateral reference system. In California PATH, a magnetic road marker-based system is being used for lateral control of passenger vehicles.[53]

Parametric uncertainties coming from the road–tire interaction, the mass of the vehicle, and the cargo shift/slosh, combined with the errors related to simplification in the control model warrant a controller that is *robust* to these changes. There exist several state-of-the-art robust techniques that have the potential to handle such problems. Advantages of these robust control algorithms are that (1) their design procedure is systematic (as opposed to manual gain tuning approach used in simple proportional-integral-derivative controllers); (2) based on the model, the system's nonlinear dynamics can be canceled; and (3) because of robustness guarantees, extensive simulations are not required. In the remaining part of this section, *robust* control algorithms for lateral control of CHVs will be emphasized.

4.2.1. Lateral Control of Single-Unit Heavy Vehicles. For single-unit vehicles, the simplified model used for control design becomes linear when the longitudinal velocity is constant. Gain scheduling can be applied to a linear controller to take care of the variation in the longitudinal velocity. This approach has been used in Ref. 17 for automatic durability testing of trucks. Linear controllers based on Frequency Shaped Linear Quadratic (FSLQ) algorithms[3] for passenger cars also used gain scheduling. FSLQ algorithms take ride comfort into consideration and are desirable for passenger vehicles. Peng[3] also applied continuous time optimal preview theory to the lateral control of passenger vehicles. These control algorithms can be applied to the lateral control problem of single-unit CHVs.

In the absence of constant longitudinal velocity assumption and in the face of large parametric changes in the model, robust nonlinear control algorithms need to be designed. Sliding Mode Control (SMC)[54] or Variable Structure Control has been popular for robust nonlinear control of passenger cars.[55,56] SMC is a special case of variable structure systems, in which the control law is switched depending on which region the state of the controlled system belongs to. The control law on each side of the "sliding surface" is selected so that the system first approaches the sliding surface. The state subsequently remains on the sliding surface and "slides" toward the desired equilibrium point. For lane following applications, the equilibrium point is the zero lateral error and zero lateral velocity point. Sliding takes place because of the high gain nature of the control law around the switching surface introduced by the switching function. The high gain is necessary for robustness against parametric uncertainties but introduces chattering. SMC approaches to steering control have been proposed by Pham *et al.*[55] and Ackermann.[57] To avoid chattering, Pham *et al.*[55] used a saturation function in place of the sign function in SMC. Furthermore, the integrated tracking error signal was used as the sliding variable, which assures zero steady state error. Ackermann[57] used steering rate as the controller command (output of the sliding mode controller). The steering angle

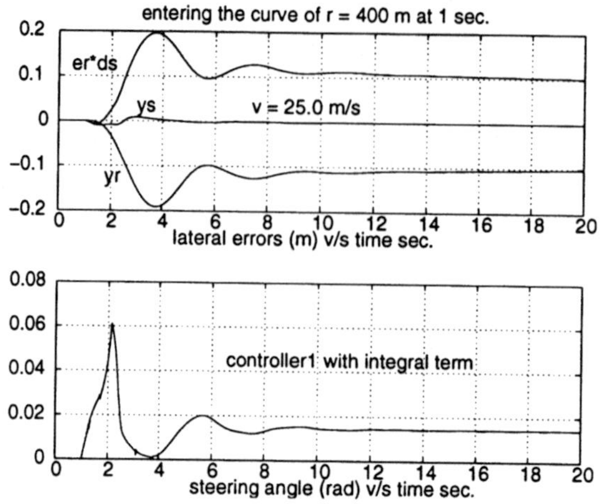

FIGURE 18. Simulation with the controller of Ref. 55 applied to the nominal plant.

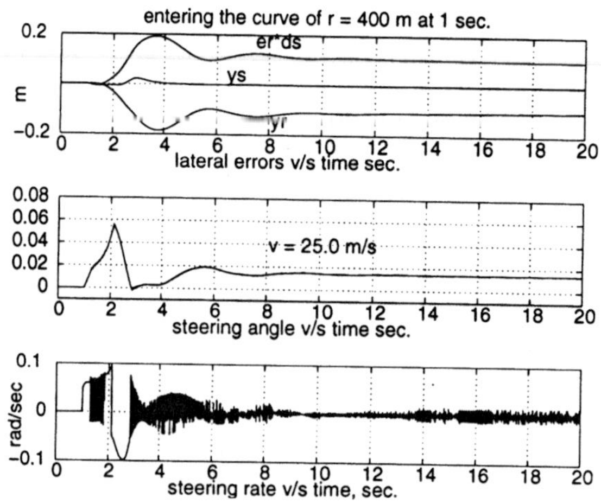

FIGURE 19. Simulation with the controller of Ref. 57 applied to the nominal plant.

command, then, was chatter free because of the filtering action of the integrator. These approaches have been studied and compared to each other for the lateral control of commuter buses (single unit-HV) in Ref. 58. Figures 18 and 19 show the result of a typical scenario. The vehicle enters a curve of 400-m radius at 1 s into the simulation. The lateral tracking error at the sensor, denoted by *ys*, is a combination of the yaw error *er* and the lateral error at the center of mass of the vehicle *yr*: $ys = yr + ds*er$, where *ds* is the distance of the sensor from the center of mass. This can be seen to be reasonably small (about 5 cm). There is significant backtracking, as indicated by the lateral tracking error as measured from the center of mass of the vehicle *yr* (≈ 20 cm). There is no chatter in the steering angle command.

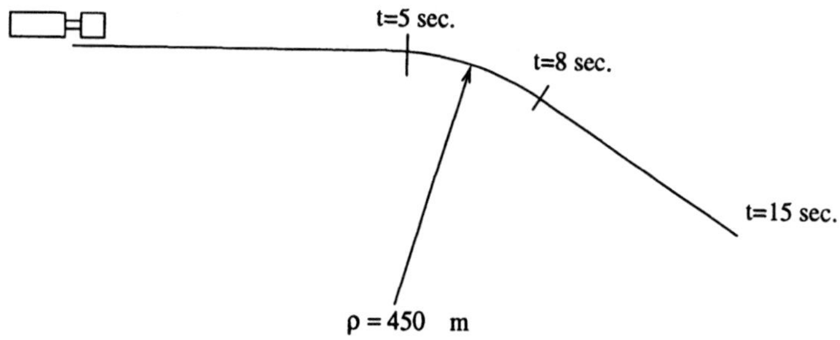

FIGURE 20. Simulation scenarios.

4.2.2. Lateral Control of Articulated Vehicles. Two linear control algorithms for articulated vehicles have been designed based on linear quadratic (LQ) optimal control and FSLQ optimal control.[59] The LQ optimal control algorithm makes a compromise between tracking performance and steering input. Perfect tracking is not attempted in these formulations. An important aspect of the frequency-shaped LQ optimal control algorithm is that it includes the lateral acceleration of the vehicle in the performance index; thus, the lateral acceleration is reduced, vehicle rollover is avoided, and driver ride quality is enhanced. Moreover, since we can penalize high-frequency components of the control input, the FSLQ control algorithm is less likely to excite unmodeled high-frequency dynamics and is more robust than LQ optimal control.

The simulation scenario we used is shown in Figure 20. The vehicle travels along a straight roadway with initial lateral displacement 15 cm and enters a curved section with radius of curvature 450 m at time $t = 5$ s. Figure 21 shows the simula-

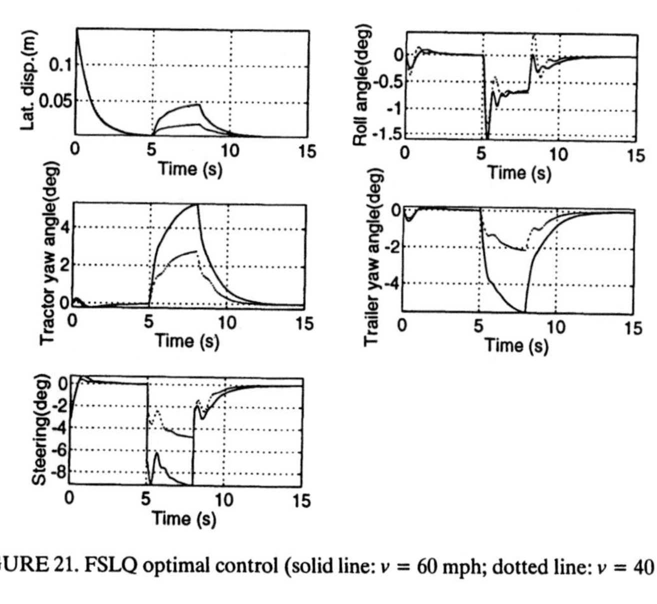

FIGURE 21. FSLQ optimal control (solid line: v = 60 mph; dotted line: v = 40 mph).

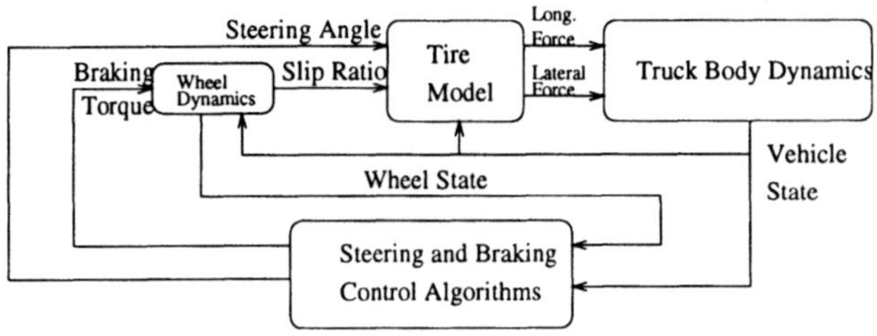

FIGURE 22. Vehicle system including wheel dynamics.

tion results of the FSLQ optimal controller at different vehicle speeds. Gain scheduling with respect to the longitudinal speed is applied in this controller. The peak lateral displacement is about 5 cm (solid line) at longitudinal speed v = 26.4 m/s (60 mph) and is about 2 cm (dotted line) at longitudinal speed 17.6 m/s (40 mph).

An input/output feedback linearization combined with the backstepping scheme[60] is designed for the lateral control of the tractor–semitrailer vehicle in Ref. 61. This control scheme uses the tractor front wheel steering as a control variable to achieve lane following in AHS, and the trailer inner and outer tire force distribution (independent braking) as the other control variable to stabilize the trailer motion. The independent braking of the trailer can be regarded as a safety augmented system (SAS) to prevent the potential yaw instability modes in an articulated vehicle. The block diagram of the closed-loop system is shown in Figure 22.

For simulation we use the same scenario as shown in Figure 20. Vehicle longitudinal speed is 26.4 m/s (60 mph). Simulation results in Figure 23 show that this scheme has better lateral tracking, while the tractor relative yaw angle, trailer

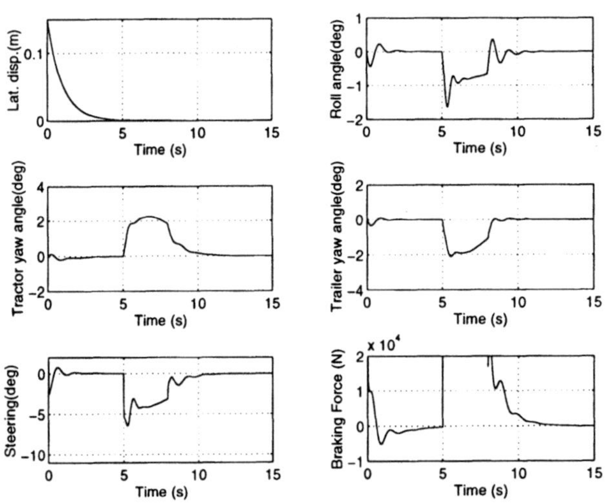

FIGURE 23. Input/output linearization control.

articulation angle, and steering control input are of the same order as those in FSLQ optimal control. The input/output linearization controller has better tracking performance because it utilizes the information of the desired tracking signal $\dot{\varepsilon}_d$ to compute the control input. Thus, the I/O linearization control scheme is a feedback-plus-feedforward controller. On the other hand, the FSLQ control scheme, a pure feedback control law, treats the desired tracking signal $\dot{\varepsilon}_d$ as a disturbance.

5. CONCLUSIONS

In this chapter, dynamic modeling and control of CHVs were presented. The need for *robust* control of these vehicles was pointed out in light of uncertainties in tire–road interaction, uncertainties in vehicle parameters such as mass, cargo shift/slosh, and nonlinearities related to varying longitudinal velocity and unmodeled dynamics. Several control algorithms for longitudinal and lateral control of CHVs were discussed.

Past and ongoing research on light passenger vehicles has shown that significant advances have been made in the technology leading to an automated highway system. However, the control issues specific to CHVs have been addressed and studied only recently. There are significant advantages associated with the inclusion of CHVs in AHS, and more work is expected and needed in the near future to complement past AVCS developments on passenger vehicles.

ACKNOWLEDGMENTS. The authors would like to thank their graduate students, Chieh Chen (UCB), Pushkar Hingwe (UCB), and Diana Yanakiev (UCLA), for their assistance with the preparation of this manuscript. The work presented here was performed as part of the California PATH Program of the University of California under MOUs 124 and 129, in cooperation with the State of California Business, Transportation, and Housing Agency, Department of Transportation; and the U. S. Department of Transportation, Federal Highway Administration.

The contents of this paper reflect the views of the authors who are responsible for the facts and the accuracy of the data presented herein. The contents do not necessarily reflect the official views or policies of the State of California. This paper does not constitute a standard, specification, or regulation.

REFERENCES

1. R. E. FENTON AND R. J. MAYHAN, Automated highway studies at The Ohio State University—An overview, *IEEE Trans. Veh. Technol.* **40**, 100–113 (1991).
2. S. SHLADOVER, C. A. DESOER, J. K. HEDRICK, M. TOMIZUKA, J. WALRAND, W. ZHANG, D. MCMAHON, H. PENG, S. SHEIKHOLESLAM, AND N. MCKEOWN, Automatic vehicle control developments in the PATH program, *IEEE Trans. Veh. Technol.* **40**, 114–130 (1991).
3. H. PENG AND M. TOMIZUKA, Preview control for vehicle lateral guidance in highway automation, *ASME J. Dyn. Syst. Meas. Control* **115**, 678–686 (1993).

4. American Automobile Manufacturers Association, *Motor Vehicle Facts & Figures '94* (Detroit, 1994).
5. *Tomorrow's Truck Symposium Proceedings,* SAE Proceedings 89-225, Detroit, 1989.
6. *Highway statistics, 1993,* 49th ed. (Federal Highway Administration, Washington, DC, 1994).
7. E. C. MIKULCIK, The dynamics of tractor–semitrailer vehicles: The jackknifing problem, *SAE Trans.* **80**, 710045 (1971).
8. R. T. BUNDORF, Directional control dynamics of automobile–travel trailer combinations, *SAE Trans.* **76**, 670099, 667–680 (1967).
9. H. DUGOFF AND R. W. MURPHY, The dynamic performance of articulated highway vehicles—A review of the state-of-the-art, *SAE Trans.* **80**, 710223, 897–906 (1971).
10. F. VLK, Lateral dynamics of commercial vehicle combinations: A literature survey, *Veh. Syst. Dyn.* **11**, 305–324 (1982).
11. F. VLK, Handling performance of truck–trailer vehicles: A state-of-the-art survey, *Int. J. Veh. Des.* **6**, 323–361 (1985).
12. P. M. LEUCHT, The directional dynamics of the commercial tractor–semitrailer vehicle during braking, *SAE Trans.* **79**, 700371, 1146–1156 (1970).
13. T. ZIMMERMANN, A. FUCHS, U. FRANKE, AND B. KLINGENBERG, VECTOR—A vision enhanced/controlled truck for operational research, *SAE Tech. Pap. Ser.* No. 942284 (1994).
14. J. M. BLOSSEVILLE, F. BLONDEEL, AND M. GRATON, Towards an automated highway system applied to freight transport in France: First considerations, *Proc. Joint University of California/PATH–France Workshop,* Richmond, CA, Oct. 1995.
15. B. FAVRE, The position of a truck manufacturer regarding intelligent transportation systems advanced technologies, *Proc. Joint University of California/PATH—France Workshop,* Richmond, CA, Oct. 1995.
16. R. D. ERVIN, IVHS and the truckmaker: Identifying the need for research, UMTRI Final Report UMTRI-92-34 (Sept. 1992).
17. R. A. BISHEL, Dual-mode truck: Automated and manual operation, *SAE Tech. Pap. Ser.* No. 931837 (Aug. 1993).
18. D. YANAKIEV AND I. KANELLAKOPOULOS, Analysis, design and evaluation of AVCS for heavy duty vehicles: Phase 1 report, California PATH Working Paper UCB-ITS-PWP-95-12 (1995).
19. D. YANAKIEV AND I. KANELLAKOPOULOS, Variable time headway for string stability of automated heavy-duty vehicles, *Proc. 34th IEEE Conference on Decision and Control,* New Orleans, LA, 1995, pp. 4077–4081.
20. D. YANAKIEV AND I. KANELLAKOPOULOS, Speed tracking and vehicle follower control design for heavy-duty vehicles, *Veh. Sys. Dyn.* **25** (1996).
21. C. CHEN AND M. TOMIZUKA, Dynamic modeling of articulated vehicles for automated highway systems, *Proc. 1995 American Control Conference,* Seattle, WA, pp. 653–657.
22. C. CHEN AND M. TOMIZUKA, Passivity-based nonlinear observer for lateral control of tractor–trailer vehicles in automated highway systems, to appear in the *Proc. 13th IFAC World Congress,* San Francisco, 1996.
23. E. HENDRICKS, Mean value modeling of large turbocharged two-stroke diesel engine, *SAE Trans.* No. 890564, 1–10 (1989).
24. J. H. HORLOCK AND D. E. WINTERBONE, *The Thermodynamics and Gas Dynamics of Internal Combustion Engines* (Clarendon Press, Oxford, 1986).
25. M. J. JENNINGS, P. N. BLUMBERG, AND R. W. AMANN, A dynamic simulation of Detroit Diesel electronic control system in heavy duty truck powertrains, *SAE Trans.* No. 891959, 5.943–5.966 (1986).
26. J. P. JENSEN, A. F. KRISTENSEN, S. C. SORENSON, AND E. HENDRICKS, Mean value modeling of a small turbocharged diesel engine, *SAE Trans.* No. 910070, 1–13 (1991).
27. J. D. LEDGER, R. S. BENSON, AND N. D. WHITEHOUSE, Dynamic modeling of a turbocharged diesel engine, *SAE Trans.* No. 710177, 1–12 (1971).
28. D. E. WINTERBONE, C. THIRUAROORAN, AND P. E. WELLSTEAD, A wholly dynamic model of turbocharged diesel engine for transfer function evaluation, *SAE Trans.* No. 770124, 1–11 (1977).
29. M. KAO AND J. J. MOSKWA, Turbocharged diesel engine modeling for nonlinear engine control and state estimation, *Proc. 1993 ASME Winter Annual Meeting,* DSC-Vol. 52, Symposium on Advanced Automotive Technologies: Advanced Engine Control Systems, New Orleans, LA.

30. D. CHO AND J. K. HEDRICK, Automotive power train modeling for control, *ASME J. Dyn. Syst. Meas. Control* **111**, 568–576 (1989).
31. D. H. MCMAHON, J. K. HEDRICK, AND S. E. SHLADOVER, Vehicle modeling and control for automated highway system, *Proc. 1990 American Control Conference,* San Diego, CA, pp. 297–303.
32. D. YANAKIEV AND I. KANELLAKOPOULOS, Engine and transmission modeling for heavy-duty vehicles, California PATH Technical Note 95-6 (1995).
33. A. J. KOTWICKI, Dynamic models for torque converter equipped vehicles, *SAE Trans.* No. 820393, 1595–1609 (1982).
34. M. W. SAYERS, Symbolic computer language for multi-body systems, *J. Guidance, Control Dyn.* **14**(6) (1991).
35. C. C. MACADAM, P. S. FANCHER, G. T. HU, AND T. D. GILLESPIE, A computerized model for simulating the braking and steering dynamics of trucks, tractor–trailers, doubles, and triples combinations—Users' manual, phase 4, Highway Safety Research Institute, Ann Arbor, MI, Report No. UM-HSRI-80-58 (1980).
36. M. SPONG AND M. VIDYASAGAR, *Robot Dynamics and Control* (Wiley, New York, 1989).
37. R. MURRAY, Z. LI, and S. SASTRY, *A Mathematical Introduction to Robotic Manipulation* (CRC Press, Boca Raton, FL, 1994).
38. D. T. GREENWOOD, *Principles of Dynamics,* 2nd ed. (Prentice Hall, Englewood Cliffs, NJ, 1988).
39. H. B. PACEJKA AND E. BAKKER, The magic formula tire model, *Proc. 1st International Colloquium on Tire Models for Vehicle Dynamics Analysis,* Deft, The Netherlands, 1991.
40. P. S. FANCHER AND Z. BAREKET, Including roadway and tread factors in a semi-empirical model of truck tires, in *Tire Models for Vehicle Dynamics Analysis* (Oct. 1991).
41. C. CHEN AND M. TOMIZUKA, Dynamic modeling of tractor–semitrailer vehicles for Automated Highway Systems, California PATH Working Paper UCB-ITS-PWP-95-8 (1995).
42. S. E. SHLADOVER, Longitudinal control of automated guideway transit vehicles within platoons, *ASME J. Dyn. Syst. Meas. Control* **100**, 302–310 (1978).
43. J. K. HEDRICK, D. H. MCMAHON, V. K. NARENDRAN, AND D. SWAROOP, Longitudinal vehicle controller design for IVHS systems, *Proc. 1991 American Control Conference,* Boston, pp. 3107–3112.
44. S. SHEIKHOLESLAM AND C. A. DESOER, Longitudinal control of a platoon of vehicles, *Proc. 1990 American Control Conference,* San Diego, CA, pp. 291–297.
45. P. VARAIYA, Smart cars on smart roads: Problems of control, *IEEE Trans. Autom. Control* **38**, 195–207 (1993).
46. W. L. GARRARD, R. J. CAUDILL, A. L. KORNHAUSER, D. MACKINNON, AND S. J. BROWN, State-of-the-art of longitudinal control of automated guideway transit vehicles, *High Speed Ground Transp. J.* **12**, 35–68 (1978).
47. C. CHIEN AND P. IOANNOU, Automatic vehicle following, *Proc. 1992 American Control Conference,* Chicago, pp. 1748–1752.
48. P. FANCHER, Z. BAREKET, AND G. JOHNSON, Predictive analyses of the performance of a highway control system for heavy commercial vehicles, *Proc. 13th IAVSD Symposium,* supplement to *Veh. Syst. Dyn.* **23**, 128–141 (1993).
49. F. BOTTIGER, H. D. CHEMNITZ, J. DOORMAN, U. FRANKE, T. ZIMMERMANN, AND Z. ZOMOTOR, Commercial vehicle and transit AHS analysis, Precursor Systems Analyses of Automated Highway Systems, Final Report, Vol. 6, Federal Highway Administration, Report FHWA-RD-95-XXX (1995).
50. K. GARDELS, Automatic car controls for electronic highways, General Motors Research Laboratory, General Motors Corp., Warren, MI, Report GMR-276 (June 1960).
51. E. DICKMANNS AND A. ZAPP, Autonomous high speed road vehicle guidance by computer vision, *Proc. 10th IFAC World Congress,* Munich, 1987.
52. J. MALIK, D. KOLLER, AND T. LUONG, A machine vision based system for guiding lane-change maneuvers, California PATH Research Report UCB-ITS-PRR-95-34 (1995).
53. W. ZHANG *et al.*, An intelligent roadway reference system for vehicle lateral guidance/control, *Proc. 1990 American Control Conference,* San Diego, CA.
54. V. UTKIN, *Sliding Modes and Their Applications in Variable Structure Systems* (MIR Publishers, Moscow, 1978).

55. H. PHAM, J. K. HEDRICK, AND M. TOMIZUKA, Combined lateral and longitudinal control of vehicles, *Proc. 1994 American Control Conference,* Baltimore, pp. 1205–1206.
56. J. ACKERMANN, J. GULDNER, W. SIENEL, AND R. STEINHAUSER, Linear and nonlinear controller design for robust automatic steering, *IEEE Trans. Control Syst. Technol.* **3**, 132–143.
57. J. ACKERMANN, Robust car steering by yaw rate control, *Proc. 30th IEEE Conference on Decision and Control,* Honolulu, HI, 1990.
58. P. HINGWE AND M. TOMIZUKA, Two alternative approaches to the design of lateral controllers for commuter buses based on sliding mode control, *Advanced Automotive Technologies,* ASME International Mechanical Engineering Congress and Exposition, 1995.
59. B. D. O. ANDERSON AND J. B. MOORE, *Optimal Control—Linear Quadratic Methods* (Prentice-Hall, Englewood Cliffs, NJ, 1990).
60. M. KRSTIC, I. KANELLAKOPOULOS, AND P. KOKOTOVIC, *Nonlinear and Adaptive Control Design* (Wiley, New York, 1995).
61. C. CHEN AND M. TOMIZUKA, Steering and independent braking control of tractor–semitrailer vehicles in Automated Highway Systems, *Proc. 34th IEEE Conference on Decision and Control,* New Orleans, LA, 1995.

CHAPTER 12

AERODYNAMIC BENEFITS FROM CLOSE-FOLLOWING

F. BROWAND, M. ZABAT, AND P. TOKUMARU

1. CLOSE-FOLLOWING AS A PART OF THE AUTOMATED HIGHWAY CONCEPT

1.1. Congestion on Urban Highways

Many believe that the highway congestion experienced in virtually all major urban areas in the United States is the most serious impediment to continued economic growth. Space limitations and cost will probably preclude the construction of significant quantities of new highway in the immediate future. As estimated from several recent freeway projects in Los Angeles, highway construction cost in Los Angeles County is approaching $12 million/lane per mile. To make a modest increase of 25% in the number of highway lane-miles in the Los Angeles County urban area would represent an investment of the order of $12 billion. (The costs are presumed to be similar for other urban areas in the United States.) It is not realistic to suppose that funding support of this magnitude would be available, or

F. BROWAND, M. ZABAT, AND P. TOKUMARU • Department of Aerospace Engineering, University of Southern California, Los Angeles, California 90089-1191. *Present address of M.Z.:* Aerodynamics and Fluid Dynamics Department, Chrysler Tech Center, Auburn Hills, Michigan 48326. *Present address of P.T.:* Aerovironment, Inc., 222 East Huntington Dr., Suite 200, Monrovia, California 91016.

Automated Highway Systems, edited by Ioannou. Plenum Press, New York, 1997.

necessarily advisable. Rather, it is the more efficient use of the presently available highway system that is the focus of the automated highway (AHS) concept. To improve the operation of the highway system, two solutions might be considered: either decrease the average vehicle width to produce more lanes of traffic, or increase the traffic flow of existing lanes. Automobiles and trucks can be made narrower but not significantly so, and only at great cost. The second choice is more compelling because large improvements in the flow (or throughput) along a highway are possible by incorporating the orderly movement of vehicles, together with a reduction in the longitudinal spacing between vehicles. Both items are contained in the close-following concept.

The mathematical argument for close-following is based on the characteristics of highway traffic flow, as illustrated in Figure 1. The data originated from the California Department of Transportation (Caltrans), and were provided and processed by Karl Petty.[1] The data, recorded in 1993, represent scatter-plotted 30-s averages for vehicle density, speed, and vehicle flux at one station on Interstate 880 in the Oakland–Berkeley area. The plots show a peak vehicle flux of approximately 2000 vehicles/lane per hour occurring at a speed of about 45 mph, followed by a decline in flux as vehicle density increases. One interesting feature is that the peak flux number, ≈2000 vehicles/lane per hour, seems to be relatively inelastic. To our knowledge, this value is not substantially different from peak fluxes measured almost 20 years earlier, in spite of significant increases in vehicle performance in the intervening years.

The vehicle density at peak flux is about 40 vehicles/mile, which translates to an average spacing between vehicles of approximately 35 m. It is easy to see that vehicle flux could be greatly increased were an artificial means employed to maintain high speed at the higher traffic densities. Such artificial control might be achieved by a coordinated motion of platoons of vehicles traveling at close spacings, or by the autonomous interaction of many single vehicles. The net effect will be the same: higher speed and shorter spacings between vehicles translates to increased traffic flow, or vehicle flux. However attractive from the operations point of view, close-following will only be possible when appropriate control systems—incorporating on-board sensors for safe high-speed travel—become available and are accepted by the public.

1.2. Reduced Fuel Expenditures and Reduced Pollution

Substantial reductions in fuel expenditure—and therefore reductions in pollutants emitted per mile traveled—arise from orderly, close-following traffic flow. These benefits accrue from two distinct sources. First, stop-and-go driving is eliminated in favor of a smoother flow at higher speed. The fuel expenditure for congested highway driving versus that for unobstructed highway travel has not been carefully studied. Roughly speaking, the difference could be approximated as the difference between a typical urban driving cycle and a freeway driving cycle. If the EPA standard driving cycles are used for this comparison, then the saving is simply the difference in fuel consumption given by the EPA estimates for a given vehicle.

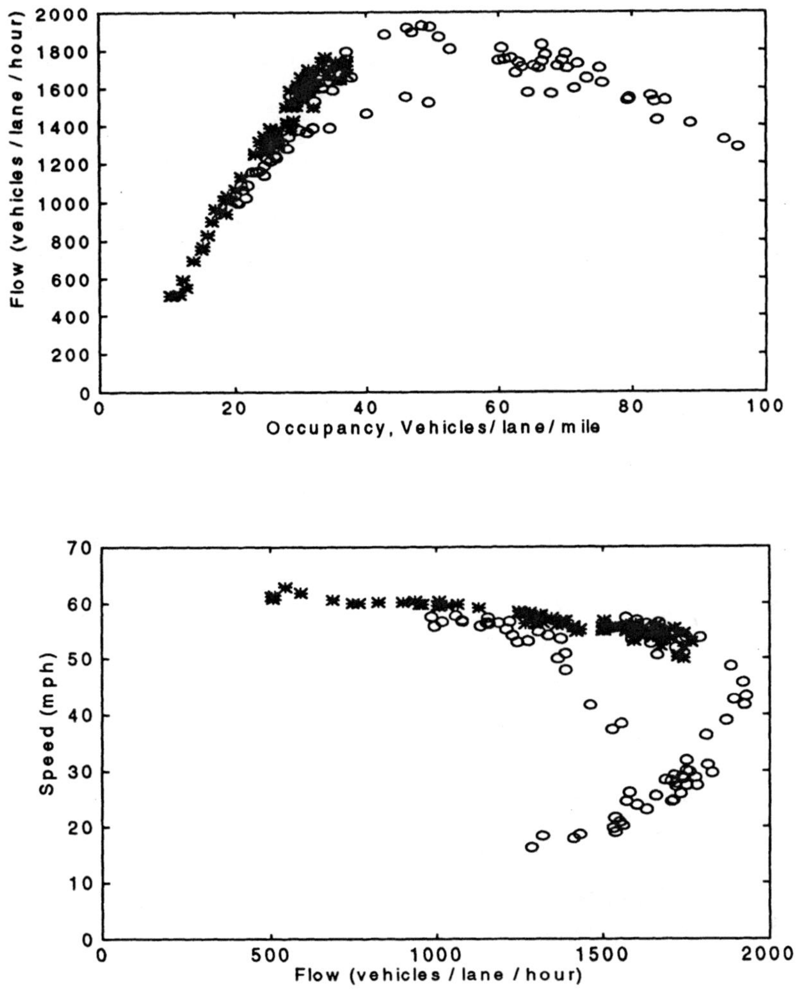

FIGURE 1. Characteristics of freeway traffic flow. Stars are morning flow, open symbols are afternoon flow.

For example, fuel consumption estimates for the 1991 Chevrolet Lumina APV are 1/17 gpm (gallons per mile) urban versus 1/24 gpm highway. Then to a first approximation, the difference between congested and uncongested highway travel becomes 0.0171 gpm, or about 30% savings.

A second and more subtle benefit arises because the average drag of N vehicles in a close-following formation is considerably smaller than the drag of the same N vehicles in isolation. Savings in drag translates directly to savings in fuel consumed and results in less pollutants emitted. An automobile traveling at constant speed consumes power—and therefore fuel—to overcome major contributions from the rolling resistance and the aerodynamic drag, as well as a residual drivetrain resistance. At 55 mph, the rolling resistance and aerodynamic drag contributions are roughly equal. Thus, at 55 mph, a drag reduction of 50% would result in a 25% reduction in fuel consumption. At higher speeds, proportionately more fuel is expended to overcome drag, since drag is quadratic in speed while rolling resistance is nearly constant.

Results presented in the following sections indicate that 50% drag reductions are indeed possible in close-following situations. These drag savings and concomitant fuel savings are discussed in detail. The present work summarizes and complements earlier results presented in Refs. 2–5.

2. AERODYNAMIC PREDICTION BASED ON NUMERICAL SOLUTIONS AND WIND TUNNEL MEASUREMENTS

The considerations of the preceding paragraphs suggest that the efficient use of the highways in the coming century will demand higher speeds. Future automobiles are likely to be smaller, relatively lighter, and as a result of close-following, vehicles will interact aerodynamically with surrounding vehicles on the roadway. All of these circumstances lead to a greater need for improved aerodynamic predictions, and for a greater understanding of aerodynamic performance.

2.1. *Defining Drag Coefficient and Reynolds Number*

When bodies under aerodynamic study are modeled numerically, or modeled in a wind tunnel test, all possible care must be taken to establish conditions consistent with full-scale operation. First, define a *drag coefficient*,

$$C_D = \frac{D}{\frac{1}{2}\rho U^2 A} \tag{1}$$

where D is the measured drag, ρ is the air density, U is the vehicle forward speed, or wind tunnel speed, and A is the maximum cross-sectional area of the body perpendicular to the flow. Couching discussion in terms of *drag coefficient* rather than in terms of *drag force* automatically accounts for differences in vehicle size and vehicle speed. It allows one to make predictions for the performance of the full-scale vehicle from wind tunnel model measurements or from numerical simulations.

In addition, the *Reynolds number* of the model flow should match the Reynolds number in full-scale operation.

$$\text{Re} = \frac{UL}{\nu} \tag{2}$$

where L is the characteristic length scale and ν is the kinematic viscosity of air. The magnitude of the Reynolds number expresses the order of magnitude of the ratio of fluid inertia force to fluid viscous force. When the Reynolds number is large, a fluid parcel moves primarily in response to differences in pressure within the flow field. Only in very thin *boundary layer* regions adjacent to surfaces is viscous shearing stress important. Because these thin boundary layer regions are retarded by viscous stress, the flow frequently leaves the surface (or separates) prematurely to form an unsteady, turbulent wake. Such separations (and possible reattachments to downstream vehicles) are difficult to predict accurately, and may in certain circumstances depend strongly on the value of the Reynolds number.

2.2. High-Reynolds-Number Flows Prove Difficult for Computer Simulation

The fundamental computational challenge at large Reynolds numbers is that of providing sufficient spatial resolution. Reynolds numbers are indicated in Figure 2 for several typical automotive applications. At these high Reynolds numbers, fluid motions in the boundary layers and within the separated flow regions are highly turbulent, and consist of unsteady motions on many scales. The ratios of largest to smallest flow scales are estimated in the last column. The body scale is denoted by L, and η signifies the smallest turbulent scale possible—some individuals would argue it is sufficient to resolve scales of the order of 10η rather than η. The cube of the scale ratio (for three space directions) is, in each case, an indication of the number of grid points required to fully describe the turbulent flow from first principles, as in a direct numerical simulation. Such numbers of grid points are well beyond present-day possibility, which is on the order of 10^7 grid points. These high-Reynolds-number flows must therefore be treated *approximately,* utilizing one of many forms of turbulence modeling. Flowfield predictions providing a basis for design are usually determined—as is done in the present chapter—by a combination of approximate numerical prediction, wind tunnel testing, and full-scale tests.

CONFIGURATION	VELOCITY	SCALE, L	Re_L	$L/10\eta$
	30 m/sec	5 m	1×10^7	1×10^4
	100 m/sec	5 m	3×10^7	2×10^4
	30 m/sec	30 m	6×10^7	4×10^4

FIGURE 2. Typical Reynolds numbers and ranges of scales for ground vehicle flows.

2.3. Description of Wind Tunnel Tests

The models are 1/8-scale replicas of the 1991 Chevrolet Lumina APV. Models and the full-scale vehicle are geometrically similar in the sense that all body dimensions have been scaled by the same ratio. The mold used to cast the models is fashioned by a stereolithographic process using computer-generated coordinates. For all practical purposes, the models are exact replicas. However, all minor external features such as window frames, side mirrors, door handles, and windshield wipers have been removed. Also, the details of the undercarriage have been left out, and the wheels are stationary.

To keep Re constant, the freestream velocity in the wind tunnel test case must be eight times higher than the full-scale vehicle velocity. This, of course, is not practical—model Reynolds numbers are lower than Reynolds numbers for full-scale operation by about a factor of eight. To account for these differences, and to judge the *sensitivity* of the drag force measurements to changes in Reynolds number, numerous tests were conducted at several Reynolds numbers. Another simulation technique is to utilize a series of small roughness elements to artificially "trip" the boundary layer on the model. Tripping ensures a turbulent boundary layer more nearly in character with a boundary layer at higher Reynolds number. Changing Reynolds number by a factor of about two, or the addition of tripping devices, typically produced drag changes of a few percent. Results of these sensitivity studies are not presented here, but can be found in Refs. 2–5.

3. TWO VEHICLES: THE CLOSE-FOLLOWING PLATOON OF MINIMUM LENGTH

3.1. Drag Coefficients from Wind Tunnel Tests

Figure 3 presents the drag of each vehicle as a function of the spacing between the two vehicles (a porous ground plane surface is employed to control unwanted boundary layer growth). The C_D of each vehicle is divided by the C_D for the same model in isolation at an identical position in the wind tunnel. This C_D *ratio* embodies the *change in drag* resulting from the unique aerodynamics of close-following. The ratio is less sensitive to possible variations in C_D resulting from small differences between models, location in the wind tunnel, and so forth. Presenting the drag coefficient ratio, rather than the drag coefficient itself, roughly accounts also for the lack of complete geometric similarity when surface detail has been ignored. Spacing is measured on the centerline from the rear bumper of the lead model to the front bumper of the trail model and is normalized with model length (0.619 m). The points shown in Figure 3 represent the average values of data taken over four separate but identical experiments. The reliability bar seen in the upper right corner is an estimate of the possible measurement error.

Several important qualitative features can be seen in Figure 3, and are contained in the three flow regions labeled as A, B, and C. In region C at spacings of

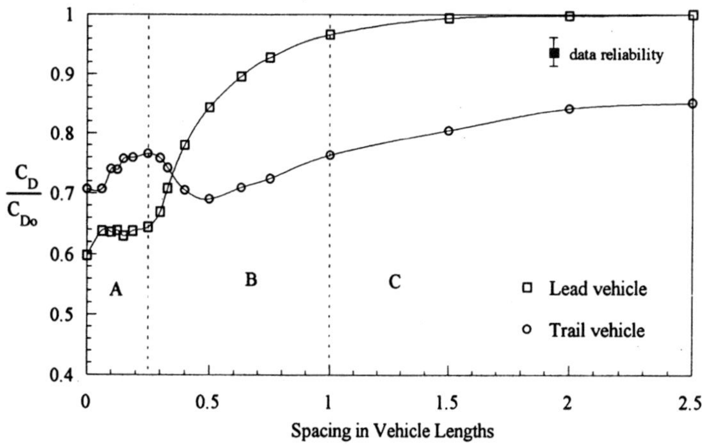

FIGURE 3. Drag coefficient ratios for two close-following vehicles. After reference 3.

the order of unity or greater, the lead vehicle is unaware of the presence of the trail vehicle (this is also true when more than two vehicles are present). The trail vehicle, which is contained in the wake of the lead vehicle, experiences a decrease in drag as expected. Extrapolation of these results to greater spacings suggests that a measurable decrease in drag will persist to a spacing of about ten. This circumstance may be termed *weak interaction* because it is one-sided—not mutual—and is entirely understandable. As spacing decreases below a value of unity in region B, the drag of the lead vehicle decreases substantially. The drag of the trail vehicle also decreases, but less rapidly. At a spacing of about one-half and below, the drag of the trail vehicle increases abruptly, and crosses the lead vehicle drag curve at about 0.35 spacing. For two vehicles of equal performance and throttle setting, the crossover position is a stable fixed point—meaning that the distance between the two vehicles will naturally equilibrate to this separation. The observation that separations of approximately this value—or slightly greater separations—are often observed in the drafting of stock cars on race tracks lends support to the wind tunnel result. Finally, at the smallest spacings of region A, the downward trend of the lead vehicle drag coefficient abruptly halts at a value of about 0.64. There is little further change in drag coefficient for the lead vehicle from spacing 0.25 to zero. The drag coefficient for the trail vehicle turns downward in region A, but remains above the lead vehicle all the way to zero spacing! Regions A and B taken together, might be termed the *strong interaction* regime.

3.2. The Intervehicle Gap Regarded as a Cavity

The reversal of the drag coefficients at close spacing is counterintuitive to the notion that shielding from the forward vehicle should always provide a lower drag coefficient for the rear vehicle. However, a counterargument can be given by regarding the flow into and out of the cavitylike region formed between the two vehicles.

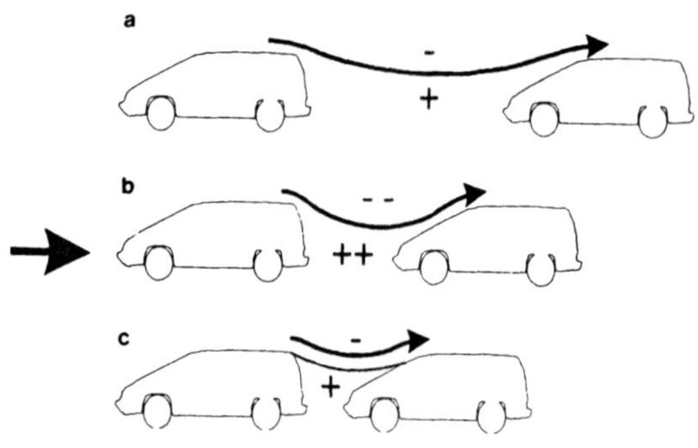

FIGURE 4. Sketch of streamlines into or out of the intervehicle gap (cavity).

Typical cavity flow streamlines are sketched for the three regions A, B, and C in Figure 4. As spacing decreases, that is, as the cavity length decreases, the inflow–outflow streamlines become more highly curved. The balance of forces requires a larger pressure difference *across* streamlines to keep fluid parcels moving *along* the more highly curved paths. This circumstance has the effect of increasing pressures along the boundaries of the cavity. Higher pressures on the boundaries of the cavity produce a *decreased* drag on the *forward* vehicle and an *increased* drag on the *trailing* vehicle. Thus, the two drag curves cross. At even shorter cavity lengths (closer spacings), flow separation from the forward vehicle and reattachment to the rear vehicle effectively cuts off most of the flow into (or out of) the cavity. Without significant cavity inflow (and outflow), the drag coefficients remain relatively insensitive to further change in cavity length.

3.3. Numerical Computations Utilizing a Simplified Model for Vehicle Shape

The previous flow model is qualitatively correct, as can be judged by computations that predict the *pressure field interaction* between two vehicles. For simplicity, the fore and aft portions of the vehicle are modeled as semiellipsoids having different fineness ratios. The upper half of the composite-ellipsoidal bodies is taken to be the vehicle, and the lower half its image below the groundplane. The geometry is shown in Figure 5. The effective body radius, b, normalized by the vehicle length, L, is given by,

$$\beta \equiv \frac{b}{L} = \sqrt{\frac{2wh}{\pi L^2}} \tag{3}$$

This choice of β ensures that the actual vehicle and ellipsoidal model have the same frontal area and effective fineness ratio. Another feature of this choice of geometry is that the volume of the ellipsoidal model, V, is independent of the choice of a, i.e.,

$$\frac{V}{L^3} = \tfrac{2}{3}\pi\beta^2 \tag{4}$$

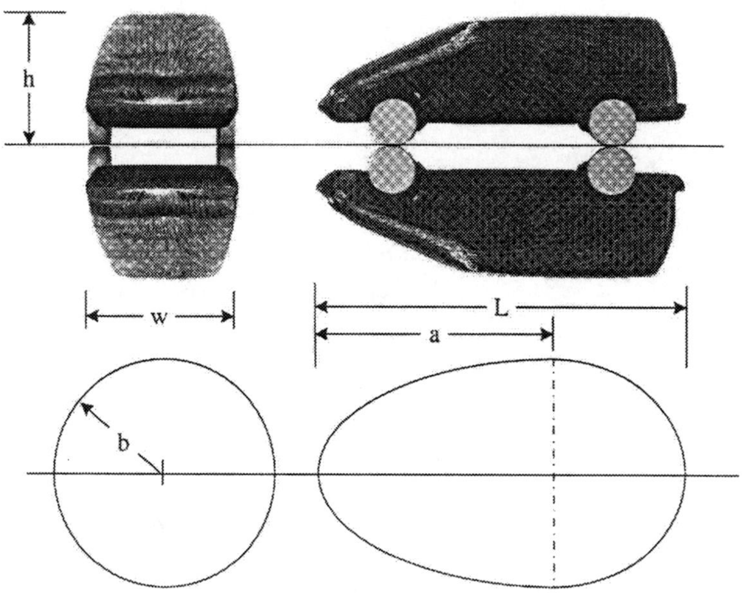

FIGURE 5. Comparison of minivan and ellipsoidal model geometries.

The effective fineness ratio for the minivans employed in the present chapter is $1/\beta = 3.5$. The parameter $\alpha \equiv a/L$ represents the fore–aft asymmetry of the vehicle shape. A value of $\alpha = 0.65$ yields a good agreement with the wind tunnel data, and is not unreasonable for the Lumina van models.

3.4. Numerical Solutions and Wind Tunnel Measurements

The force exerted by the pressure field interactions on the lead vehicle of a two vehicle platoon is estimated using an axisymmetric panel method.[(6)] (In this case the flowfield approximation ignores boundary layer regions altogether.) Pairs of vehicles are examined for a range of spacings and in four configurations corresponding to the wind tunnel experiments: Figure 6a, both vehicles face forward, and both vehicles face backward; Figure 6b, front-to-front arrangement; Figure 6c, back-to-back arrangement. In all of these panels, the drag coefficient ratio is plotted for the lead vehicle. The symbols are the experimental results, and there are at least two separate but identical experiments for each geometry. The solid and dashed curves are constructed by adding the pressure force coefficient computed using the panel method to the experimental value for the isolated vehicle, and then normalizing by the drag of the isolated vehicle. Thus, the calculation provides the predicted drag coefficient behavior—for the lead vehicle—with the influence of the *interaction pressure field* taken into account. No comparisons of trail vehicle drag have been included. According to the theory, the trail vehicle experiences a pressure field interaction of equal magnitude but opposite sign, so no additional information is generated.

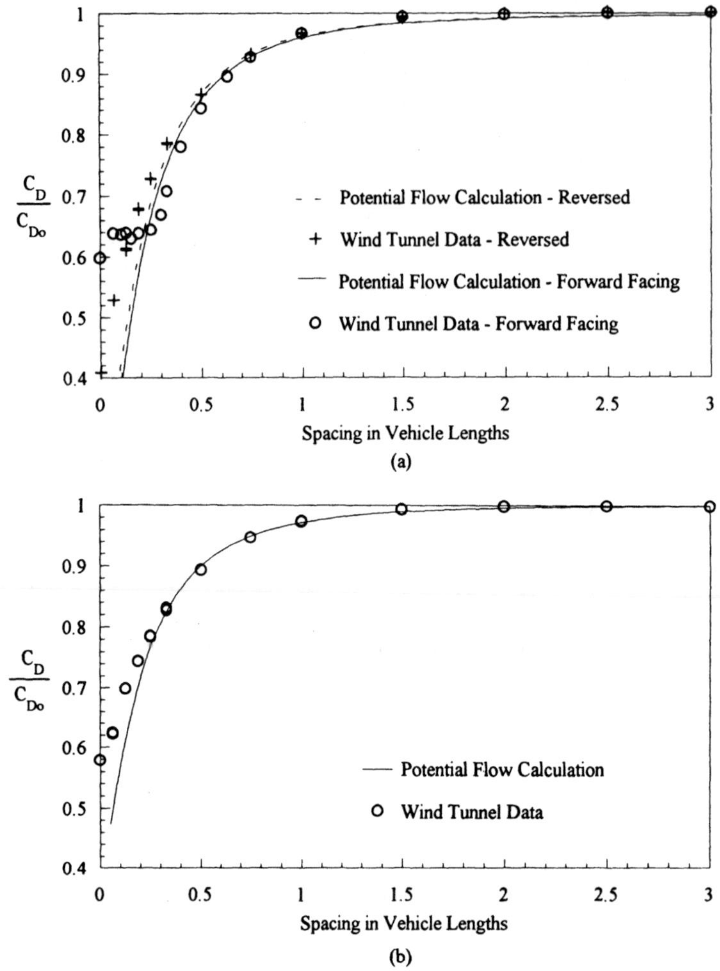

FIGURE 6. Comparisons between potential flow calculations and wind tunnel data. C_{Do} is taken from wind tunnel data for an isolated vehicle traveling in the appropriate direction. (a) Vehicles in forward facing and reversed orientation; (b) front-to-front orientation; (c) back-to-back orientation; (d) intercomparison of potential flow solutions for all four previous orientations.

The broad conclusion from the comparisons is that for vehicle spacings greater than about a quarter of a vehicle length, the drag reduction of the lead vehicle in the platoon—in any of the four configurations considered—is mostly attributable to the pressure field interaction between the lead and following vehicle. In fact, the drag decrease is surprisingly well predicted by the potential flow theory. The only adjustable factors in the comparison are the above-mentioned shape and fineness ratio values which are matched in a reasonable manner to the actual geometry, as discussed in Section 3.3. Figure 6d presents all four theoretical curves on the same plot. This intercomparison is useful as a measure of the

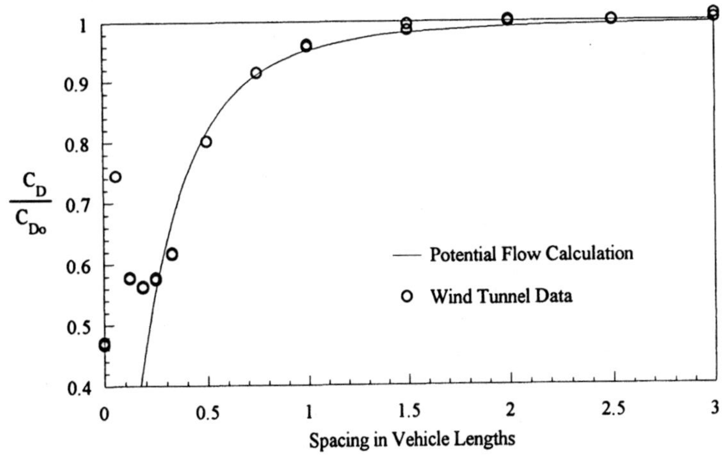

(c)

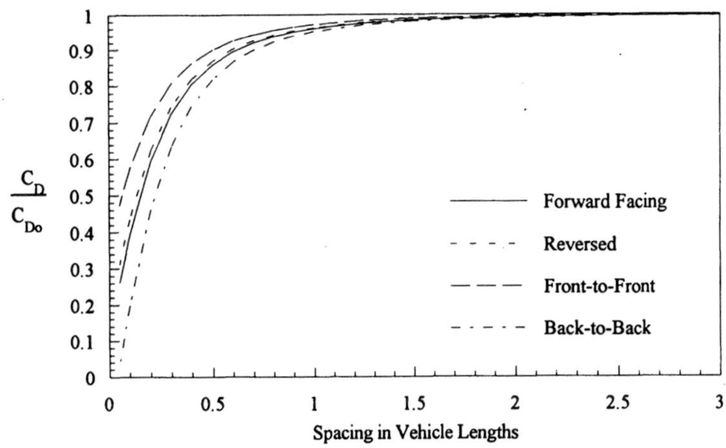

(d)

FIGURE 6. (*continued*).

possible range of behaviors for vehicles of various geometries—at least for spacings greater than one-quarter.

At spacings shorter than one-quarter, the drag measurements tend to diverge from the simple theory. This is not surprising, since flow separation for the forward vehicle must become important at short spacings, and flow separation—when it occurs—is not addressed at all by the theory. However, differences in drag behavior can still be related in a qualitative manner to the types of geometry provided by the two vehicles in tandem. When the forward vehicle faces forward as in Figure 6a (circles) and in Figure 6c, it presents a bluff afterbody shape (see Figure 5). It is surmised that a strong *separation* from the afterbody closes the intervehicle gap abruptly, and the drag coefficient levels off. When the forward vehicle faces backwards, as in Figure 6a (crosses) and in Figure 6b, it becomes a vehicle having a slanted rear deck with an angle of approximately 27 degrees. The flow now remains attached over the deck, and a further substantial decrease in the drag coefficient occurs with decreasing gap length because of the presence of the trail vehicle. Attached flow is probably maintained by the presence of strong longitudinal vortices forming along the sides of the deck—as described in Ref. 7.

Finally, Figure 6c illustrates an additional complicated behavior resulting from the particular geometry of the back-to-back arrangement. At very short spacings ≈0.1, a sharp peak in the drag of the lead vehicle occurs. The peak represents an increase in drag coefficient from 0.56 to 0.75, an increase of about 35%. As spacing is further decreased, the drag falls again. We believe this local peak represents a *resonance condition* for which the turbulent vortices in the separating shear flow have a spacing that is an exact multiple of the gap length. This circumstance (somehow) allows the flow to separate from the bluff base of the lead vehicle, and to attach smoothly near the forward edges of the trail vehicle. In this circumstance, a relatively low pressure is maintained within the gap. The relatively low pressure over the rear of the forward vehicle results in an increased drag coefficient. In close-following operations, such resonances are to be avoided. We believe similar flow field effects have been observed and discussed in simpler geometrical contexts in Refs. 8 and 9.

4. THREE AND FOUR VEHICLES IN TANDEM

The results of drag measurements as a function of spacing on three- and four-vehicle platoons using the porous surface ground plane are presented in Figures 7

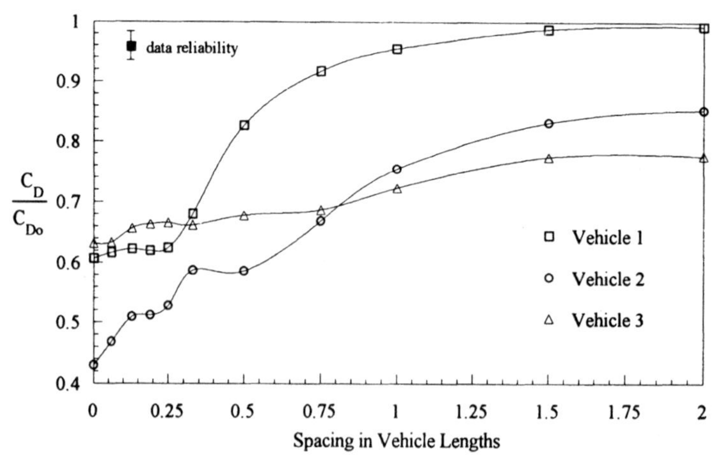

FIGURE 7. Drag coefficient ratios for a three-vehicle platoon. After reference 3.

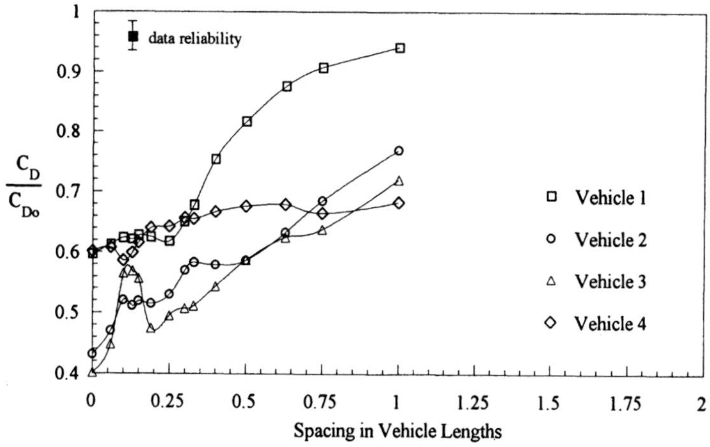

FIGURE 8. Drag coefficient ratios for a four-vehicle platoon. After reference 3.

and 8. The intervehicle spacing within the platoon is uniform and is measured in the same manner as previously. The points presented here are averaged over four separate data sets. Again note the data reliability bar, which is an estimate of the root mean square data scatter. At first glance the results may appear "noisy" because of the numerous small and unexpected variations—particularly at short spacings. Much of this variation is in fact greater than our estimated error and reflects the actual physical changes taking place in the flow field at short spacings. Many of the details show consistent trends as the number of vehicles within the platoon increases.

As the number of vehicles in the platoon increases, one observes more complex behaviors in the *strong interaction* regime. As expected, the interior vehicles generally have the lower drags, but there are also curious small plateau regions that form. For example, the drag of the second vehicle in a three-vehicle platoon contains two small plateau regions—one extending from 0.1 to 0.2 and one from 0.3 to 0.5. The second vehicle experiences similar behavior in the four-vehicle platoon, and the third vehicle exhibits even more dramatic drag variation—a plateau at 0.6–0.7, a decrease to a relative minimum (plateau) at 0.2–0.3, then a dramatic rise (to a plateau) at 0.1–0.2 before finally decreasing to a value of about 0.4 at zero spacing. The narrowness of the plateau peak in drag at 0.1–0.2 again suggests a flow-induced *resonance* that is quite sensitive to vehicle spacing and to vehicle position within the platoon.

5. THE MANY-VEHICLE LIMIT BY EXTRAPOLATION

5.1. Defining an Average Drag Coefficient for Close-Following

An overall measure of platoon drag performance may be obtained by defining an average drag coefficient ratio:

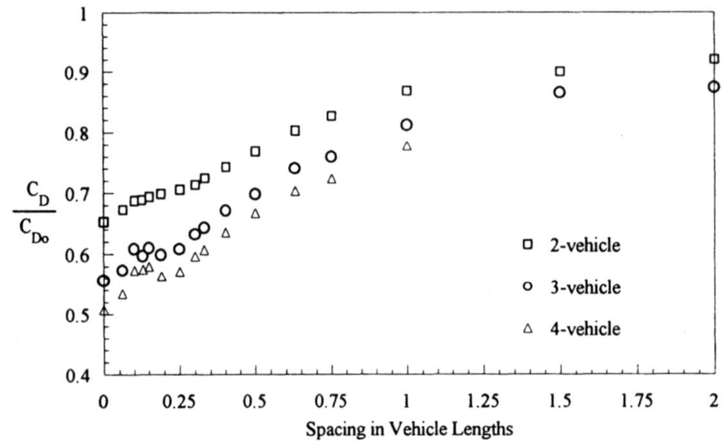

FIGURE 9. Average drag ratios for two-, three-, and four-vehicle platoons. After reference 3.

$$\frac{(C_D)_{av}}{C_{Do}} = \frac{1}{N}\sum_{i=1}^{N}\frac{C_{Di}}{C_{Do}} \tag{5}$$

This ratio is shown in Figure 9. Platoon-averaged drag coefficient ratios possess much smoother behavior than the drags of individual vehicles. In the platoon containing four vehicles, the average drag has been reduced by a factor of 1/2 at zero spacing. More importantly perhaps, the average drag coefficients are dramatically lowered, and vary only moderately, in the range of spacings 0.1–0.3 where close-following could be implemented with the help of control systems. Note that the addition of each succeeding vehicle decreases the average platoon drag, but as the number of vehicles increases the increments become smaller. One would expect the average drag ratio to approach a limit as the number of platoon members increases indefinitely.

An extrapolation to larger platoon size may be obtained by the following procedure. The platoon-averaged drag coefficient ratios are plotted for a series of separations as a function of the inverse of the number of vehicles in the platoon. Continuing the drag coefficient ratio into the domain between one-quarter ($1/N = 1/4$) and zero gives the desired behavior. This is accomplished using a linear fit and a quadratic fit to the four data points at $1/N = 1/1, 1/2, 1/3, 1/4$. The resulting two-dimensional combined extrapolation/interpolation is shown in Figure 10, expressing platoon-averaged drag coefficient ratio as a function of separation distance and number of vehicles. The difference in the results between the linear fit and the quadratic fit can be used to estimate the range of uncertainty in the predictions. Of course the extrapolation procedure assumes a well-behaved drag dependence as platoon size increases, but this is not an unwarranted assumption—at least for the *platoon-averaged* behavior. Contours of constant drag savings show a diminished

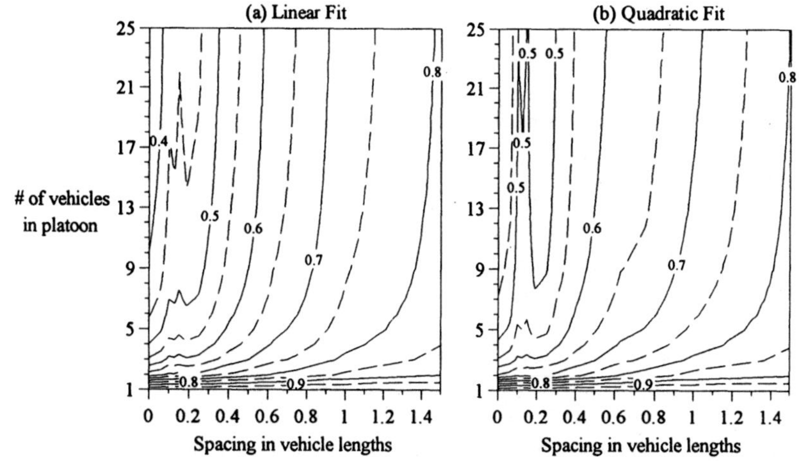

FIGURE 10. Estimated reduction in drag coefficient for large platoons. Contours are constant values of drag coefficient ratio. After reference 3.

dependence on platoon size beyond six or seven vehicles, and suggest that the four-vehicle platoon results are probably sufficient to make a reasonable extrapolation.

5.2. The Fuel Consumption Estimates of Sovran

Sovran[10] suggested a general method for determining the influence of aerodynamic drag on fuel consumption. The method accounts for engine performance during breaking and idling, and utilizes the standard Environmental Protection Agency (EPA) driving cycle model for 1983. A determination of the actual rate of fuel consumption requires detailed information about the engine and drivetrain for a particular vehicle. Sovran, in his paper, limits his analysis to changes in the fuel consumed per unit distance, Δ(liters/km) or Δgpm, as a result of specific changes in the vehicle drag.

The result of Sovran's analysis applied to platoon operation is the linear relationship

$$\left[\frac{(\text{liters}/\text{km})_\infty - (\text{liters}/\text{km})_{\text{platoon}}}{(\text{liters}/\text{km})_\infty}\right] = \left[\frac{\text{gpm}_\infty - \text{gpm}_{\text{platoon}}}{\text{gpm}_\infty}\right] = \xi_A \left[\frac{C_{D\infty} - (C_{D\text{platoon}})_{\text{av}}}{C_{D\infty}}\right] \tag{6}$$

where, for the EPA highway driving schedule,

$$(\xi_A^*)_H = \frac{0.89}{1 + \{0.0310 r_0 + 0.000126\} \Big/ \dfrac{C_{D\infty} A}{M}} \tag{7}$$

TABLE 1
Vehicle Specifications for 1991 GM Lumina APV

Nominal Coefficient of Drag, C_D	0.32[a]
Vehicle mass, M (based on manufacturer's curb weight)	1736 kg
Vehicle length, L	4.93 m
Cross-sectional area, A	2.816 m^2
Fuel consumption per unit distance, highway, $(\text{liters/km})_H$	0.098 (1/24 gpm)
Tire rolling resistance coefficient, r_0	0.012

[a]Obtained from General Motors Corporation.

The rolling resistance coefficient, r_o, and vehicle mass, M, are given in Table 1. ξ_A is the influence coefficient relating percentage changes in C_D to changes in fuel expenditure. It depends upon the particular vehicle chassis, the engine, the fuel, the drivetrain, and the accessories. These items are reflected in the various coefficient values in the above expression. ξ_A^* represents the highest attainable value of ξ_A. When a vehicle traveling at its cruising speed is subjected to a smaller load, say because of a change in drag, the driver can use the extra power available to speed up or to maintain the original speed by reducing the throttle setting. In opting for the latter, the engine is subsequently operated at a slightly less than optimum condition (lower efficiency) where its specific fuel consumption is slightly higher. This will be reflected in the above equations by a value of ξ_A which will be slightly lower than ξ_A^*. To maintain ξ_A^*, the optimum operating condition should be recovered.

5.3. *Fuel Consumption Calculations*

Table 1 lists information on the 1991 GM Lumina APV that is pertinent to this discussion. The tire rolling resistance is nominal for 205/70SR15 tires. Based on values listed in Table 1, the influence coefficient for highway driving conditions is found to be

$$(\xi_A^*)_H = 0.454 \tag{8}$$

This value can now be used to determine the percentage change in fuel consumption using equation (6).

The result of using the experimental data for two- and four-vehicle platoons in this relationship is shown in Figure 11. Also shown is the linearly extrapolated result for $N = 8$ and $N = 20$. As anticipated, there is a considerable savings in fuel as a result of the reduction in drag related to close-following—as much as a 26% reduction from the isolated vehicle consumption is possible at spacings of 0.1–0.25 for large platoons on the highway.

6. CONCLUSIONS

From the wind tunnel measurements and calculations presented here, and from reasonable extrapolations of these results, the following conclusions can be drawn regarding drag behavior at short spacings, and on anticipated fuel savings.

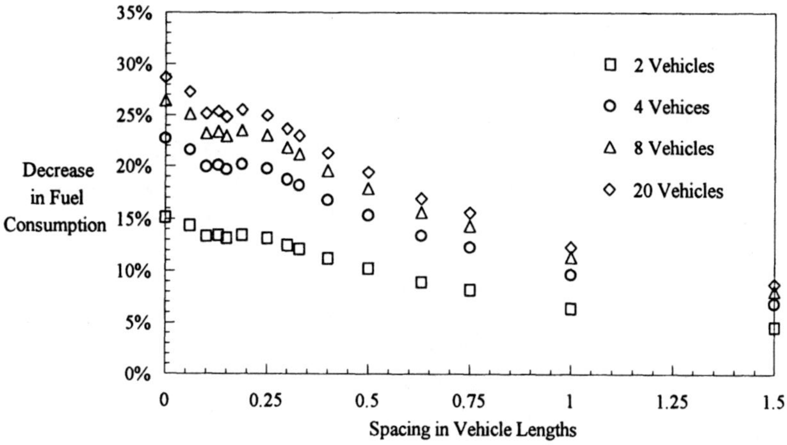

FIGURE 11. Decrease in fuel consumption for close-following vehicles in highway operation. After reference 4.

1. All vehicles in a close-following platoon share the drag savings benefits. It is important that all members share this benefit—otherwise drivers would be too concerned about relative position within the platoon. Vehicles following the lead vehicle(s) are shielded by the forward vehicle(s) and experience lower drag. But, quite surprisingly, a vehicle in a more forward position benefits from the presence of a following vehicle as a result of an induced pressure field. The pressure field *decreases* the drag of the forward vehicle and contributes an incremental *increase* in drag on the trailing vehicle.
2. Interior vehicles have less drag than vehicles at either end of the platoon. Based on the wind tunnel results, it is likely that the most downstream interior vehicle will have the least drag.
3. The benefit to the lead vehicle can be predicted by considering the potential flow interaction—provided the spacing is not less than about one-quarter vehicle length.
4. Vehicle interactions at spacing of less than one-quarter vehicle length are complicated by the details of separation and reattachment of the turbulent shear flow. These processes are, in turn, strongly dependent on vehicle shape. At certain short spacings, the drag may increase sharply, indicating a *resonance* phenomenon which, we believe, occurs when the local turbulence scale matches an integral multiple of the separation distance.
5. The average drag coefficient for the platoon can be as low as one-half the value for all members in isolation, leading to a decrease in fuel consumption of about 25%. Since all members of the platoon participate individually in drag savings, all members experience fuel savings. Extrapolation of these drag results indicates that little additional benefit will be gained by having platoons larger than six to eight vehicles.

These conclusions are based primarily on model tests in a wind tunnel. The conclusions are further supported by potential flow calculations, which attest to the correctness and generality of the results. Unfortunately, we know of no measurements at full scale which could be used as documentation. We are presently designing a field test in which the forward vehicle will tow trailing vehicles. Drag values will be determined utilizing a tow bar of special design featuring a low-pass dynamic response to limit unwanted road noise, and containing load cells for force measurement.

ACKNOWLEDGMENTS. We gratefully acknowledge the financial support of the California Department of Transportation through the California PATH Program of the University of California. Special mention also goes to S. Frascaroli and N. Stabile—former students who contributed greatly to the project.

REFERENCES

1. K. PETTY, Freeway service patrol (FSP) 1. 1: The analysis software for the FSP project, PATH Research Paper No. UCB-ITS-PRR-95-20 (1995).
2. M. ZABAT, S. FRASCAROLI, AND F. BROWAND, Drag measurements on 2, 3 & 4-car platoons, *SAE Paper* No. 940421 (1994).
3. M. ZABAT, N. STABILE, S. FRASCAROLI, AND F. BROWAND, Drag forces experienced by 2, 3 & 4-vehicle platoons at close spacings, *SAE Paper* No. 950632 (1995).
4. M. ZABAT, N. STABILE, AND F. BROWAND, Estimates of fuel savings from platooning, *Proc. 5th Annual ITS America Meeting,* Washington, DC, 1995.
5. M. ZABAT, N. STABILE, S. FRASCAROLI, AND F. BROWAND, The aerodynamic performance of platoons, PATH Research Paper No. UCB-ITS-PRR-95-35 (1995).
6. J. L. HESS AND A. M. O. SMITH, Calculation of potential flow about arbitrary bodies, *Prog. Aeronaut. Sci.* **8**, 1–138 (1966).
7. W.-H. HUCHO AND G. SOVRAN, Aerodynamics of road vehicles, *Annu. Rev. Fluid Mech.* **25**, 485–537 (1993).
8. A. ROSHKO AND K. KOENIG, Interaction effects on the drag of bluff bodies in tandem, in *Aerodynamic Drag Mechanisms of Bluff Bodies and Road Vehicles,* edited by G. Sovran, T. Morel, and W. T. Mason (Plenum Press, New York, 1978), pp. 253–286.
9. M. GHARIB AND A. ROSHKO, The effect of flow oscillations on cavity drag, *J. Fluid Mech.* **177**, 501–530 (1987).
10. G. SOVRAN, Tractive-energy-based formulae for the impact of aerodynamics on fuel economy over the EPA driving schedules, *SAE Paper* No. 830304 (1983).

CHAPTER 13

THE EFFECTS OF AHS ON THE ENVIRONMENT

MATTHEW J. BARTH

1. INTRODUCTION

As a key culmination of ITS technology, an Automated Highway System (AHS) will substantially improve the safety and efficiency of highway travel. The general AHS concept consists of instrumented vehicles operating on instrumented roadways without operator intervention.[(1)] Various sensors and communication devices will link the vehicles and roadway, virtually eliminating driver error and maximizing traffic performance. Drivers equipped with a vehicle instrumented for AHS operation will be able to enter an AHS through a check-in area and proceed onto a transition lane, where control of the vehicle will be assumed by the AHS. The system will then move the vehicle onto one of the automated lanes where it then merges with other traffic. When the driver's destination is reached, the system will move the vehicle back to the transition lane where the driver will be able to resume control of the vehicle.[(1)]

Some of the key benefits of an AHS are[(2–4)]:

- Improved safety, since the human error factor will be eliminated

---

MATTHEW J. BARTH • College of Engineering, Center for Environmental Research and Technology, University of California, Riverside, California 92521.

Automated Highway Systems, edited by Ioannou. Plenum Press, New York, 1997.

- Increased roadway flow rates, since vehicle spacings can be safely reduced, resulting in higher traffic densities traveling at higher speeds
- Reduced congestion, since traffic flow variances related to slow human reactions will be eliminated, and the number of accidents will be reduced
- Increased user comfort, since there will be less strain on the drivers

In addition, there are several potential benefits for the environment. Individual vehicle fuel consumption and emissions will likely be reduced, primarily related to smoother traffic flow (i.e., fewer accelerations/decelerations) and less congestion, resulting in shorter trip times. Further, vehicles in an AHS will likely operate at very close spacings, which will significantly reduce the aerodynamic drag on the vehicles. As a result, fuel consumption and emissions will further be reduced.

There is concern, however, that a reduction in congestion and the enhanced mobility and access associated with an AHS will lead to an increase in travel demand, resulting in an increase of total vehicle kilometers traveled (VKT) and a concomitant increase of vehicle emissions. If an AHS allows smoother flow and higher speeds on the roadways, people may choose to live farther away from work while still commuting in the same amount of time—thereby increasing the VKT. Farther, attractive trip destinations will become reachable, again increasing VKT. Although an indirect effect, this induced traffic demand issue is important when evaluating the environmental impact of an AHS.

An AHS will also have an impact in other environmental areas, such as acoustic noise pollution, land use, aesthetics, and electromagnetic radiation.[4] Acoustic noise levels will vary as a function of vehicle type, traffic density, traffic speed, acceleration profiles, and tire designs. AHS roadway infrastructure such as soundwalls and elevated structures will also have an influence on noise. Further, if an AHS is implemented using existing highways and rights of way, the demand for building new highways will be reduced and any detrimental impact on the land will be small. It will also be important to minimize the visual impacts of AHS-related roadway infrastructure. Lastly, an AHS should be designed so that electromagnetic radiation associated with any power transfer or communications should not increase significantly relative to ambient fields.

In this chapter, the emphasis of discussion is on the direct effects an AHS may have on vehicle emissions. Although individual vehicle emissions have decreased significantly over the years, the vehicle–highway system is still one of the largest contributors to air pollution in urban areas as a result of increases in VKT and associated congestion. In order to estimate the direct effects an AHS may have on vehicle emissions, preliminary research has been carried out using unique simulation models and appropriate emission modeling tools (see, e.g., Refs. 5, 6). A short description of this research is given first, with an emphasis on the modeling methodology and corresponding results. A brief review is then given of recent work addressing the induced demand issue.

2. THE DIRECT EFFECTS OF AHS ON VEHICLE EMISSIONS

Research has shown conceptually that an AHS can provide safe, efficient movement of vehicles on the highway (e.g., Refs. 7–10). However, an AHS is a complex system and must be capable of performing a wide range of operations, such as network traffic management, route planning and guidance, coordination of vehicle movements, and automated vehicle maneuver control. Each of these operations will have an effect on vehicle emissions. Other factors that will influence vehicle emissions on an AHS are weather conditions (e.g., inclement weather will likely reduce AHS speeds) and the types of vehicles that use the AHS.

In order to determine the impact of AHS on vehicle emissions, traffic models capable of simulating various AHS concepts must be closely integrated with vehicle emission models. Unfortunately, existing traffic, emissions, and planning models have been developed independently of each other and are difficult to integrate together when determining accurate air quality impacts. Current emission models (i.e., MOBILE, EMFAC[11–13]) simply relate vehicle emissions to average traffic densities and speeds on a specific network, and are not adequate for analyzing traffic at the microscale level required for AHS evaluation. What is required is a *modal* emissions model, i.e., a model that directly relates emissions at high time resolutions (e.g., second by second) to vehicle operating modes such as idle, steady-state cruise, various levels of acceleration/deceleration, and so forth. In a related research project, the author and colleagues are developing a comprehensive modal emissions model that can easily be combined with ITS simulations.[14]

Other modeling efforts are under way to predict emissions for various ITS scenarios, such as a Volpe National Transportation System Center project to evaluate the environmental impact of ITS by combining existing emissions and transportation models.[15] In related research, Guensler *et al.* have qualitatively evaluated various ITS technology bundles with respect to vehicle emissions.[16,17] Recently, Washington has developed and applied a modal emissions model to evaluate the emissions impact of electronic tolling operations.[18] For this research, a power demand modal emission model has been developed and effectively combined with an AHS simulation model called SmartPath. These models are described below.

2.1. *Power Demand-Based Modal Emission Model*

A power demand modal emission model has been developed based on second-by-second vehicle operation data and simultaneously measured emissions data.[19] The emission data are registered with vehicle dynamic operating mode, e.g., acceleration, deceleration, steady-state cruise, idle, and so on. Using this modal type of emission model is in sharp contrast to the current driving cycle-based emission inventory techniques [i.e., the Federal Test Procedure (FTP)[20]], where emissions are collected in bags over long periods of time (on the order of 500 s), and then analyzed as a whole.

The second-by-second emission data are combined with an analytical model of the instantaneous power requirements placed on a vehicle's engine. The instanta-

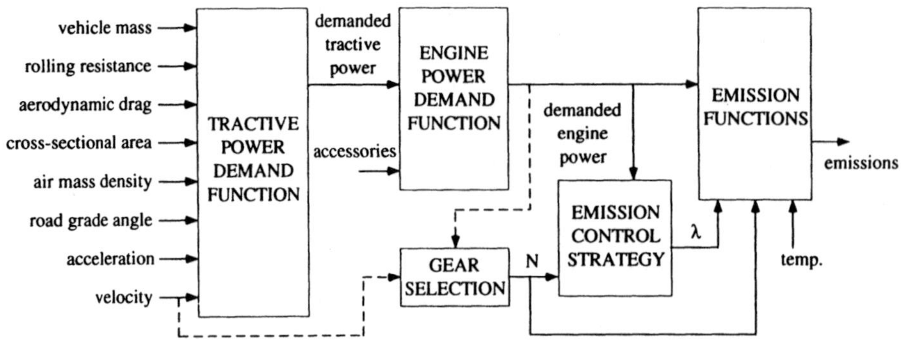

FIGURE 1. Power demand-based modal emissions modeling methodology.

neous power requirements placed on a vehicle (at the wheels) depend on three types of factors: (1) *environmental factors* (e.g., mass density of air, road grade), (2) *static vehicle parameters* (e.g., vehicle mass, rolling resistance coefficient, aerodynamic drag coefficient, cross-sectional area), and (3) *dynamic vehicle parameters* (e.g., commanded acceleration, velocity). Given these parameters, it is possible to first determine the demanded tractive power, as shown in Figure 1. Further, the power demand on the engine can be determined by modeling the power transfer through the vehicle's drive system and incorporating other variables such as use of accessories (e.g., air conditioning, power steering).

One of the most important aspects of this power demand-based method for estimating emission output is modeling the vehicle's emission control system. Modern vehicles have complex emission control systems that include as the primary component a catalytic converter. An electronic engine controller regulates the air–fuel ratio to the engine so that the ratio is as near as possible to the stoichiometric ratio where the catalytic converter operates most efficiently. During normal operation of the vehicle, the air–fuel ratio is kept at the stoichiometric ratio (lambda = 1). However, there can be cases when the conversion efficiency of the emission control system is reduced. One example is during cold-start events, when the catalytic converter is not at its proper operating temperature and thus is not operating at its peak performance. Another important example includes power enrichment events. When a vehicle has a high engine power demand (which may be induced by a hard acceleration or steep grade), it has been shown that the emission control system can go "open-loop" and the air–fuel ratio is commanded rich for peak demand power and protection of engine components. Recent studies have shown that power enrichment events can contribute significantly to overall emission production (e.g., Refs. 21–23).

As a first approximation to modeling power enrichment events, a simple thresholding technique is used. When the demanded engine power exceeds a particular threshold, the emission control system goes into an open-loop state, and the air–fuel ratio is rich, producing significant emissions. When the demanded power is below that threshold, the system maintains the air–fuel ratio at stoichiometry, resulting in low emissions. The emission output of carbon monoxide (CO), hydrocarbons (HC), and oxides of nitrogen (NO_x) can be measured and correlated to demanded engine power induced under numerous operating conditions. The emission output can then be approximated by a function that relates emission species output to demanded power.

For use in the AHS simulation experiments, the equations corresponding to vehicle dynamics and load-based emissions have been calibrated to a 1991 Ford Taurus. This electronic fuel injection vehicle has state-of-the-art emission control equipment. Automated vehicles used in an AHS will likely have similar emission control systems that give the same or better performance; therefore, modeling this type of vehicle will serve as an outside bound for AHS emissions production.

2.2. *SmartPath Simulator*

The California PATH (Partners for Advanced Transit and Highways) program has developed over the years an AHS simulator called SmartPath.[24–26] SmartPath is a microscale AHS simulator where vehicles and highways are modeled having appropriate sensors, communication devices, and actuators. SmartPath was designed around the AHS control and communication architecture described in Refs. 8, 27, and 28.

Other AHS concepts and architectures exist (see, e.g., Ref. 29); however, the architecture implemented in SmartPath is believed to be the most developed at this time. SmartPath can be used as a tool to determine AHS performance measures such as highway capacity, traffic flow, and other measures of interest to transportation system planners and users. Because SmartPath is a microsimulator, each system element and control policy is individually modeled. These elements and control policies are parametrically specified, thus the user can alter these parameters to study the effect on AHS performance.

SmartPath is highly modular, so that users can change individual control laws in all of the layers. The user needs to set up configuration files that specify the highway network, the link layer, coordination, and regulation layer parameters. There are two types of highway lanes featured in SmartPath: automated lanes and nonautomated or transition lanes. Platoons of vehicles under full automated control occupy the automated lanes and vehicles that are manually driven occupy the transition lanes.

Communications in SmartPath are modeled as one-way or point-to-point links. Within a platoon, there are two-way links between vehicles, as well as two-way links between the platoon leader and all followers. It is also assumed that there are two-way links between the leaders of neighboring platoons. Finally, there is a link between the highway link layer and every vehicle.

By knowing each vehicle's acceleration, velocity, and position (and thus grade) at every simulation time interval, a program module was integrated into the simulation to calculate total emissions of CO, HC, and NO_x based on the power demand modal emissions model.

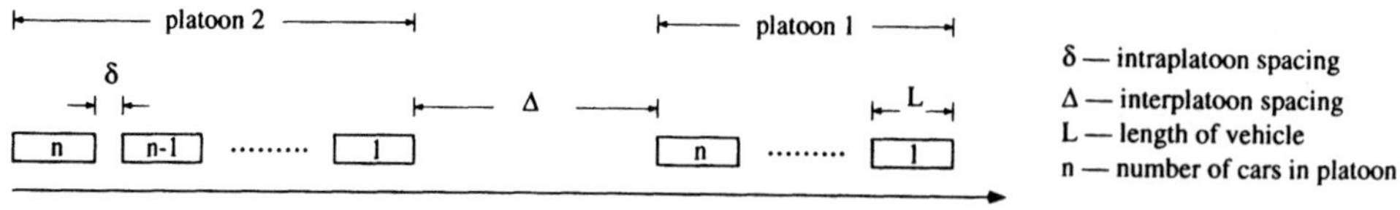

FIGURE 2. Platoons of vehicles on a highway.

2.3. AHS Operation

ITS technology in the form of Advanced Vehicle Control Systems (AVCS) can be applied to control vehicle motion so that vehicles can operate in platoons, i.e., follow each other very closely at high speeds, while still maintaining a high safety margin. The notation associated with platooned vehicles is given in Figure 2.

Traveling in platoons has a number of implications: (1) traffic flow will increase dramatically over current highway conditions as a result of denser traffic traveling at higher speeds; (2) congestion should decrease since the stop-and-go effect caused by relatively long human reaction delays will be eliminated and accidents will be minimized.

Current highway traffic (i.e., uninterrupted traffic flow) can be characterized by the traffic volume (v), average vehicle speed (S), and vehicle density (D). These terms are generally related by the product $v = S \times D$.[30]

Further constraints operate on these parameters which restrict the type of flow conditions on a highway link. The general form of these constraints is shown in Figure 3, which illustrates some key points of uninterrupted traffic flow:

- Zero rate of flow occurs in two distinct cases: (1) when there are no vehicles on the roadway and (2) when the density is so high that all vehicles are stopped and cannot move. In the first case, the density is zero, thus the flow rate is zero, and the speed in this case is assumed to be the driver's desired speed (i.e., vehicle free speed). In the second case the density is at its maximum and the vehicle speed is zero. The density at which this occurs is called the jam density.

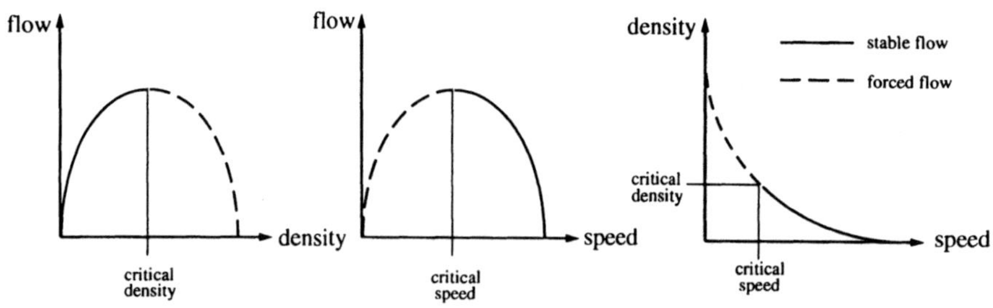

FIGURE 3. Flow, density, and speed relationship of uninterrupted traffic flow.

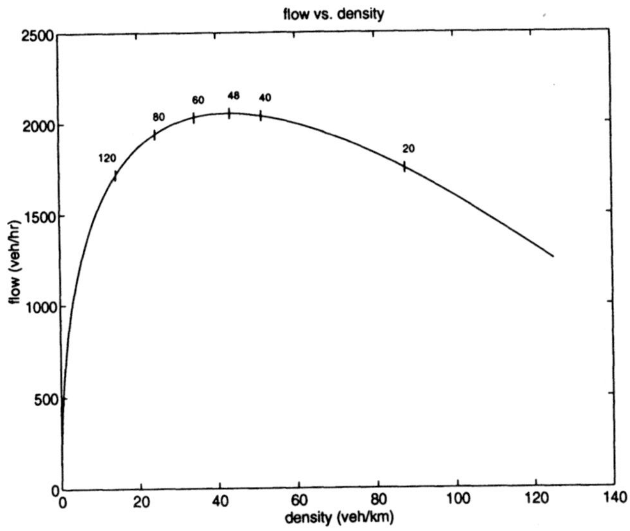

FIGURE 4. Flow–density relationship for manual traffic. Velocity values (km/h) are annotated on the curve.

- As density increases from zero, the traffic flow increases because of the increased number of vehicles. The average vehicle speed is reduced to maintain safety during higher density conditions.
- Traffic flow is maximized at a specific critical density. As density increases above the critical density point, speed drops off at a faster rate. Traffic flow tends to become unstable in this region as a result of perturbations from lane change maneuvers, merging, or any external variables (e.g., debris in roadway, accident in adjoining roadway). These perturbations can create disturbances that are not damped or dissipated in the flow. These unstable, forced flow regions in the curves are characterized by stop-and-go congestion.

Nonautomated flow–density–speed relationships can be approximated mathematically by specifying the spacing, or gap, between vehicles required for safe stopping if one car suddenly brakes, and after a time lag, the second car also brakes without collision. The flow–density curve in Figure 4 was produced for the case when the first car brakes at $0.9g$ (8.82 m/s^2) and the second car brakes at $0.6g$ (5.88 m/s^2) after a 1-s time lag. This curve (after Ref. 9) is for a single lane and is similar to curves predicted by the *Highway Capacity Manual*.[30]

A similar mathematical formulation can be developed for the flow–density–speed characteristics of an AHS. Within a platoon of vehicles, the spacings are much smaller and closely regulated by automated controls. Therefore, platooned vehicles can travel faster at higher densities, thus improving the traffic throughput. If we consider a single lane of platooned traffic as shown in Figure 2,

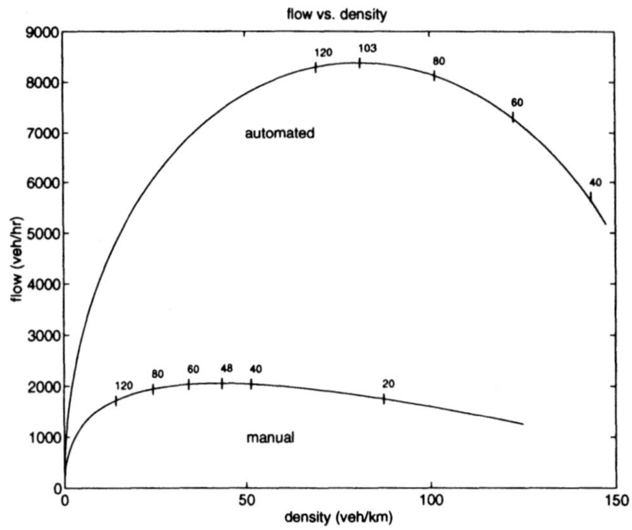

FIGURE 5. Flow–density relationship of traffic for both manual driving and automated driving. Velocity values (km/h) are annotated on the curve.

we can mathematically approximate the flow–density–speed characteristics. Using the notation given in Figure 2, the vehicle density for an automated lane is given as:

$$D = \frac{n}{\Delta + n(L+\delta) - \delta} \tag{1}$$

The interplatoon spacing is determined as described before, i.e., requiring safe stops if one platoon suddenly brakes, and after a time lag, the second platoon (leader) also brakes. In the automated scenario, the time lag is much shorter than that for human drivers. The resulting flow–density curve in Figure 5 (upper curve) was produced for the more restrictive case when the first platoon brakes at $2g$ (19.6 m/s^2), and after a 0.3-s time lag, the second platoon (leader) brakes at $0.3g$ (2.94 m/s^2). It is assumed that the intraplatoon spacing is precisely controlled and can also perform safe stops under these specified stopping conditions. In the mathematical formulation, the intraplatoon spacing is set to 1 m, the car length is 5 m, the number of vehicles in each platoon is 20, and the vehicle free speed is 120 km/h. The difference between the flow for manual driving and automated driving is substantial. The maximum traffic flow for this automated case is roughly four times that of the manual driving case. The maximum flow for the automated case occurs at an average speed of 103 km/h, and for the manual case it occurs at 48 km/h.

In the AHS architecture that serves as the basis for the SmartPath simulator, a vehicle enters the AHS and announces its destination exit. The system then assigns a route for the vehicle, consisting of various path changes along the route. A path consists of a sequence of platoon maneuvers, where the three key maneuvers for a

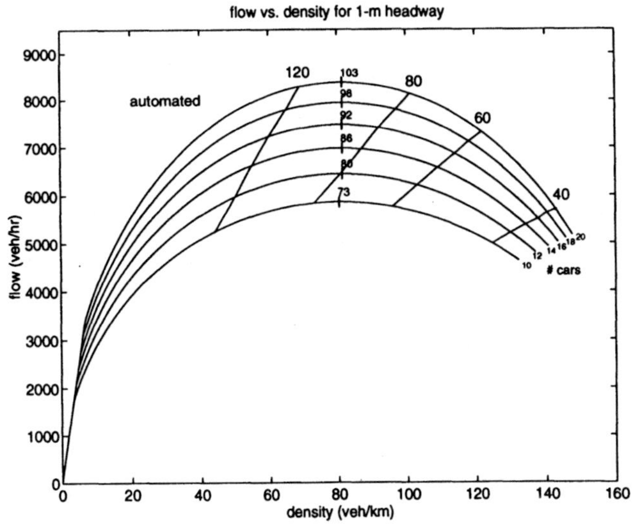

FIGURE 6. Flow–density relationship of automated traffic for different platoon lengths.

platoon are *merging, splitting,* and *lane changing.* These maneuvers are selected and implemented by the platoon leaders (and free agents, defined as one-vehicle platoons), after communication takes place within the platoon. Because of these maneuvers, platoons will dynamically vary in length as vehicles travel from their specific origins to destinations. Thus, the high flow rate for automated traffic shown in Figure 5 will be somewhat reduced, since the average platoon length per lane will be less than the specified maximum (in this case, 20 vehicles). Figure 6 shows the reduction of flow with shorter platoon lengths. Further, because of the maneuvers described above, the average speed of automated traffic will be less than the assigned free speed of the platoons. This also reduces the average flow rate per lane.

2.4. Steady-State Velocity Emissions

Emissions produced by traffic (manual or automated) will depend on several factors. In general, large variations in velocity (i.e., numerous acceleration, deceleration events) lead to higher emissions, therefore to minimize total emissions, traffic should be kept as smooth as possible. We first consider vehicle emissions at steady-state vehicle velocities.

Using the AHS simulation models coupled with the power demand emissions model, platoon simulations were carried out to determine average emission rates at different steady-state velocities. The motion parameters, engine power demand, and associated emission rates were calculated for each modeled vehicle in the simulation. Because the follower vehicles within a platoon have very small intraplatoon spacings (e.g., on the order of 1 m), the aerodynamic drag coefficient of each follower is significantly reduced in accordance with the "drafting effect."

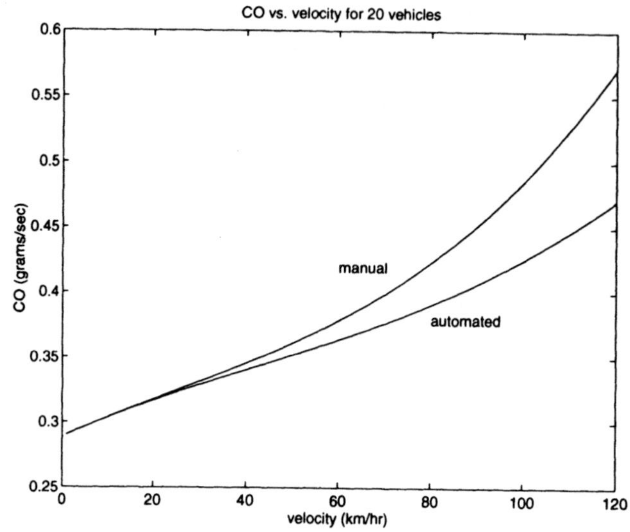

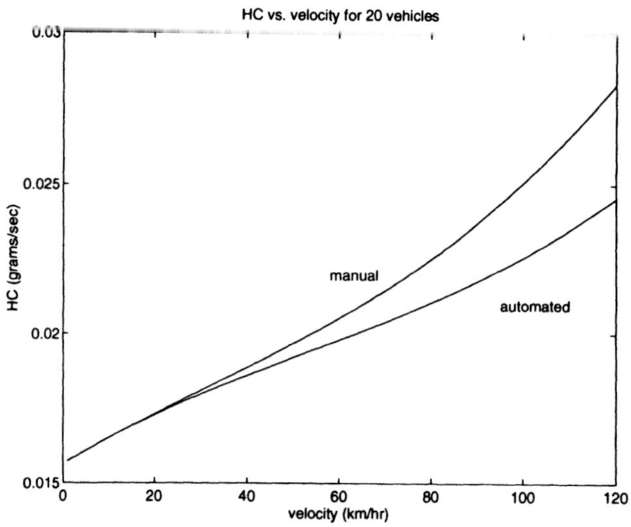

FIGURE 7. Constant-velocity carbon monoxide (a), hydrocarbon (b), and NO_x (c) emission rates for 20 vehicles platooned and nonplatooned.

Using preliminary aerodynamic drag reduction data for vehicles in platoons produced by the University of Southern California,[31,32] the calculated power demand on the engine is significantly smaller at higher speeds. Based on the data, even the lead vehicle of a platoon has its aerodynamic drag coefficient reduced via the vehicle following closely behind.

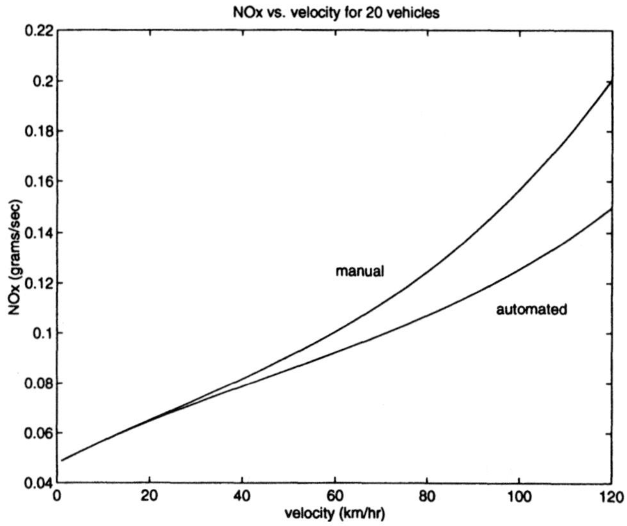

FIGURE 7. *(continued)*

Emission rates were calculated at constant speeds ranging from 5 km/h to 120 km/h. A third-order polynomial was then used to fit the data to give continuous emission values at any prescribed speed. Comparisons between constant-velocity CO, HC, and NO_x emission rates for 20 vehicles that are platooned and nonplatooned are shown in panels a, b, and c, respectively, of Figure 7. In both cases the traffic density is low, and the vehicles are traveling near their free speeds. At lower speeds, vehicle emission rates of the two cases are roughly the same. However, at higher speeds, the platooned vehicles benefit from the drafting effect, which results in less engine load and thus less emissions output. Similar reductions for fuel consumption have been predicted in Ref. 32. In order to illustrate the benefits of drafting, Figure 8 shows the constant-velocity CO emission rate for 20-vehicle platoons for two specific cases, namely, when the platoons are spaced 1 or 3 m apart. It is seen that when the vehicles in the platoon are closer together, lower emission rates can be achieved.

In order to determine total steady-state emissions of an automated lane within an AHS, these emission data were applied to the flow–density curves shown in Figure 5. It is important to note that the curves in Figure 5 reflect traffic density and flow associated with specified safe spacings. Thus, to generate flow values at lower densities, vehicle speeds greater than the free speed (i.e., the maximum speed a driver will go on the freeway without interference from other traffic) were used in the calculations. For purposes of generating total link emissions at lower densities, the flow values were adjusted so that the vehicle velocities at low densities were at the constant free speed.

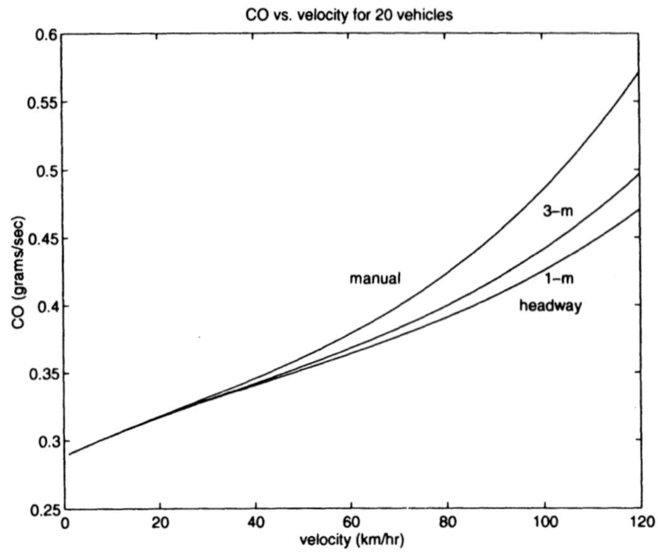

FIGURE 8. Constant-velocity carbon monoxide emission rates for 20 vehicles nonplatooned and platooned for the cases of 1- and 3-m spacings.

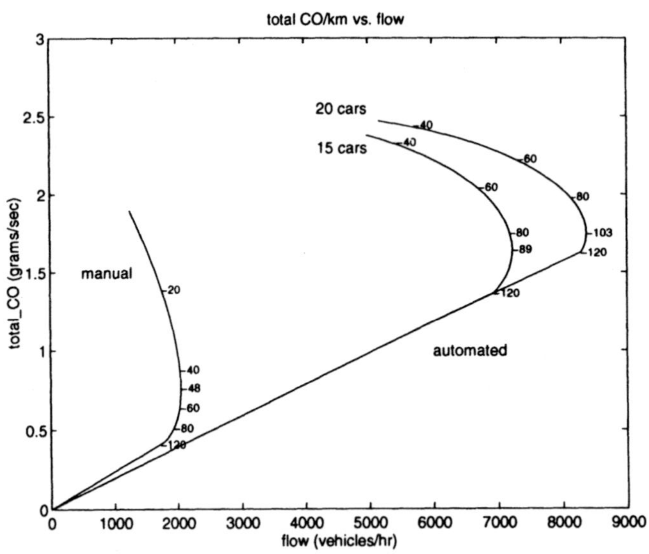

FIGURE 9. Total CO (a), HC (b), and NO_x (c) emissions versus traffic flow for manual and automated traffic, for a 1-km lane. Velocity values (km/h) are annotated on the curve.

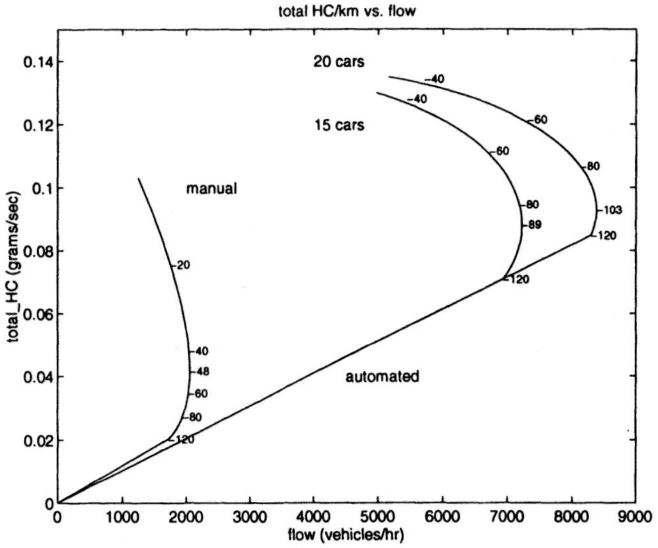

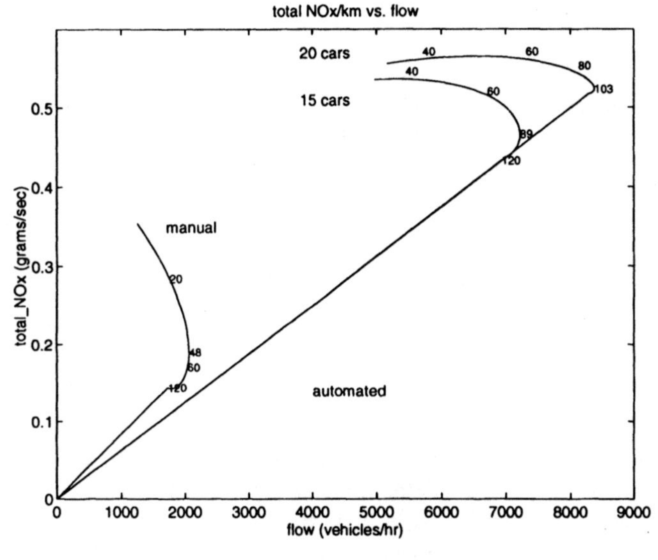

FIGURE 9. *(continued)*

Total CO, HC, and NO_x emission rates for a 1-km lane are shown as a function of traffic flow for both the manual and automated (platooning) cases in panels a, b, and c, respectively, of Figure 9. There are several key points to note in these plots:

1. The maximum traffic flow for a manual lane is 2053 vehicles/h at an average vehicle speed of 48 km/h. At the same traffic volume, the automated lane produces roughly *half* as much emissions as in the manual case, regardless of whether the platoon size is 15 or 20 vehicles:

	CO (g/s)	HC (g/s)	NO_x (g/s)
Manual	0.76	0.0415	0.1882
Automated	0.40	0.0210	0.1190

2. Given the emissions rate for maximum manual traffic volume, roughly *twice* the traffic volume can occur in the automated lane (regardless of whether platoon size is 15 or 20 vehicles) to produce the same amount of emissions:

	Flow		
	CO	HC	NO_x
Manual	2053	2053	2053
Automated	3899	4005	3005

3. The maximum traffic flow for an automated lane is 8286 vehicles/h at an average speed of 103 km/h for a constant platoon size of 20 vehicles. The associated emissions at this point are roughly *twice* that of the maximum flow rate of manual driving:

	CO (g/s)	HC (g/s)	NO_x (g/s)
Manual	0.76	0.0415	0.1882
Automated (20)	1.75	0.0911	0.5214
Automated (15)	1.64	0.0868	0.4610

It is important to point out that the emissions associated with higher traffic densities and lower average speeds for the manual case are underestimated in these curves. Remember that these emissions are calculated based on steady-state velocities, and the negative slope region of the flow–density curve is inherently unstable, leading to stop-and-go traffic. The accelerations associated with stop-and-go traffic will lead to a greater amount of emissions.

2.5. *Vehicle Emissions for AHS Maneuvers*

The emission rates for automated traffic calculated in the previous section were based on steady-state traffic flow with platoons traveling at constant velocities.

However, this idealized traffic flow will be difficult to achieve since the entrance and egress of vehicles in the AHS will cause traffic stream disturbances, inducing variations in speed (i.e., accelerations and decelerations). At the heart of these velocity variations are several vehicle/platoon maneuvers such as a vehicle joining a platoon, one platoon merging with another, or a vehicle leaving a platoon. Using the SmartPath simulator and emissions model, we have investigated several of these maneuvers that can have a large impact on emissions production.

2.5.1. Free-Speed Accelerations. When a vehicle enters the AHS and switches to automated control, it must first accelerate in the transition lane to reach the prescribed highway velocity. If using a modern vehicle with an emissions-control strategy as described in Section 2.1, this free-speed acceleration rate can have a dramatic impact on cumulative AHS emissions. If the commanded acceleration is sufficiently high, the power demand on the vehicle's engine will be large and the vehicle will enter a state of power enrichment. During power enrichment, the air–fuel ratio is commanded rich for peak power performance, and as a result, prodigious amounts of emissions are produced. For the modeled vehicles in our simulations, CO emissions increase by two orders of magnitude when in the power enrichment state.

In the current SmartPath implementation, vehicles entering the AHS are commanded to a constant free-speed acceleration rate of 1 m/s^2. At low speeds, the power demand at this acceleration rate is small. However, at high speeds, the power demand on the vehicle's engine is already large because of the aerodynamic drag factor (which increases with the square of velocity). A 1 m/s^2 acceleration rate places further power demand on the engine, which may cause it to enter power enrichment. The vehicles modeled in the current simulation enter power enrichment at a velocity of 34 m/s (122 km/h), when accelerating at 1 m/s^2 on level ground.

2.5.2. Platoon Merge. As part of the AHS architecture implemented in SmartPath,[28] two neighboring platoons in the same lane will merge to create a longer platoon if they are less than the optimal platoon length (specified by the link layer, see Ref. 28). When the lead platoon falls within the detection range of the following platoon, communication is initiated between the platoon leaders, and the merge maneuver proceeds. As with the free-speed acceleration, the acceleration rate of the merging, upstream platoon can have a significant effect on the production of vehicle emissions. If the acceleration rate is high enough, not only will the leader of the upstream platoon enter power enrichment mode, but every vehicle that belongs to the upstream platoon as well.

In the current SmartPath version, platoons perform the merge maneuver at a constant 2 m/s^2 acceleration. When traveling at a modest 20 m/s (72 km/h) and accelerating at this rate, the modeled vehicles easily enter the power enrichment state when traveling on level ground. In fact, at that acceleration rate, a leader of a platoon will enter power enrichment if it is traveling at 12 m/s (43 km/h) or greater, and a follower vehicle will undergo power enrichment at 13 m/s (46 km/h) or greater, getting a slight benefit from the aerodynamic drafting effect.

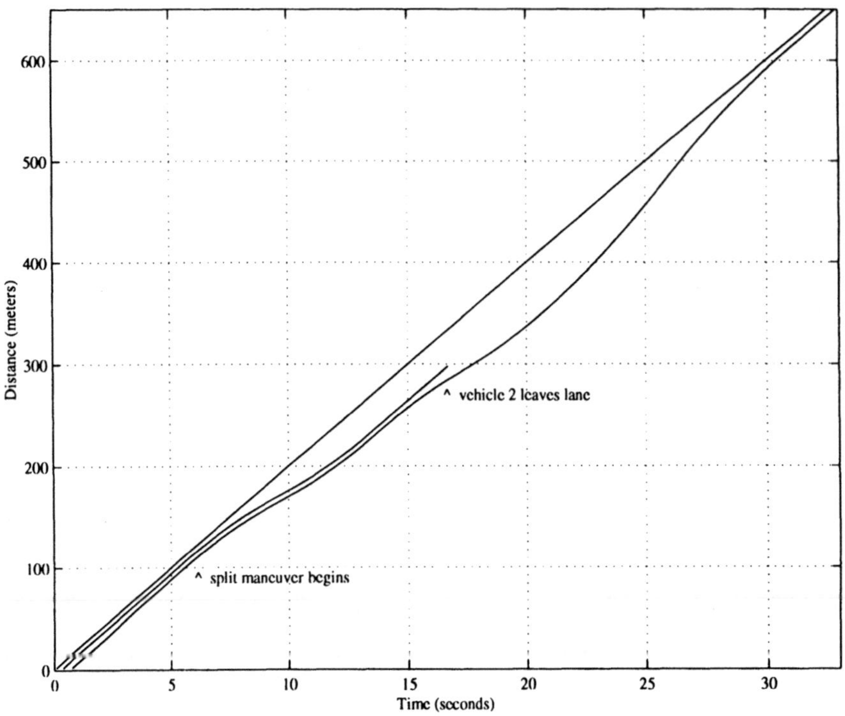

FIGURE 10. Time–distance diagram of a three-vehicle platoon, split maneuver.

2.5.3. Platoon Split. When a member of a platoon needs to exit the AHS, a platoon split maneuver is performed. In a platoon split, the platoon is divided at the vehicle that needs to exit. The upstream segment of the platoon decelerates as a whole, then accelerates back up to the nominal highway speed, leaving a sufficiently large safety gap between the two platoons. The leader of the upstream platoon (the vehicle that needs to exit) then changes lanes, and the second vehicle in the platoon becomes the new leader. If the downstream platoon is still within sensing range of the upstream platoon, then the upstream platoon will perform a merge maneuver, rejoining the two segments.

An example split maneuver is illustrated for a three-vehicle platoon as a time–distance diagram in Figure 10. At approximately 6 s into the simulation, the split is initiated and vehicles 2 (middle line) and 3 (bottom line) decelerate at a constant rate of -2 m/s^2 for approximately 4 s. The two vehicles then accelerate at 2 m/s^2 until they are back at the nominal highway velocity of 20 m/s. Vehicle 2 then changes lanes, maintaining an approximate 20 m/s velocity. Vehicle 3 decelerates while vehicle 2 is changing lanes, then begins to accelerate at 2 m/s^2 in order to rejoin vehicle 1 (top line), which has not changed its velocity throughout the maneuver. This merge acceleration lasts for approximately 9 s, after which vehicle 3 decelerates and joins up with vehicle 1.

The corresponding velocity, acceleration, power demand (normalized to maximum rated power), and CO emissions for vehicles 2 and 3 are shown in panels a and b, respectively, of Figure 11. It can be seen that during the period when vehicles 2 and 3 accelerate after their initial deceleration, the power demand on the two vehicles is greater than the enrichment threshold, and the corresponding CO emission rate is far greater than that produced during constant velocities. The power demand is also very high when vehicle 3 performs its merge maneuver with vehicle 1, and the CO emissions rate for vehicle 3 is again at its maximum.

2.5.4. Constant Power Acceleration Approach. Instead of using constant accelerations when performing these platoon maneuvers, we have modified the acceleration strategy used in the AHS simulation. In its initial design, SmartPath assigns accelerations that are held constant for different components of each maneuver. Not only is this detrimental when considering vehicle emissions, in some cases it is impossible to perform the commanded accelerations given limitations in engine power. For example, when traveling at high speeds, a commanded acceleration of 2 m/s^2 will be impossible to perform since the power demand on the engine will exceed what it can actually produce.

Based on equations of vehicle dynamics, we have devised a constant power approach when vehicles accelerate in the simulation. The inertial power requirements for a vehicle (in kW) are given in simplest form as:

$$P_{\text{inertial}} = \frac{M}{1000} \cdot V \cdot (a + g \cdot \sin\theta) \tag{2}$$

where M is the vehicle mass (kg), V is the vehicle velocity (m/s), a is the vehicle acceleration (m/s^2), g is the gravitational constant (9.81 m/s^2), and θ is the road grade angle. The power requirements related to the drag components are given in simplest form as:

$$P_{\text{drag}} = \left(M \cdot g \cdot C_r + \frac{\rho}{2} \cdot V^2 \cdot A \cdot C_a \right) \cdot \frac{V}{1000} \tag{3}$$

where C_r is the rolling resistance coefficient, ρ is the mass density of air (1.225 kg/m^3, depending on temperature and altitude), A is the cross-sectional area (m^2), and C_a is the aerodynamic drag coefficient. Thus, the total tractive power requirements placed on the vehicle (at the wheels) are given as:

$$P_{\text{tractive}} = P_{\text{inertial}} + P_{\text{drag}} \tag{4}$$

If the maximum power of the engine is limited such that the vehicle does not go into power enrichment mode, acceleration can be determined as:

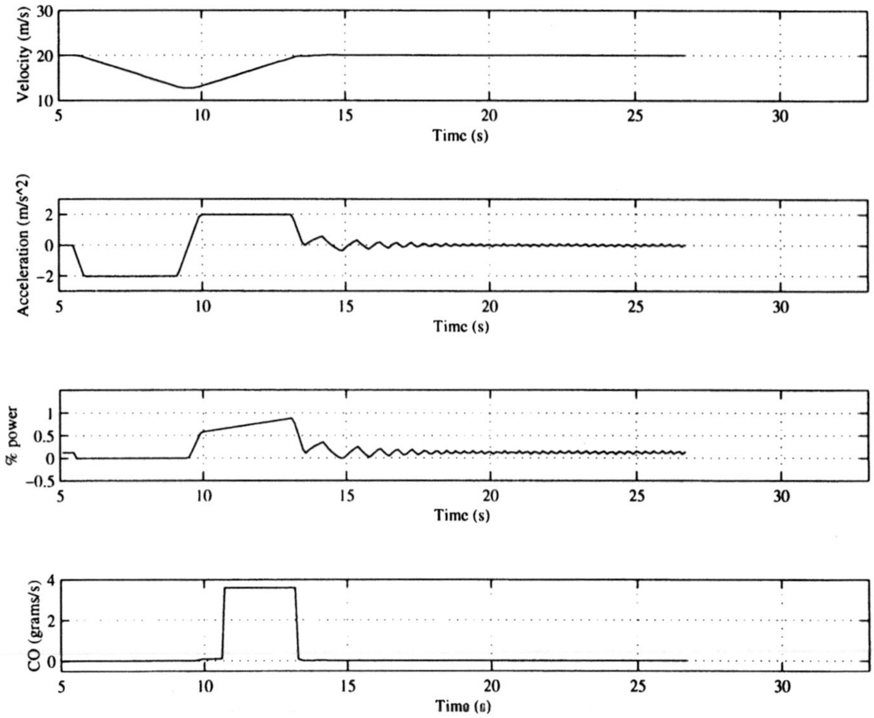
Velocity (m/s)
Acceleration (m/s^2)
% power
CO (grams/s)
Time (s)

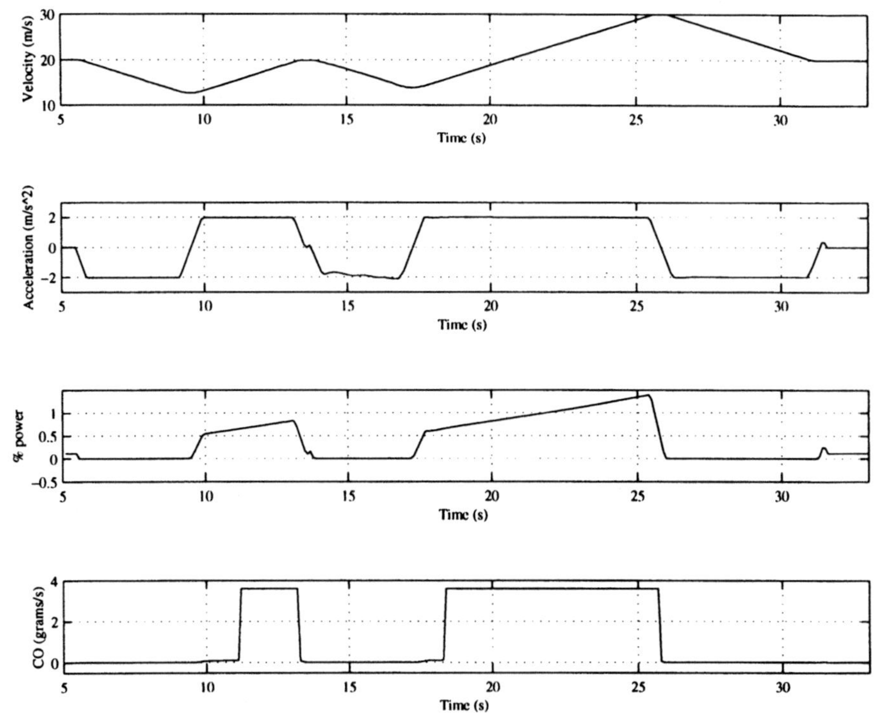
Velocity (m/s)
Acceleration (m/s^2)
% power
CO (grams/s)
Time (s)

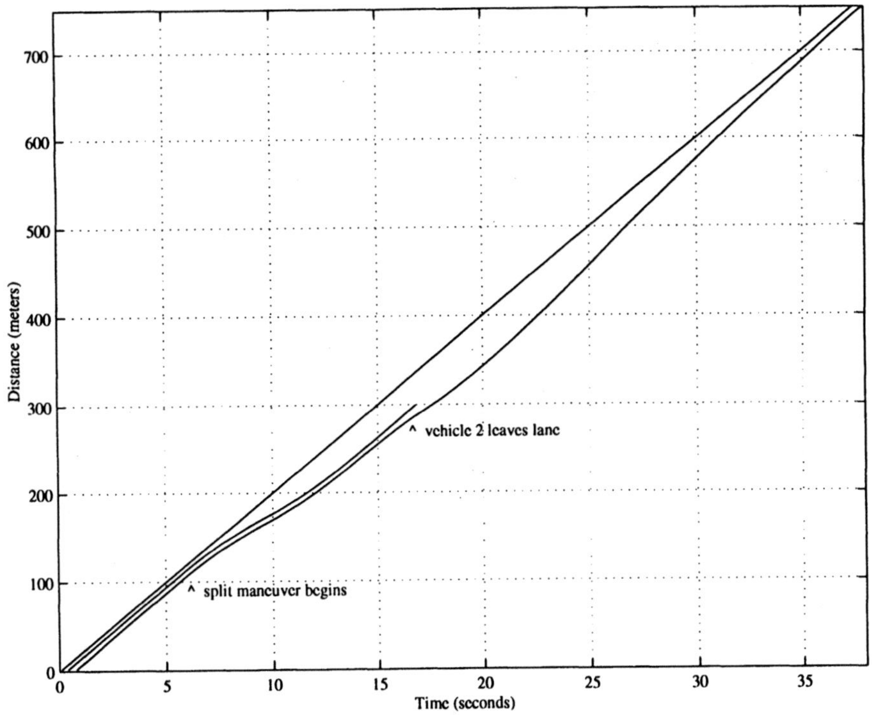

FIGURE 12. Time–distance diagram of a three-vehicle platoon, constant-power split maneuver.

$$\text{Acceleration} = \frac{1000 \cdot (P_{\text{maximum}} - P_{\text{drag}})}{M \cdot V} - g \cdot \sin\theta \tag{5}$$

Using this methodology, each vehicle in the simulation will achieve its highest possible acceleration performance without going into the power enrichment mode. This is substantially different than simply using constant acceleration values during the platoon maneuvers.

This acceleration methodology was applied to the same split maneuver described in the previous section. Figure 12 shows the time–distance diagram of the three vehicle trajectories, and panels a and b of Figure 13 show the velocity, acceleration, power, and CO emissions of vehicles 2 and 3, respectively. Again, the leader vehicle (vehicle 1, top line in Figure 12) does not change speed at all, and thus appears as a straight line. Vehicles 2 and 3 still perform the same maneuver, but because they do not accelerate as quickly as before, the maneuver itself takes longer to perform (the maneuver is finished after 32 s compared to 26 s shown in Figure 10). The savings in emissions, however, is significant. Cumulative CO emis-

FIGURE 11. Vehicle 2 (a) and vehicle 3 (b) velocity, acceleration, normalized power demand, and CO emissions for maneuver.

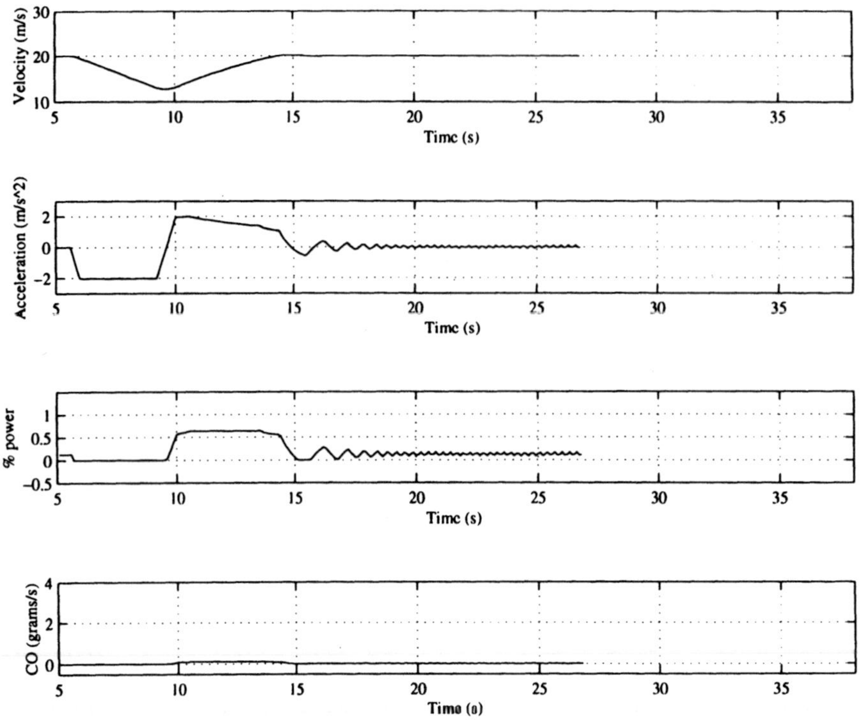
Velocity (m/s)
Time (s)
Acceleration (m/s^2)
Time (s)
% power
Time (s)
CO (grams/s)
Time (s)

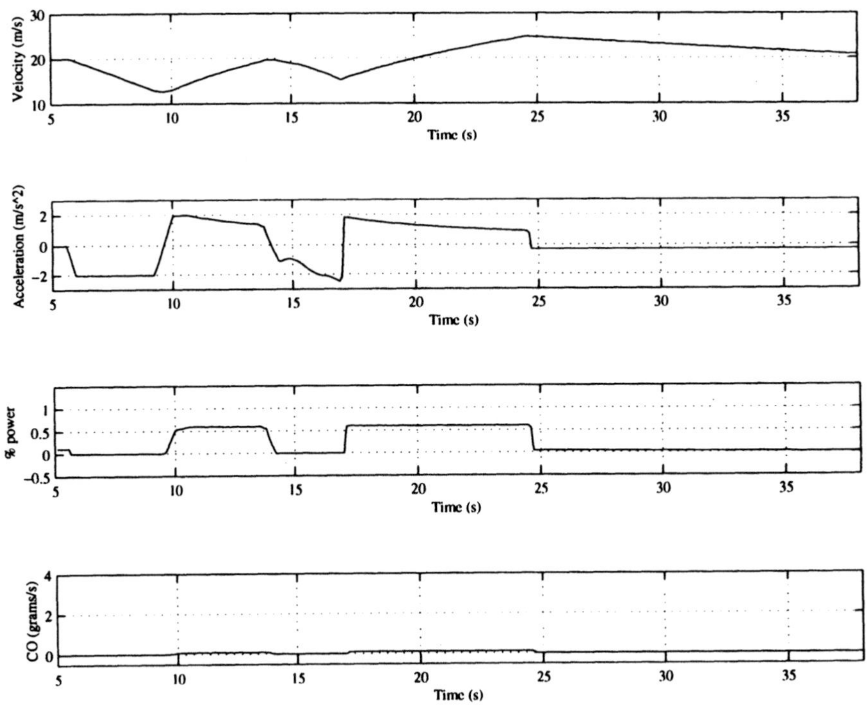
Velocity (m/s)
Time (s)
Acceleration (m/s^2)
Time (s)
% power
Time (s)
CO (grams/s)
Time (s)

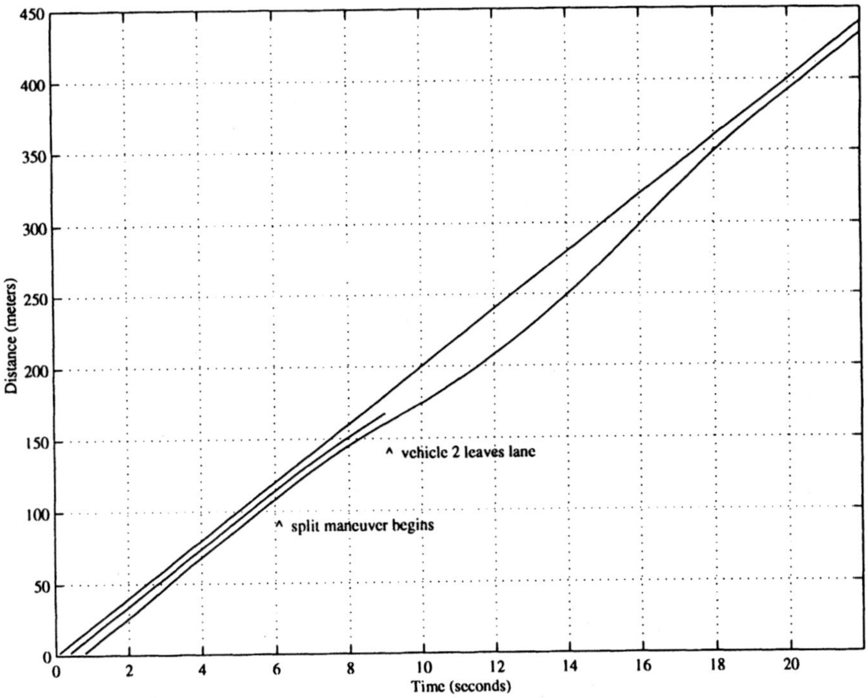

FIGURE 14. Time–distance diagram of a three-vehicle platoon, modified split maneuver.

sions for the constant-acceleration split are 60.4 g, while the cumulative CO emissions for the constant-power split are 5.9 g, an order of magnitude less (10.2).

2.5.5. *Modified Split Protocol.* Another method for reducing cumulative AHS emissions is to modify the protocols of the various vehicle/platoon maneuvers. In the initial implementation of SmartPath, the protocols were designed primarily from a safety standpoint. However, it is possible to design the protocols so that vehicle emissions are minimized while still maintaining a sufficiently high degree of safety.

As an example, we modified the protocol of the same split maneuver described in the previous sections. Using the original constant-acceleration approach, the protocol was modified such that when the split begins, vehicle 2 simply changes lanes. Instead of having both vehicles 2 and 3 slow down and then speed back up in order to create a gap between the front of vehicle 2 and the back of vehicle 1, vehicle 2 initiates a lane change immediately while operating within the platoon. The safety

FIGURE 13. Velocity 2 (a) and vehicle 3 (b) velocity, acceleration, normalized power demand, and CO emissions for maneuver.

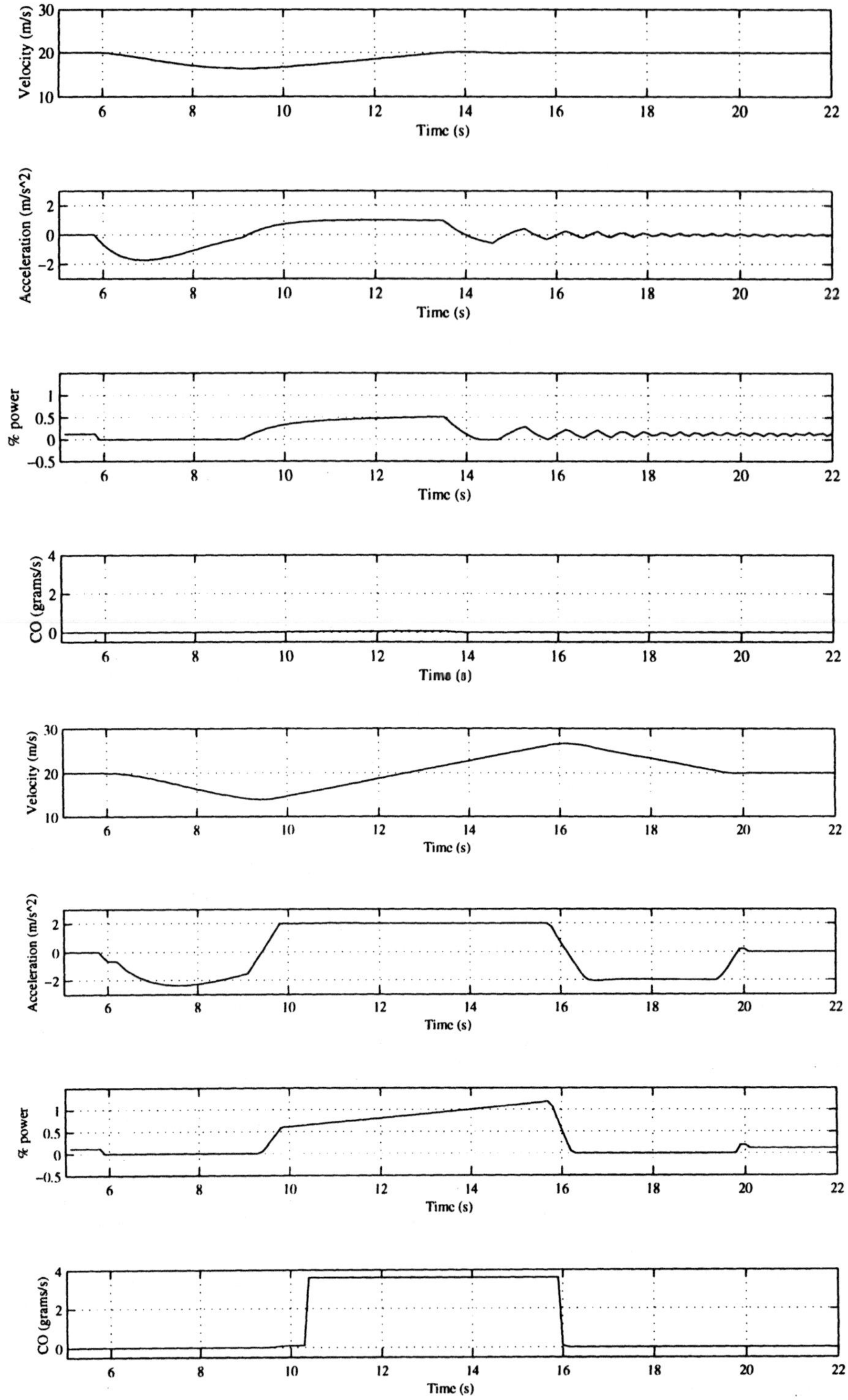

Velocity (m/s)
Acceleration (m/s^2)
% power
CO (grams/s)
Time (s)

of this maneuver modification should still remain high, since the spacings between vehicles remain small, and any collision will have only a small impact. After vehicle 2 changes lanes, vehicle 3 again performs a merge with vehicle 1, but in this case, the distance between the two vehicles is shorter than before. The trajectories of the vehicles are shown in Figure 14. The corresponding velocities, accelerations, power, and CO emissions are shown for vehicles 2 and 3 in panels a and b, respectively, of Figure 15.

With this maneuver modification, the maneuver is finished after 14 s, compared to the 26 s of the original scenario. Cumulative CO emissions for the modified split are 23.6 g, compared to the 60.4 g of the original split maneuver. Emissions could be further reduced by combining the two methods described in this and the previous section.

3. INDUCED TRAVEL DEMAND

Any ITS technology that serves to reduce congestion and increase the efficiency and reliability of roadway travel has the potential to make driving a more attractive alternative. It is likely, then, that an AHS will induce additional travel demand, resulting in more trips, more vehicles in use, more kilometers traveled, and a modal preference for personal automobiles.[4] The environmental benefits (particularly reduced emissions and energy consumption) gained through smoother traffic flow and reduced congestion can easily be offset by additional traffic demand.[33]

Induced travel demand associated with ITS has been the subject of several studies. In particular, Shladover addressed both the supply and demand of ITS, and the potential contributions to reducing transportation's greenhouse gas production.[34] Ostria *et al.* have developed an analytical framework based on microeconomic principles for evaluating the potential impacts of ITS on the quantity of travel, with emphasis on the resulting changes in consumers' surplus and external costs.[35,36] Another objective of this work was to identify the policy implications of the potential impacts. Further, Vaughn and Murphy have also performed a qualitative analysis of induced travel demand associated with IVHS technology.[37] Also, Meyer, Mohaddes Associates have recently incorporated the benefits of IVHS technology into regional travel demand models.[38]

To offset potential increases in travel demand, Travel Demand Management (TDM) and Travel System Management (TSM) policies should be implemented concurrently with the deployment of ITS technology, such as AHS.[4] TDM measures will influence individual traveler behavior by discouraging personal travel and encouraging drivers to switch travel modes to car pools, vanpools, or transit during

◄

FIGURE 15. Velocity 2 (a) and vehicle 3 (b) velocity, acceleration, normalized power demand, and CO emissions for maneuver.

congested travel times. Examples of TDM mechanisms include real-time rideshare matching, congestion pricing, and telecommuting. The role of TSM policies is to increase the performance of the existing transportation network through mechanisms such as incident management, traffic signal coordination, ramp metering, and HOV lanes.

4. CONCLUSIONS AND FUTURE WORK

Based on microscale simulation models and modal emissions data for a modern, closed-loop emission-controlled vehicle, steady-state (i.e., constant velocity) emission rates have been estimated for both manual and automated lanes. An automated lane using platooning can potentially improve the traffic flow by a factor of four, and at maximum flow values, the total emissions (grams per second per kilometer) increase is by a factor of two (for the modeled Ford Taurus). If only half of the automated lane capacity is used, the traffic flow improves by a factor of two, and the associated emission rates are roughly the same as the full-capacity manual case. If the automated lane carries the same traffic volume as in the manual case, the emissions are reduced by a factor of two. These results are related to reduced engine power demand on vehicles that are platooning and on higher vehicle densities achieved in an automated scenario.

Further, preliminary analysis has been carried out to evaluate vehicle emissions associated directly with the AHS maneuvers of free-speed accelerations, platoon merging, and platoon splits. The current version of SmartPath uses a constant-acceleration strategy in these maneuvers. This can be problematic at high speeds since a constant-acceleration constraint can cause a modern emission-controlled vehicle to enter a power enrichment state, in which very high emissions are produced. We have devised a constant-power approach that limits the accelerations of automated vehicles, eliminating power enrichment states, and greatly reducing emissions. Emissions can also be reduced by developing "emission-friendly" protocols, which do not require high-power episodes while still maintaining system safety. Also, by keeping the size of platoons as large as possible, traffic density and highway capacity will both increase, and the number of vehicles benefiting from the aerodynamic drafting effect increases.

This analysis assumed a constant platoon size of 20 vehicles, however, platoons will vary in length as a result of vehicles dynamically entering and leaving platoons as they travel from their specific origins to destinations. Shorter platoons will lead to lower automated lane capacities and higher average vehicle emissions.

In order to minimize emissions in an AHS, several points should be observed:

- The size of platoons should be kept as large as possible; this increases the density of traffic and increases the highway capacity, and the number of vehicles benefiting from the aerodynamic drafting effect increases.
- During AHS maneuvers, vehicles should be commanded so that they operate at power levels below the threshold where they enter a power enrich-

ment state. By eliminating power enrichment events, significantly lower emissions result.

- The number of maneuvers in general should be kept to a minimum, which will keep the traffic flow smoother, resulting in less accelerations and decelerations that lead to higher emissions.

Further research must be carried out to better characterize vehicle emissions associated with stop-and-go traffic in the unstable traffic flow–density regions. If congestion is to be avoided, the traffic should be kept in the positive slope region of the flow–density curve (see Figure 5). When in the positive slope region, interaction between vehicles in traffic is minimal, leading to smoother traffic flow. It can be seen that the extent of the positive slope region is much greater for the automated lane than for the manual lane. For the automated case, the network and link layer controllers in the AHS will attempt to keep traffic in the stable operating regime at all times.

Another area that can be explored in the future is an analysis of the problems associated with the end points of automation, i.e., the dumping of high flow rates onto off-ramps, arterials, and collectors. In effect, automation causes peak period compression, meaning that higher flow rates occur throughout the system. In some cases this could cause congestion on off-ramps, arterials, and collectors at automation egress points. The emissions associated with these cases should also be examined.

In addition, macroscale modeling techniques need to be developed for estimating system-level emissions of an AHS. Instead of modeling every vehicle at high time resolutions, aggregate traffic parameters for an AHS can be used as input to a macroscale emissions model. Examples of traffic parameters indexed over different AHS roadway sections may be vehicle density, entrance/exit density, and ratio of average traffic speed over assigned free speed. This aggregate model would then allow the accurate estimation of cumulative tailpipe emissions in system-level applications.

Again, it is important to point out that the emission rates used in this analysis were for a single vehicle. For current manual driving, the vehicle population is quite varied, and to more accurately predict total emissions, emission rates for different vehicle classes must be incorporated. For an automated scenario, however, the vehicle population will be somewhat more restricted. Vehicles that have automated platoon technology will tend to be newer passenger vehicles with closed-loop emission control systems, similar to the vehicle modeled here. It may be that by the time AHS technology is in place in our transportation systems, vehicle emission control technology will have improved to the point where the potential problems outlined in this report do not apply.

Lastly, it is imperative to couple the deployment of AHS with appropriate TDM and TSM strategies in order to avoid potential increases in travel demand, which could negate the direct environmental benefits of AHS.

ACKNOWLEDGMENTS. This work was performed as part of the California PATH Program of the University of California, in cooperation with the State of California

Business, Transportation, and Housing Agency, Department of Transportation; and the U. S. Department of Transportation, Federal Highway Administration.

The contents of this report reflect the views of the author who is responsible for the facts and the accuracy of the data presented herein. The contents do not necessarily reflect the official views or policies of the State of California. This report does not constitute a standard, specification, or regulation.

REFERENCES

1. U.S. Department of Transportation, *National Program Plan for Intelligent Transportation Systems (ITS),* 1st ed. (ITS America, 1995).
2. L. SAXTON, Automated Control–Cornerstone of Future Highway Systems, *IVHS Rev.* pp. 1–16 (Summer 1993).
3. N. CONGRESS, Automated Highway System—An Idea Whose Time has Come, *Public Roads* **58**(1) (1994).
4. NAHSC, Automated Highway System (AHS)—System Objectives and Characteristics, Technical Report, National Automated Highway System Consortium (1995).
5. M. BARTH AND J. NORBECK, Transportation Modeling for the Environment; Year 1 Interim Report, Research Report No. UCB-ITS-PRR-94-27, California PATH Program (1994).
6. M. BARTH AND J. NORBECK, Transportation Modeling for the Environment; Year 2 Final Report, Research Report, California PATH Program (1995).
7. U. KARAASLAN, P. VARAIYA, AND J. WALRAND, Two Proposals to Improve Freeway Traffic Flow, Research Report No. UCB-ITS-PRR-90-6, California PATH Program, ITS, University of California (1990).
8. P. VARAIYA AND S. SHLADOVER, Sketch of an IVHS Systems Architecture, Research Report No. UCB-ITS-PRR-91-3, California PATH Program, ITS, University of California (1991).
9. Rockwell, Potential Payoffs from IVHS: A Framework for Analysis, Research Report No. UCB-ITS-PRR-92-7, California PATH Program, University of California (1992).
10. W.-B. ZHANG, S. SHLADOVER, AND R. HALL, A Functional Definition of Automated Highway Systems, Research Report No. UCB-ITS-PRR-94-9, California PATH Program, ITS, University of California (1994).
11. H. MALDONADO, Methodology to Calculate Emission Factors for On-Road Motor Vehicles, Technical Report, California Air Resources Board (1991).
12. H. MALDONADO, Supplement to Methodology to Calculate Emission Factors for On-Road Motor Vehicles July 1991, Technical Report, California Air Resources Board (1992).
13. D. EISINGER, Preview of MOBILE, in *Proc. Transportation Modeling: Tips and Trip-Ups,* San Mateo, CA, 1993).
14. M. BARTH *et al.*, Modal Emissions Modeling: A Physical Approach, in *1996 Transportation Research Board Annual Meeting,* Washington, DC, 1996.
15. C. LITTLE AND J. WOOSTER, IVHS and Environmental Impacts: Implications of the Operational Tests, in *National Conference on Intelligent Transportation Systems and the Environment,* Arlington, VA, 1994, pp. 315–338.
16. R. GUENSLER, D. SPERLING, AND S. WASHINGTON, IVHS Technologies and Motor Vehicle Emissions, in *Proc. IVHS America Annual Meeting,* Washington, DC, 1993, pp. 226–235.
17. S. WASHINGTON, R. GUENSLER, AND D. SPERLING, Air Quality Impacts of Intelligent Vehicle Highway Systems, in *Transportation Planning and Air Quality II,* edited by P. Benson (American Society of Civil Engineers, New York, 1993).
18. S. P. WASHINGTON, Carbon Monoxide Impacts of Electronic Tolling Operations: Two Conflicting Assessments of a Promising Intelligent Transportation Technology, in *88th Annual Meeting of the Air & Waste Management Association,* San Antonio, TX, 1995.

19. M. J. BARTH AND J. M. NORBECK, A Power-Demand Approach to Estimating Vehicle Emissions, in *Fourth CRC-APRAC On-Road Vehicle Emission Workshop,* San Diego, CA, 1994, pp. 5-51–5-72.
20. FTP, Code of Federal Regulations. Title 40. Parts 86-99 (portion of CFR which contains the Federal Test Procedure), Office of the Federal Register (1989).
21. S. H. CADLE *et al.*, CRC-APRAC Vehicle Emissions Modeling Workshop Summary, *J. Air Waste Manage. Assoc.* **41**, 817–820 (1991).
22. N. A. KELLY AND P. J. GROBLICKI, Real-World Emissions from a Modern Production Vehicle Driven in Los Angeles, *J. Air Waste Manage. Assoc.* **43**, 1351–1357 (1993).
23. M. MEYER *et al.*, A Study of Enrichment Activities in the Atlanta Road Network, in *Proc. International Specialty Conference on Emission Inventory Issues,* Durham, NC, 1992.
24. F. ESKAFI, D. KHORRAMABADI, AND P. VARAIYA, SmartPath: An Automated Highway System Simulator, Technical Report No. UCB-ITS-PTM-92-3, ITS-PATH (1992).
25. B. HONGOLA, J. TSAO, AND R. HALL, SmartPath Simulator—Version MOU62, Technical Report No. UCB-ITS-PWP-93-8, ITS-PATH (1993).
26. F. ESKAFI AND D. KHORRAMABADI, SmartPath User's Manual, Technical Report Department of Electrical Engineering and Computer Sciences, University of California, Berkeley (1993).
27. S. E. SHLADOVER *et al.*, Automated Vehicle Control Developments in the PATH Program, *IEEE Trans. Veh. Technol.* **40**(1), 114–130 (1991).
28. A. HSU *et al.*, The Design of Platoon Maneuver Protocols for IVHS, Research Report No. UCB-ITS-PRR-91-6, California PATH Program, ITS, University of California (1991).
29. W. B. Stevens, The Use of System Characteristics to Define Concepts for Automated Highway Systems (AHS), in *Transportation Research Board, 73rd Annual Meeting,* Paper No. 940990, Washington, DC, 1994.
30. *Highway Capacity Manual, Special Report 209* (Transportation Research Board, Washington, DC, 1985).
31. M. ZABAT, S. FRASCAROLI, AND F. K. BROWAND, Drag Measurements on 2, 3, & 4 Car Platoons, in *SAE Tech. Pap.* No. 940421 (1994).
32. M. ZABAT, N. S. STABILE, AND F. BROWAND, Estimates of Fuel Savings from Platooning, in *ITS America Annual Conference,* Washington, DC, 1995, pp. 1202–1208.
33. P. WINTERS AND A. POLK, Transportation Demand Management and Intelligent Vehicle–Highway Systems: The Need for Mutual Cooperation, in *National Policy Conference on Intelligent Transportation Systems and the Environment,* Arlington, VA, 1994, pp. 309–313.
34. S. E. SHLADOVER, Potential Contributions of Intelligent Vehicle/Highway Systems (IVHS) to Reducing Transportation's Greenhouse Gas Production, Technical Research Report No. 91-4, California PATH Program (1991).
35. S. OSTRIA, M. LAWRENCE, AND D. PICKRELL, Capacity-Induced Increases in the Quantity of Travel with Special Reference to IVHS, in *National Policy Conference on Intelligent Transportation Systems and the Environment,* Arlington, VA, 1994.
36. S. J. OSTRIA, ITS-Induced Increases in the Quality of Travel with Special Reference to Emission Impacts, in *88th Annual Meeting of the Air & Waste Management Association,* San Antonio, TX, 1995.
37. K. VAUGHN AND D. MURPHY, Air Quality Impacts of IVHS, in *ITS America Annual Conference,* Washington, DC, 1995, pp. 1195–1201.
38. Meyer, Mohaddes and Associates, Incorporation of the Benefits of IVHS Technology into Regional Travel Demand Models, Technical Report, Meyer, Mohaddes Associates (1994).

CHAPTER 14

Regional Mobility Impacts Assessment of Highway Automation

Mark A. Miller, Anne Bresnock, Steven E. Shladover, and Edward H. Lechner

1. INTRODUCTION

Urban traffic congestion and air pollution are crucial issues in most metropolitan areas, but are more acute in Southern California than in most other North American regions. The PATH Program at the Institute of Transportation Studies, University of California, Berkeley and the Southern California Association of Governments (SCAG) have investigated some of the long-term regional impacts that could result from implementation of advanced highway technologies (automation and electrification) in the Los Angeles area. This chapter focuses on the evaluation of mobility impacts of highway automation technologies applied to portions of the Southern California freeway network in 2025.

Mark A. Miller • California PATH Program, University of California, Berkeley, California 94804. Anne Bresnock • California State Polytechnic University, Pomona, California 91768. Steven E. Shladover • California PATH Program, University of California, Berkeley, California 94804. Edward H. Lechner • Science Applications International Corporation, Systems Control Technology Group, Los Altos, California 94022; *present address:* Jakes Associates, Inc., San Jose, California 95112.

Automated Highway Systems, edited by Ioannou. Plenum Press, New York, 1997.

Highway automation, or the Automated Highway System (AHS), appears to be just about the only option available to make a substantial contribution toward relieving congestion without imposing serious mobility restrictions. The basic idea behind AHS is that by use of automatic control technology it should be possible to overcome the performance limitations of human drivers that significantly limit both the capacity and the safety of the present-day surface transportation system. The automated system should enable road vehicles to travel substantially closer together than they do today, producing significant increases in the throughput of the roadway infrastructure.

The AHS would represent a dramatic change from the operation of today's road system in terms of capacity (or productivity) and safety as well as in the driver's perception of the travel experience. Such dramatic changes give rise to many uncertainties, making it difficult to predict the consequences of AHS implementation with high confidence. However, the initial explorations of possible AHS impacts have appeared promising enough to justify further explorations. This is, however, clearly an area of high potential risks as well as high potential paybacks.

The uncertainties surrounding AHS cover a wide range of subject areas:

- Technical performance
- System-level reliability and safety
- Direct transportation system impacts (congestion)
- User acceptability of the automated driving experience
- Capital and operating costs
- Indirect transportation impacts (such as energy and air quality)
- Long-term transportation/land use interaction effects
- Liability and legal issues

Of these, primarily the issues of technical performance have been addressed in most previous research on AHS. Some preliminary studies of the reliability and safety issues are currently under way, while this chapter addresses the direct transportation system impacts. A major study of the human factors and user acceptability issues is also under way. The other topics appear to be even more difficult to address, requiring extensions beyond the scope of validity of currently available modeling tools and data.

In 1993, the Federal Highway Administration (FHWA) initiated a series of 1-year studies under its Precursor Systems Analyses (PSA) of Automated Highway Systems to promote research into the issues and risks associated with AHS. The PSA studies investigated both technical and nontechnical issues: technical areas included lateral and longitudinal control, automated check-in and check-out requirements, malfunction management, roadway infrastructure deployability, and roadway and vehicle operational analysis; nontechnical areas included legal and liability issues, environmental issues, equity, funding, and user acceptance.

Since the conclusion of the PSA set of research projects, the next phase of AHS work has begun within the collaborative framework of the National Automated

Highway System Consortium (NAHSC). The NAHSC has been established to (1) analyze alternative AHS concepts, (2) demonstrate the feasibility of AHS in 1997, (3) develop and test a prototype AHS, and (4) provide specification for a future operational development. These uncertainties, as well as others, will be investigated during the course of the NAHSC work.

These uncertainties are of course indications of some of the risks associated with AHS. However, they are also fruitful opportunities for research that can confer benefits on the entire transportation community. Because AHS represents such a large stretching of the frontiers of transportation system performance, it requires reexamination of the available transportation system analysis tools and methodologies. The enhancements that will be needed to effectively represent AHS can lead to improvements in the state of the art of modeling used to predict the impacts of more traditional transportation system changes.

The research reported here is an initial attempt to quantify some of the direct transportation system impacts of widespread use of a fully mature AHS technology in a large metropolitan area (Los Angeles). The fidelity of the results is of course limited by the available data and modeling tools, which were not designed with the intent of analyzing such a radical transportation system change. However, in spite of such limitations, the results appear to be quite dramatic and worthy of serious consideration. The principal questions to be addressed here are:

1. How much of an urban freeway network would need to be set aside for exclusive AHS use in order to make a significant impact on congestion within the region?
2. Could such an exclusive AHS network be accommodated within existing rights-of-way without requiring extensive new construction or imposing severe congestion on the rest of the freeway network?
3. What effect would the AHS have on the levels of congestion on the conventional (nonautomated) freeways and arterials?
4. How would the congestion relief benefits be distributed among the drivers of the AHS-equipped vehicles and nonequipped vehicles?
5. How would the AHS have to be connected to the rest of the road system (access and egress ramp arrangements)?

The results derived in the course of this study have produced significant answers to all of these questions, without necessarily offering definitive quantifications of the specific impacts that would follow from long-term implementation (such as land use changes).

Very few prior studies have attempted to answer questions such as these. The two most ambitious such studies are now quite old, having been completed in 1971 and 1981, respectively. Given the changes in technology and the condition of the transportation system that have occurred since then, the results of those studies have limited applicability today. The earlier study, by the Transportation Systems Center (TSC),[1] was a comparative evaluation of several advanced technology "dual mode" options for the Boston metropolitan area, using a 20-year planning horizon. The options that were studied as alternatives to new freeway construction were an AHS serving both buses and automobiles, an automated pallet system, and a system of completely new, small automated guideway vehicles. The AHS option was in the middle range of both costs and benefits in this study, with the automated guideway [or Personal Rapid Transit system (PRT)] being the most ambitious. A 249-mile AHS network was estimated to cost $1.6 billion, and the overall ratio of benefits to costs was estimated to exceed 2. The impacts included an arterial speed increase of 18% and an areawide travel speed increase of 57% compared to the baseline scenario using only conventional technology and the planned highway expansions. Induced demand was assumed to be 20% for purposes of the study, but this was acknowledged to be purely an *a priori* estimate for purposes of sensitivity evaluation rather than a predicted outcome.

The 1981 study was a broad-based AHS concept design and evaluation conducted by General Motors under contract to FHWA.[2,3] Case studies of implementation were conducted for Philadelphia as a sample metropolitan area and the Northeast Corridor as a sample intercity region. The Philadelphia study assumed a 320-mile AHS network with 51 "stations" (access/egress points), which would serve about 17% of the interzonal vehicle trips within the region (represented by 71 travel analysis zones). The mode split model used to predict this level of usage was adopted from the earlier TSC study.[1] Although system costs were estimated in considerable detail, under a variety of assumptions, no estimates of regional transportation benefits were included in the final report.[2]

There have been two recent studies that have investigated the mobility impacts of highway automation. One study[4–6] modeled the use of automation technologies on the freeway network in the Sacramento (California) urban region. Important differences exist between that study and the study reported herein with respect to: (1) size and level of congestion within the region, (2) automation scenario design assumptions, (3) methodologies utilized to develop the automation scenario, (4) configuration of the regional roadway network, (5) size of the automated network compared to the entire regional freeway system, and (6) representation of the automation technology by the regional transportation model. These differences make it extremely difficult to compare the results of the two studies.

The other study was performed as part of the PSA research endeavor by the Calspan Corporation.[7] A performance evaluation was conducted for a simulated AHS implementation on traffic corridors in urban, suburban, and rural settings. While mobility benefits were obtained in terms of speed and travel time on the simulated AHS facility, as in the case above, there were numerous differences between the Calspan work and the study reported herein, making an extensive comparison of the two studies difficult.

2. HIGHWAY AUTOMATION SCENARIO DESCRIPTION

The impacts analysis compared baseline (no advanced technology) mobility predictions with those for the automated system for 2025. Baseline projections were compiled for transportation demand, population, and employment for the Los Angeles region. SCAG's regional transportation model was used to determine the 2025 transportation demand. A 2025 transportation network consisting of the existing highway network, currently funded new highway construction and reconstruction specified in SCAG's Regional Mobility Plan for 2010, and long-range corridors identified to assist future transportation needs, was constructed for the purposes of this study.[(8)] The SCAG 2025 population and employment projections were used to update the socioeconomic component of the regional transportation model.

Two automation applications were considered to address the issue of access and egress in the automation scenario design. The *base network ramps* application contained only access ramps that were shared with vehicles traveling in conventional lanes, also referred to as mixed flow or manual lanes. Access and egress points in the base network ramps case occurred at approximately 5-mile intervals or less. The *additional ramp facilities* application included separate ramps at the designated access and egress locations for automated vehicle use. These additional ramps allowed for improved access to the automated lanes since the automated vehicles no longer had to weave through mixed flow lanes.

To guide selection of the specific configuration for the 2025 automated scenario network, combinations of alternative automated VMT (vehicle miles of travel) market penetration and automated network size were simulated for the AM-peak period in the SCAG region. Market penetrations of 5, 15, 30, 45, and 60% of AM-peak VMT were modeled on networks of 234, 431, and 657 automated freeway center-line miles. A sensitivity analysis was performed to determine the number of automated lanes required for each market penetration/network size combination. For each such combination, link traffic volumes in each automated freeway corridor were examined. The automation scenario was chosen to be of the largest size automated network (657 center-line miles) assuming a 45% market penetration. A complete description of the scenario selection process and its methodological development and outputs of the sensitivity analyses is contained in Refs. 9 and 10.

The automation scenario chosen to estimate the regional impacts consisted of the characteristics given in Table 1. It was assumed that freeway trips less than 4.0 miles would not utilize the automated facility. Figure 1 shows the 2025 automation scenario network.

3. TRANSPORTATION TECHNOLOGY ASSIGNMENT METHODOLOGIES

At the beginning of the scenario development process, trip assignments were performed for the 2025 AM-peak for each of the three potential automation networks (see Section 2), and on- and off-network trip length distribution tables were produced. Each of these tables grouped the system's origin–destination (O-D) pairs by on- and off-freeway network mileage. These trip length matrices enabled identification of short freeway trips, which were eliminated from the pool of potential trips that could use the automation technology.

TABLE 1
Automation Scenario Characteristics

Automated freeway network size	2165 lane miles (20% of regional freeway network)
Automated vehicle market penetration	23.6 million VMT 45% of AM-peak VMT for region 1.1 million AM-peak trips 19.3% of AM-peak trips for region
Technical characteristics	Automated lanes segregated from manual Special access and egress studied but not required Lane capacity = 6000 vehicles/lane/hour (based on 15-vehicle platoons traveling at 55 mph)

For the chosen automation scenario's network size and market penetration, the number of automated trips designated for its trip assignment was based on a random selection of trips within the set of O-D pairs identified for potential automation use. This set consisted of all trips with an on-network component greater than 4 miles. It was assumed that this bias for trips with a longer on-network com-

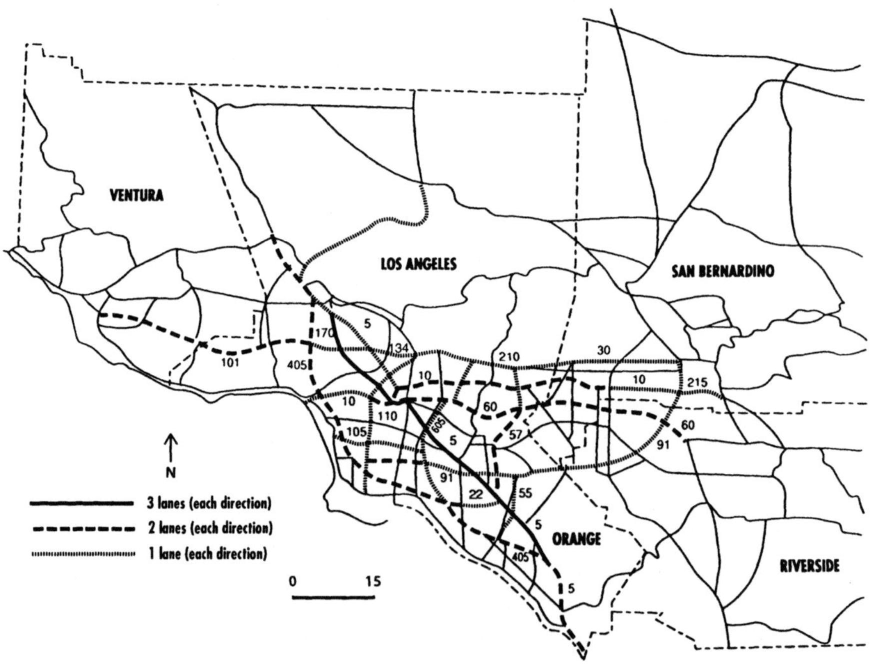

FIGURE 1. Automation scenario 2025 regional highway network.

ponent would realistically depict the behavior of the systems' users. For the automation scenario, 19.3% of the potential automated trips (or 45% of the VMT) were selected to travel on the automated facility.

It was not possible to assign mixtures of conventional trips and trips equipped with the automation technology to the freeway system simultaneously, given the limitations of the transportation model. As a result of this constraint, alternative trip assignments were modeled to study the effect of trip priority on mobility. As expected, the traffic volume estimates on the automated and nonautomated freeway links resulting from loading automated trips first and last were significantly different. For the impacts analysis, automated trips were assigned after the conventional trips since this ordering was assumed to be a more realistic representation of expected travel behavior. That is, by assigning the automated trips last, a small time penalty was attached to automated travel that could serve as a surrogate for the time spent traversing congested mixed flow lanes to enter or exit the automated facility.

Because of the technical requirement that the volume/capacity (V/C) ratio not exceed 1.00 on any of the links of the automated facility, it was necessary to develop further adjustments to the standard trip assignment procedure. Initially the standard speed–volume relationship developed by the U. S. Bureau of Public Roads as

$$\text{Speed} = \text{Speed}_0 \,/\, [1 + 0.15(V/C)^4]$$

(where Speed_0 = 55 mph) was utilized in the trip assignment. After first loading the conventional (nonautomated) vehicle trips to the network's mixed flow lanes, 30% of the automated trips were assigned to the automation network. A review of link traffic volumes produced by this assignment determined if congestion had developed on any of the links of the automated facility. Congestion was evident if reported link V/C ratios were above 1.00. Then, the following adjustments were made to the traditional speed–volume relationship for further trip loadings. For automated links that indicated

$$V/C < 1.00, \qquad \text{Speed} = 55 \text{ mph}$$

$$V/C > 1.00, \qquad \text{Speed} = 1 \text{ mph}$$

for the next trip loading stage to discourage assignment of trips in excess of those that the technology could accommodate.

The remaining automated trips were loaded in 10% increments. After each automated trip increment was assigned, V/C ratios on each link of the automated facility were evaluated to determine which links had V/C ratios above 1.00. Again, links indicating V/C ratios greater than 1.00 were assigned speeds of 1 mph and those with V/C ratios less than 1.00 were assigned speeds of 55 mph. This procedure continued until all automated trips were assigned. Speeds on all automated network links, including those previously assigned a 1 mph speed for trip loading purposes, were maintained at 55 mph to calculate mobility statistics. This method was as-

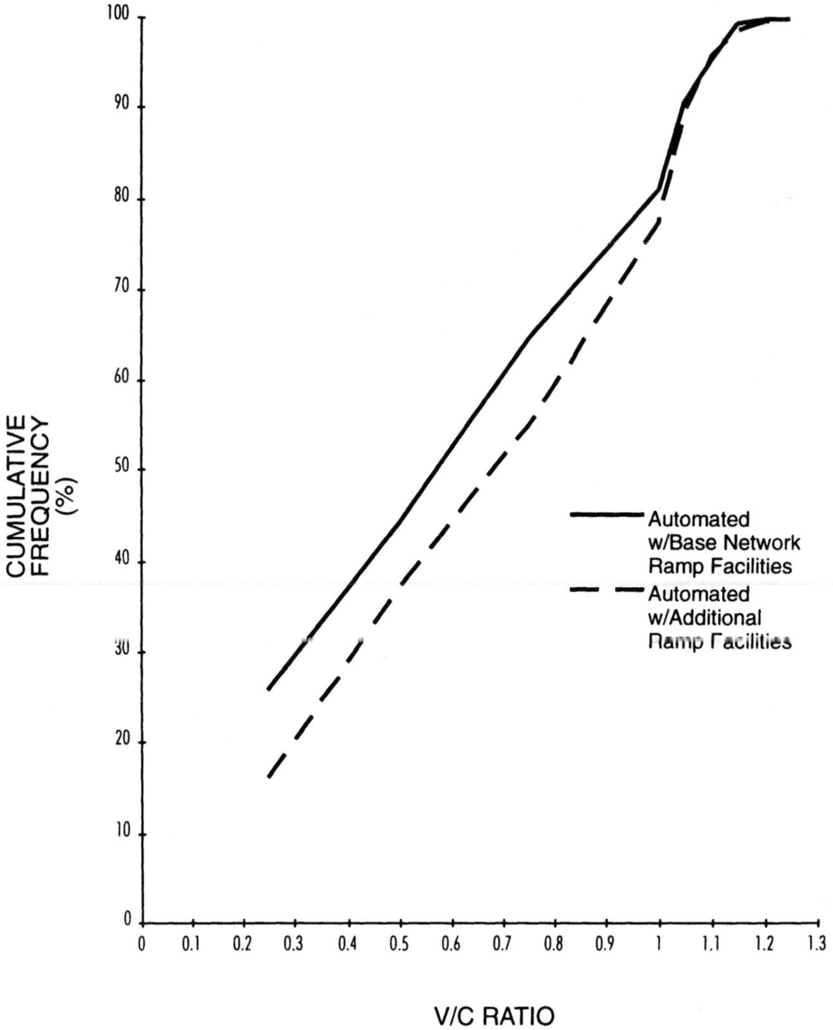

FIGURE 2. Highway automation cumulative V/C ratio frequency distribution—automated facility links.

sumed to most closely approximate the characteristics of the automation technology given current transportation modeling capabilities.

Figure 2 depicts the V/C ratios for each automation scenario after the trip assignments were completed.

For the base network ramps automation application, 81.2% of the automated links had V/C ratios less than or equal to 1.00 and all were less than 1.19. For the additional ramps automation application, 77.5% of the automated links had V/C ratios less than or equal to 1.00 and 98.5% of the automated links had V/C ratios less than 1.20.

The results just cited were derived assuming 6000 vehicles/lane per hour for the automated lanes. Altering this condition would obviously modify the number of automated links indicating congestion. An automated lane capacity definition of 7000 vehicles/lane per hour would produce fewer automated links with V/C ratio greater than 1.00, while 5000 vehicles/lane per hour would generate more automated links with V/C ratio greater than 1.00. The results were accepted, however, for analysis of the regional mobility impacts since transportation model constraints prevented a perfect representation of the automation technology, that is, prohibition of V/C ratios from exceeding 1.00 on all automated links.

Trip assignment results indicated that 25.6% of regional VMT, or 13.4 million VMT, occurred on the automated freeway in the base network ramps application. For the additional ramps case, 28.9% of VMT for the regional highway system, or 15.1 million VMT, traveled in the automated lanes. Total VMT attributed to the automated vehicles for operations *both* on and off the automated network were 23.6 million for the base network ramps application, and 23.5 million for the additional ramps case. The automated lanes in the base network ramps application carried 56.8% of the VMT driven by automation equipped vehicles, and in the additional ramps case, the automated lanes served 64.1% of the automated vehicles' VMT. The larger percentage associated with the additional ramp facilities application occurred since the separate set of on-and off-ramps made access and egress to and from the automated facility more direct. The result was less VMT on the mixed flow lanes, and more VMT on the automated lanes. The remaining VMT driven by vehicles equipped for automated operation occurred on other highway facilities. Mobility impacts detailed in Sections 4 and 5 cover various levels of aggregation and facility types.

4. REGIONAL MOBILITY IMPACTS

Comparisons of VMT, VHT (vehicle hours of travel), VHD (vehicle hours of delay), and average vehicle speeds were developed for the baseline trip assignment (no advanced technology), and automated base network and additional ramp facilities cases. Of the travel statistics studied, VHD and speed were considered the most appropriate indicators of congestion. Two types of VHD statistics are reported. Actual percentage change in VHD is first shown, followed by the percentage change in VHD *per vehicle mile of travel*. The latter mobility statistic is shown since part of the delay reduction on nonautomated facilities is attributed to fewer vehicle miles traveled on these facilities, and the effect of this can readily be factored out. Mobility indicators were derived for each scenario at several levels of aggregation, such as the freeway corridor, county, and region (see Figure 1). Complete mobility results for the automation trip assignments may be found in Refs. 11 and 12.

Mobility results were compiled for several facility types: automated freeway lanes, mixed flow freeway lanes, major and minor arterials.

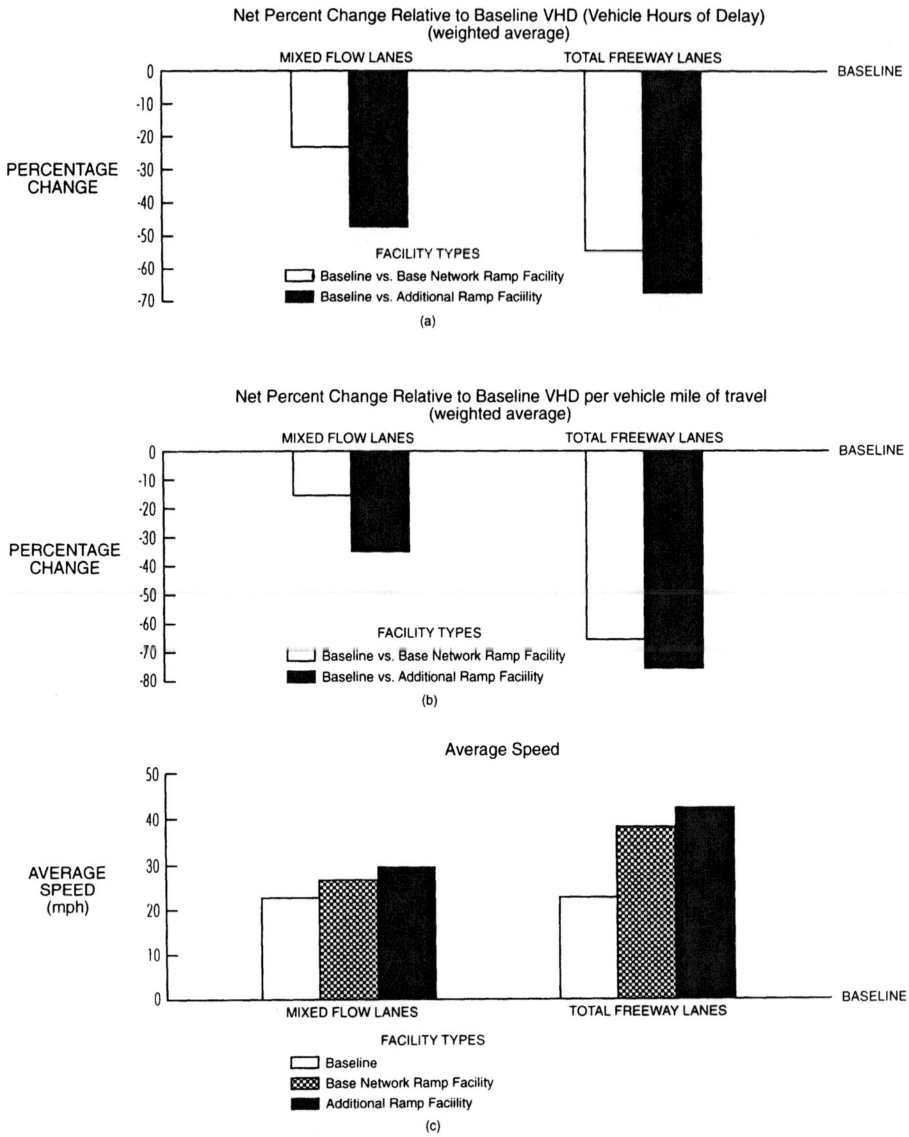

FIGURE 3. Highway automation/automated freeway corridors.

Panels a and b of Figure 3 identify changes in congestion levels over all automated freeway corridors on a per lane basis. The additional ramp facilities application reduced freeway congestion more than the base network ramps case. For example, the average per lane reductions in total VHD for the mixed flow lanes were 23.0 and 47.5%, respectively, for the base ramps and additional ramp facilities automation cases relative to the baseline. Corresponding per lane percentage

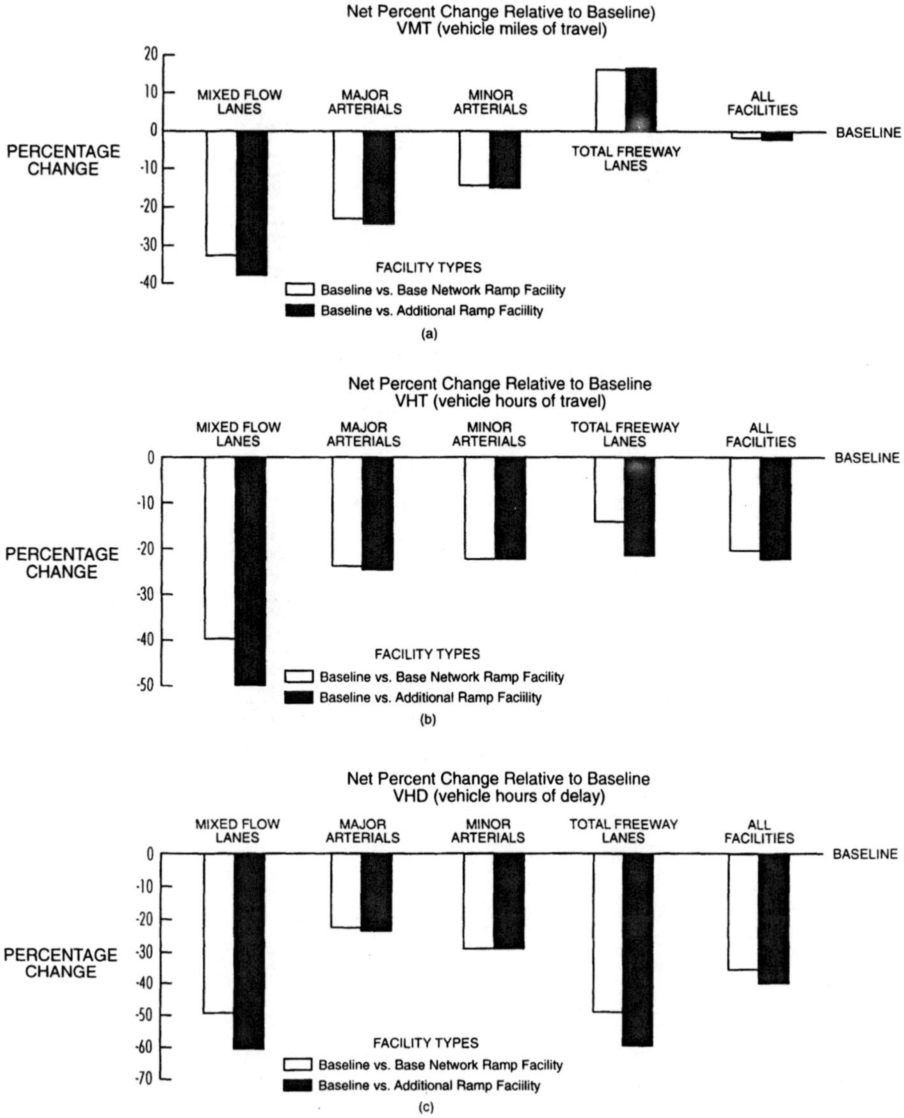

FIGURE 4. Highway automation 2025 regional highway network (am-peak period).

reductions in VHD per vehicle mile of travel were 16.5 and 35.5%, respectively, for the mixed flow lanes. The difference between the base ramp network and the additional ramp facilities cases was expected since the additional ramp facilities case offered more access and egress opportunities to the automated vehicles and hence resulted in fewer vehicles on the mixed flow lanes than in the base ramp facilities case. Similarly, the results normalized per VMT were as expected because the automation cases increase the total VMT on the freeway while reducing

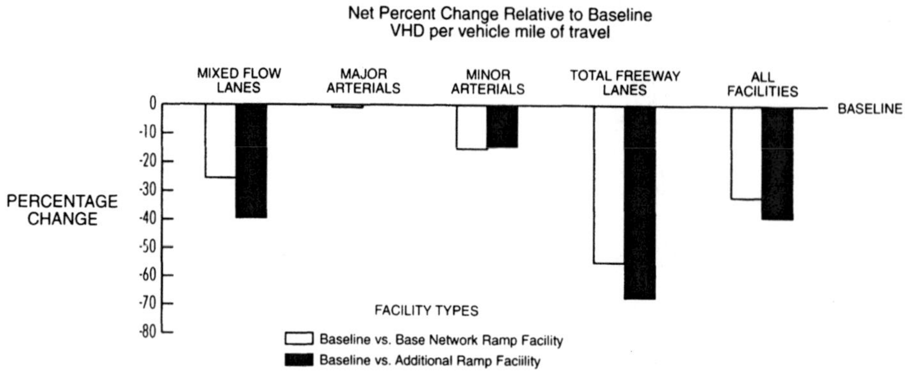

(d)

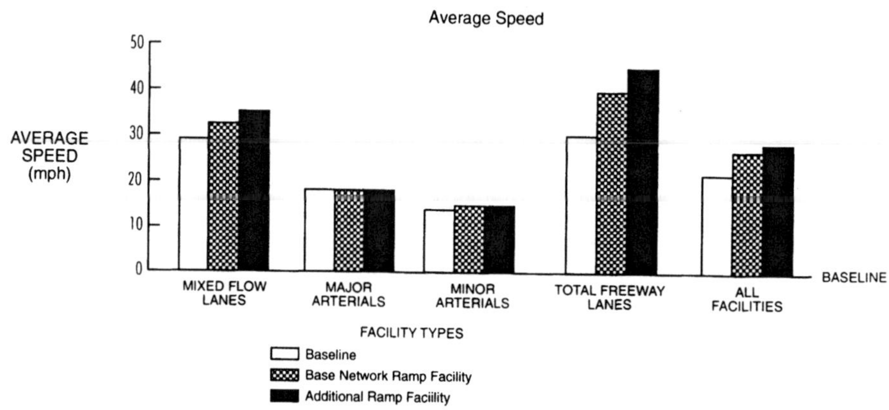

(e)

FIGURE 4. (*continued*)

the VMT in the mixed flow lanes. Figure 3c depicts corresponding changes in average speed between the baseline scenario and each automation scenario.

Figure 4 presents performance measure totals across the entire region for each automation case compared to the baseline. The results indicate that while the volume of freeway traffic increased relative to the baseline, VHD decreased considerably. For example, aggregated over all facilities, for the base ramps and additional ramp facilities cases respectively, total VHD declined by 34 and 40% compared to the baseline; corresponding results for VHD per vehicle mile of travel indicated 33 and 39% delay reductions, respectively. Figure 4e depicts average speeds for the baseline and both automation scenarios. Congestion mitigation on arterials suggested that vehicle trips were drawn to the automated freeway lane(s) as well as mixed flow lanes from the arterials when automated trips enabled freeway mobility conditions to improve. For example, major and minor arterials showed an average percentage VHD reduction of 23 and 28%, respectively, for the base ramp applica-

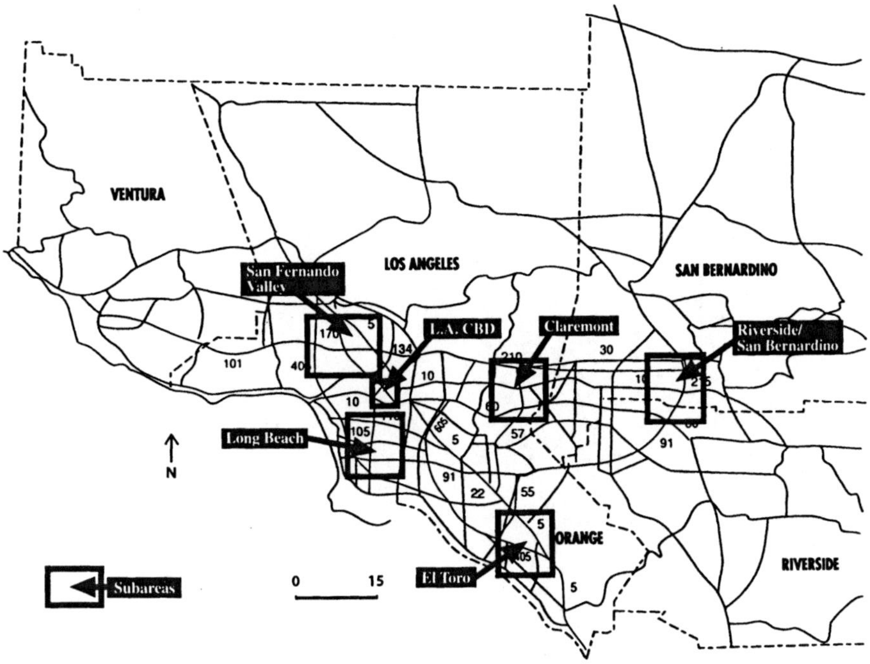

FIGURE 5. 2025 regional highway network.

tion compared to the baseline, and 24 and 28% for the additional ramp facilities case relative to the baseline (Figure 4c). Corresponding percentage changes for VHD per vehicle mile of travel may be found in Figure 4d. A more detailed analysis of arterial congestion reduction is discussed in Section 5.

5. SUBAREA ASSESSMENT

In order to obtain a more detailed understanding of the interactions between the automated freeway system and the remainder of the transportation system, six geographically diverse subareas of the SCAG region were selected for closer study. Figure 5 shows the chosen subareas, which covered approximately 100 square miles each, except for the Los Angeles central business district (CBD), which spanned 25 square miles. The six subareas were Claremont, El Toro, Los Angeles CBD, Long Beach, Riverside/San Bernardino, and the San Fernando Valley.

Analysis of the mobility results began with compilation of the V/C ratios for each arterial link located within approximately 1 mile of an automated facility. For each subarea, the link V/C ratio cumulative frequency distribution was derived for the baseline, automation base ramps, and automation additional ramp facilities trip assignments.

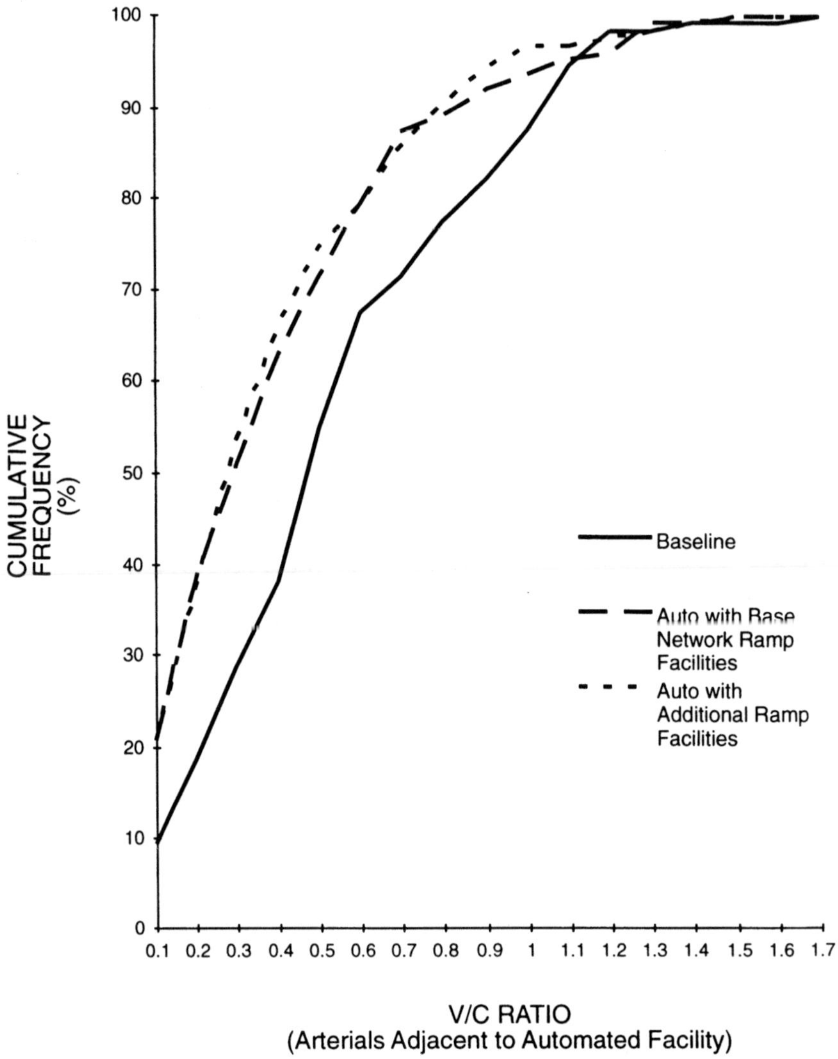

FIGURE 6. Highway automation cumulative V/C ratio frequency distribution for Los Angeles CBD subregion.

Figure 6 reports these findings for the Los Angeles CBD, which was fairly representative of all six subareas. As shown in Figure 6, for any given V/C ratio threshold value, a larger percentage of links had V/C ratios less than this threshold in both automation frequency distributions than the baseline frequency distribution. These results indicated that arterial travel was less congested when automation technology was applied than when it was not. For example, in the baseline case, 87.6% of the arterial link V/C ratios were less than 1.0 compared to 93.7 and 96.8% for the automation base network ramps and automation additional ramp facilities assignments,

TABLE 2
Highway Automation Ramp Traffic Comparisons
(Average Ramp Traffic Volume Percentage Change)

Subarea	Baseline versus base ramps	Baseline versus added ramps
Claremont	+5.3	+9.7
El Toro	+5.5	+13.6
LACBD	+32.8	+46.5
Long Beach	+14.4	+19.9
Riverside/San Bernardino	+20.5	+22.6
San Fernando Valley	+9.6	+13.1

respectively. In general, there was a leftward shift of the automation cases' cumulative arterial link V/C ratio frequency distribution compared to that for the baseline.

Individual link traffic volumes were compiled for all three scenarios for arterial links adjacent to freeways and for freeway on- and off-ramp links. For the arterial link case, the average percentage changes in total VHD, VHD per vehicle mile of travel, and speeds for each automation scenario compared to the baseline for each subarea were computed (Figure 7). For the case of ramp links, the average percentage change in traffic volume for each automation scenario relative to the baseline was derived (Table 2).

The reductions in total VHD on arterials across subareas comparing the baseline to the automation cases ranged from 30% to 75%. The corresponding reductions in VHD per vehicle mile of travel ranged from 20% to 60%, while the increases in speed were from 1% to 8%. These results support the earlier conclusion that arterial travel was less congested when automation technology was applied than when it was not.

The traffic volumes on ramps relative to the baseline case increased by 5–33% for the automation base ramp scenario and by 10–47% for the automation additional ramp facilities scenario. Because the number of ramps available to all vehicles in both the automation base ramp scenario and the baseline case were the same, these results suggest that freeway ramps became more congested when automation technology was applied than when it was not. For the additional ramp facilities scenario, even though traffic volume increased, the separate set of ramps for automated vehicles suggests that congestion is reduced compared to the baseline. Ramp facilities doubled in number, whereas traffic volume increased at a much smaller rate (Table 2). Modeling restrictions precluded an accurate estimation of ramp delays to verify the conjecture about congestion reduction.

6. CONCLUSIONS

This study has provided some initial answers to some of the major questions about the impacts that large-scale AHS implementation could have on transporta-

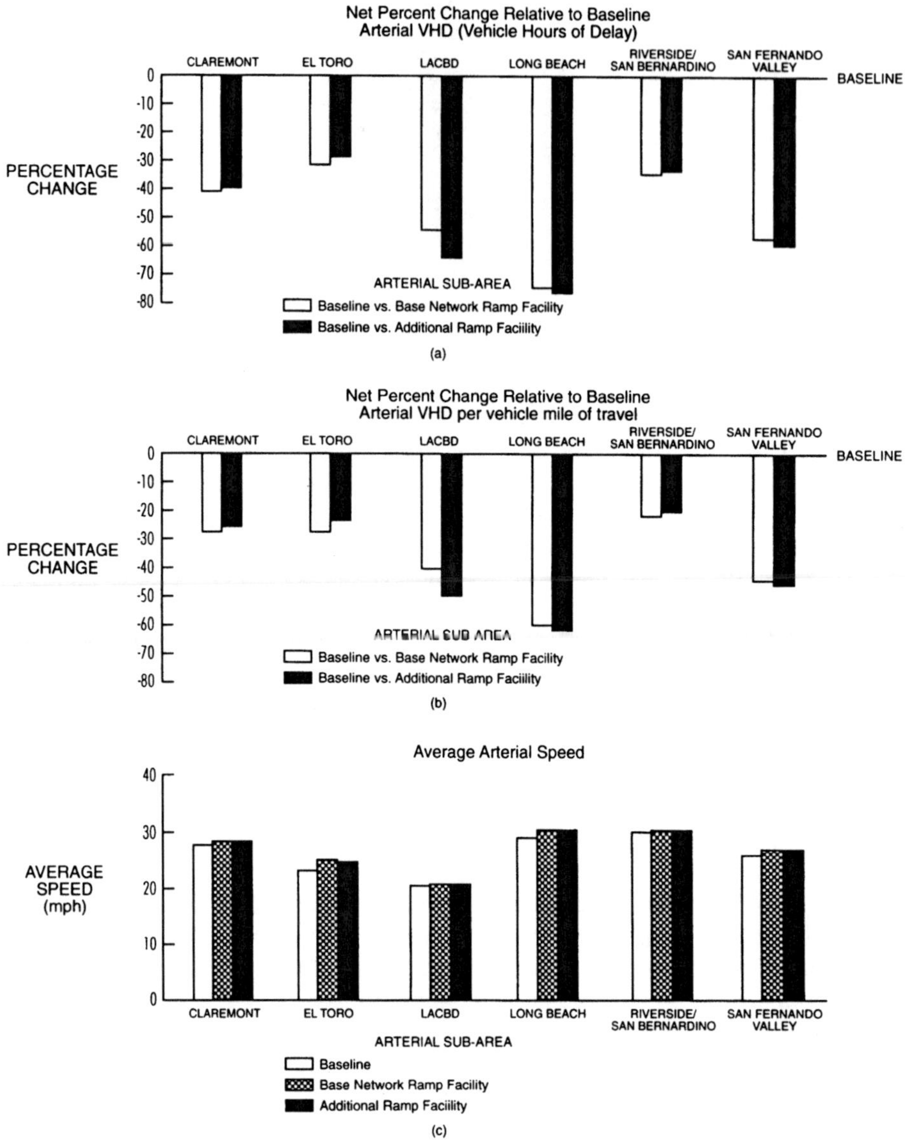

FIGURE 7. Highway automation arterial mobility comparisons.

tion service in a major metropolitan area. Given the paucity of hard data and the limitations of the modeling tools that are available today, the results should be viewed as only roughly quantitative rather than definitive. Nevertheless, these results shed important light on the general scale of transportation benefits that could be gained from AHS.

In the absence of a mode split model that could provide a realistic representation of the mode-specific attributes of the demand for AHS service, the mode split had to be treated exogenously. Similarly, in the absence of a model of transportation–land use interactions for the region, the effects of latent and induced demand could not be incorporated. These would tend to reduce the magnitude of the transportation improvements predicted by the model that was used. However, it should be noted that the additional trips would confer mobility (or accessibility) benefits on the travelers that would more or less balance the loss of VHD savings that could not be represented here.

Reviewing the answers that have been developed to the questions that were posed in the Introduction:

1. Within the Los Angeles region, an AHS network representing about 20% of the regional freeway lane miles should be able to produce dramatic transportation improvements.
2. This AHS network was assumed to be made available by converting some of the existing freeway infrastructure for AHS operation, *not* by construction of any additional roadway. This is assumed to rely in part on the potential for automatic steering control to enable use of narrower lanes than on the present-day freeways. Some additional construction is likely to be needed for access and egress ramps and barriers to segregate automated traffic from manual.
3. The level of congestion on the remaining (manually operated) freeway lanes was still reduced significantly, representing 23 and 47.5% reductions in freeway VHD from the baseline case for the two AHS scenarios, and 16.5 and 35.5% reductions in freeway VHD normalized for the volume of traffic remaining in the manual lanes.
4. Not only is the delay experienced by manually driven vehicles on the freeway reduced, but the arterial traffic delays are also reduced by 23% to 28%, essentially because of the diversion of substantial traffic from the arterials to the high-capacity AHS lanes. The arterial speeds are not substantially improved by this, but the arterial volumes are reduced. Both of these results indicate that very substantial benefits are gained by those who choose not to (or cannot afford to) use the AHS. This result casts serious doubt on the accusation that AHS is an elitist system that benefits the wealthy at the expense of the rest of the population.
5. The comparison between the two AHS scenarios that were studied indicates the desirability of increasing the access ramp capacity to accommodate the increased AHS travel, particularly by use of additional ramps that could provide access directly to the automated lanes without requiring AHS travelers to merge across all of the lanes of manual traffic. Considerable improvements in most of the measures of effectiveness are achieved by this, although it should be noted that even with the ramp delays observed in the absence of additional ramp capacity, the overall system improvements remain very substantial.

Much research is still needed so that the data and modeling tools can be made available to produce higher-fidelity estimates of the impacts that AHS can have. These impacts estimates will need to include not only refinements of the preliminary mobility estimates reported here, but also estimates of the safety, environmental, energy consumption, and land use impacts of AHS. The available knowledge base needs to be expanded considerably in order that the AHS technology development and city and regional planning and policy development can move forward in harmony rather than discord. Research in this area has begun within the context of the work of the NAHSC.

ACKNOWLEDGMENTS. The authors gratefully acknowledge funding provided by the Federal Highway Administration, the California Department of Transportation (Caltrans) through the California PATH Program of the University of California, and the Southern California Association of Governments (SCAG), and technical and administrative assistance provided by William Boyd, Teresa Wang, and Hong Kim of SCAG, Dan Empey and Constance Klein of Systems Control Technology, Inc., the members of the project's Project Advisory Group, Robert Parsons of Parsons Transportation Associates, Charles Price and Roy Bushey of Caltrans, and Professor Brett Trueman of the University of California, Berkeley.

The contents of this report reflect the views of the authors who are responsible for the facts and the accuracy of the data presented herein. This report does not constitute a standard, specification, or regulation.

REFERENCES

1. P. BENJAMIN, Analysis of Dual Mode Systems in an Urban Area. Volume I: Summary, DOT-TSC-OST-73-16A, I (1973).
2. J. G. BENDER, L. S. BONDERSON, R. E. SCHMELZ, J. F. THOMPSON, T. R. BENYO, D. MILLER, AND D. STUART, Systems Studies of Automated Highway Systems, Final Report to FHWA (1981).
3. J. G. BENDER, An Overview of Systems Studies of Automated Highway Systems, *IEEE Trans. Veh. Technol.* **40**(1) (1991).
4. R. A. JOHNSTON AND D. L. PAGE, A Preliminary Systems-Level Evaluation of Automated Urban Freeways, *Proc. Second International Conference on Applications of Advanced Technologies in Transportation Engineering,* Minneapolis, MN, 1991.
5. R. A. JOHNSTON AND R. CEERLA, A Continuing Systems-Level Evaluation of Automated Urban Freeways, UCB-ITS-PRR-93-15, University of California, Berkeley (1993).
6. R. A. JOHNSTON AND R. CEERLA, A Continuing Systems-Level Evaluation of Automated Urban Freeways, *J. Transp. Eng.* **120**(6) (1994).
7. Calspan Corporation, *AHS Roadway Analysis,* Vol. 3, FHWA-RD-95-121 (1995).
8. Southern California Association of Governments, Regional Mobility Plan, Los Angeles (1989).
9. Southern California Association of Governments and Partners for Advanced Transit and Highways, Highway Electrification and Automation Technologies—Regional Impacts Analysis Project, Phase II: Scenario for Advanced Highway Technologies, UCB-ITS-PRR-93-20, Los Angeles (1993).
10. A. BRESNOCK, M. A. MILLER, E. H. LECHNER, AND S. E. SHLADOVER, Highway Automation: System Modeling for Impacts Analysis, *Proc. Second International Conference on Vehicle Navigation and Information Systems,* Dearborn, MI, 1991.

11. Southern California Association of Governments and Partners for Advanced Transit and Highways, Highway Electrification and Automation Technologies—Regional Impacts Analysis Project, Phase III—Appendices: Mobility Statistics, UCB-ITS-PRR-93-22, Los Angeles (1993).
12. Southern California Association of Governments, Highway Electrification and Automation Technologies—Regional Impacts Analysis Project, Phase III: Impacts Analysis Results, UCB-ITS-PRR-93-21, Los Angeles (1993).

CHAPTER 15

Institutional and Societal Issues Associated with Automated Highway Systems

An Environmental Perspective

Mark A. Miller

1. INTRODUCTION

Automated Highway Systems (AHS) have the potential for providing considerable improvements over current roadway conditions in the areas of vehicle throughput, traffic safety, user comfort, as well as numerous other areas of significance. At the present time, however, issues exist about numerous aspects of AHS deployment and operation as well as the risks involved if these issues are not resolved. AHS research and development is still at a stage that precludes the ability to completely answer such concerns now. The success of AHS, that is, the realization of its potential in all areas of importance, begins with an inquiry into and the eventual resolution of these issues and concerns.

Issues may be classified as technical, nontechnical, or a combination of these two types. The nontechnical issues generally fall under the heading of institutional

Mark A. Miller • California PATH Program, University of California, Berkeley, California 94804.

Automated Highway Systems, edited by Ioannou. Plenum Press, New York, 1997.

and societal concerns including, for example, privacy, liability, equity, funding, and environmental issues. In fact, environmental issues straddle both the technical and nontechnical domains. Environmental issues have generated a considerable level of intense discussion and debate, and may, in fact, pose overall a greater challenge to resolve than the more technical issues. In 1993, the Federal Highway Administration (FHWA) initiated a series of 1-year studies under its Precursor Systems Analyses (PSA) of Automated Highway Systems to promote research in this and numerous other AHS-related areas.[1] The research results reported in this chapter are an outgrowth of the PSA work that the California PATH Program at the University of California at Berkeley performed in the area of institutional and societal issues, with a particular emphasis on environmental concerns. The work performed by the California PATH Program was under a subcontract to Delco Systems Operations under Purchase Order 67204 to the University of California at Berkeley. This research project concluded in November 1994 and the results were presented at the 5th Annual Meeting of the Intelligent Transportation Society of America held in Washington, D. C. in March 1995 and have been published in its Proceedings.

Since the conclusion of the PSA set of research projects, the next phase of AHS work has begun within the collaborative framework of the National Automated Highway System Consortium (NAHSC) consisting of organizations representing the automotive, vehicle electronics, aerospace, and transportation infrastructure development industries, as well as a state department of transportation and academia; the California PATH Program at the University of California at Berkeley is a member of the NAHSC and the author is currently working on several NAHSC- related projects. The NAHSC has been established to (1) analyze alternative AHS concepts, (2) demonstrate the feasibility of AHS in 1997, (3) develop and test a prototype AHS, and (4) provide specification for a future operational development.

The phrase "environmental issues" in the context of AHS is open to varying interpretations, ranging from the more restricted view of air quality impacts and vehicle fuel use only, to a broader picture that goes beyond these two characteristics to include certain attributes of urban form or quality of life relative to the local or regional roadway transportation system, such as (1) the degree of dependence on the automobile for mobility, (2) community or neighborhood cohesion, and (3) level of induced travel demand. The broader perspective was adopted here to derive as comprehensive an examination of such issues vis-à-vis AHS as possible.

The research into environmental issues associated with AHS consisted of the identification and analysis of potential issues that have emerged or are likely to emerge and recommendations for an approach for their resolution. Such recommendations would include action items that could be taken to resolve a problem directly or further research that needs to be conducted to increase understanding to the level needed to produce a well-supported recommendation for direct action. This chapter consists of a discussion of the PSA research project results supplemented with the current status of work on these issues to reflect NAHSC-conducted work that began after the conclusion of the PSA studies.

2. SOURCES OF INFORMATION

A significant feature of environmental issues that could help explain part of the intensity of debate surrounding the subject and the barriers to resolving these issues is the vast array of interest in this subject from groups of individuals from diverse backgrounds with frequently conflicting interests in this subject. Such interests are probably more diverse than for purely technical problems which at least give the perception of a higher degree of "cut-and-dried" quality about them and thus more readily attained and less controversial solutions. Such groups include (1) governmental organizations (local and regional departments of transportation, metropolitan planning organizations, councils of governments, and transit agencies), (2) professionals in engineering, planning, economics, and related fields (transportation engineers and planners, urban planners, land use planners, academic researchers), (3) other professionals (public policy analysts, administrators, lawyers, and politicians), (4) interest groups, primarily the environmental community, (5) the media, and (6) the general public, both traveling and nontraveling. A complete analysis and resolution of environmental issues associated with AHS should draw on all of these groups for input to provide analyses, both quantitative and qualitative, from different perspectives.

The sources of information for this research were principally governmental organizations (MPOs), professionals in engineering, economics, and planning, and members of the environmental community. Data were gathered by means of a focus group and through individual interviews.

A focus group was conducted prior to an Institute of Transportation Engineers meeting in the San Diego, California, area in March 1994. The meeting was a collaborative effort between George Mason University (George Mason University was a participant in the PSA set of research projects under a subcontract to BDM, Federal Inc. Dr. Tom Horan, then of the Institute of Public Policy at George Mason University, is gratefully acknowledged for his planning and facilitating of the focus group) and the California PATH Program. The overall purpose of the focus group was to investigate user acceptance issues of AHS, with an emphasis on environmental issues. The focus group consisted of 15 professionals attending the conference from throughout the United States in the transportation engineering, urban planning, transportation planning, and land use planning areas as well as members of the environmental community. The conference itself had an environmental theme to it, and a short list of candidate focus group participants was culled from the full list of conference registrants, with the aim of achieving at least a modicum of geographical and professional diversity among the participants. Initial introductory telephone discussions were held to ascertain the interest level in focus group participation. It was felt that a group of no more than approximately 20 individuals would be the most suitable and appropriate to use for this focus group setting. Each participant received prior to the meeting a brief written introduction to AHS that also described the AHS federal program under the direction of the FHWA. Prior to the beginning of the focus group session, additional material on AHS was presented that described possible AHS evolutionary deployment and operational scenarios.

While the focus group was very structured in format, the individual interviews were conducted in a more open, less structured way. Ten interviews were conducted, for which each participant received in advance the same write-up on AHS that was distributed to the focus group participants. A few specific questions were asked of the interviewees in combination with a rather structure-free format of receiving input from them.

Both the focus group participants and the individual interviewees possessed different levels of knowledge of AHS, differing in accuracy and sophistication. The purpose of providing AHS-related information to them was to at least ensure that all participants had a common information base with which to form opinions and provide input, while obviously not their entire knowledge base. The participants' different levels of accurate information on AHS and how it may lead to reinforcing certain beliefs and attitudes about AHS and the federal AHS program is an issue that will be discussed in detail in the following section.

3. ISSUES ANALYSIS AND RECOMMENDATIONS

A review of the issues raised during the focus group and the individual interviews revealed three common discussion themes under which these issues were subsequently grouped. These three categories are: travel-related issues, infrastructure and urban form, and institutional, from which numerous other issues were derived. This divisional structure serves to facilitate the process of understanding and integrating the issues on a macrolevel.

3.1. *Travel-Related Issues*

The travel-related issues stem from concerns over the consequences of AHS implementation and operation on how much more travel is made, by what means, and its secondary impacts such as vehicle emissions and fuel usage. The issues presented below have linkages and thus are not mutually exclusive.

The current transportation paradigm or model of "how things are done" or "the way the urban transportation system works" may be described succinctly as people driving alone in their cars on vast networks of urban freeways that over time have led to urban sprawl. The issue is that AHS, as currently envisioned, would only encourage the *continuation of this same type of "business as usual" behavior*, i.e., more driving, more single-occupancy-vehicle (SOV) driving, and more sprawl. AHS, it is feared, would also emphasize the further development of highways and make only SOV driving more attractive by increasing its convenience and comfort at the expense of other modes of travel, such as public transit and high-occupancy-vehicle (HOV) driving. In the opinion of the interviewees, a considerable amount of resources have already been invested in public transit and HOV travel modes and this investment should be continued and even strengthened, not reduced or abandoned.

The encouragement of more SOV driving could then mean an *increase in trips, trip length, and volume of drivers,* even above what such increases might be over time without AHS. These effects could be the result unless there were in place strong measures to counteract them, such as (1) transportation demand management, (2) congestion pricing, (3) parking pricing, and (4) land use planning and management. The AHS program has not been viewed as having mitigating measures such as these or others at a central and important place in its research and development effort.

The encouragement of more SOV driving could also mean an emphasis on a *single mode of travel,* when the emphasis should be placed on the multimodal nature of the transportation system. AHS research and development work should not exclusively focus on the automated mainline portion of a trip, but should consider more holistically the entire trip from origin to destination, which could involve the use of more than one mode of travel.

Associated with the potential encouragement of longer trips with AHS is the concern over the *decrease in use of other travel modes used primarily for shorter trips,* such as walking and bicycling, possibly leading to a degradation of what is referred to in the environmental community as the short-trip network or infrastructure in local neighborhoods.

The concern that AHS would encourage and eventually lead to an increase in driving is also referred to as the induced or latent demand effect. Recall that induced demand may be moderated by the advent of measures such as transportation demand management or land use planning and management. Associated with induced demand is a concern for the potential increase in *vehicle emissions*. Several variables, however, play a role in the determination of the actual net impact on emissions. Understandably and expectedly, concerns over numerous possible negative emissions impacts dominated the views and opinions gathered during the focus group and individual interviews. Awareness and expression of the potential emissions benefits were, in general, minimal.

Emissions may be affected through changes in the (1) operation of the vehicle, (2) number of drivers, (3) volume of trips, and (4) trip length. The most prevalent environmental concerns expressed about AHS involved the potential for leading to increases in volume of drivers, trips, and trip length, which on an aggregate basis in terms of total emissions tonnage, would mean an emissions increase. Moreover, certain pollutants, such as, oxides of nitrogen, tend to increase with increases in speed. Thus, with more freely flowing traffic and greater speeds associated with AHS, amounts of such pollutants could increase. Also, if vehicles accessing or egressing the AHS develop into lengthy queues, buildups of additional emissions could result at the on- and off-ramps.

These concerns notwithstanding, AHS research and development will proceed along with other technological advances in areas such as emission control technologies, clean fuels, electric vehicles, and other areas that could have the effect of reducing emissions on a per mile basis. Moreover, AHS have the potential for smoothing out the flow of traffic, removing or at least reducing stop-and-go, idling, and sharp acceleration and deceleration driving modes which are known to contribute to vehicle emissions. Moreover, in the context of automated vehicles traveling with much smaller headways than presently possible, that is, in platoons, preliminary research has indicated there are emission reductions for all vehicles, including the lead vehicle.[2] The true net effect on vehicle emissions is unknown at the present time.

Also associated with induced demand is a concern for the potential increase in *vehicle fuel usage*. However, as in the case for vehicle emissions, several variables play a role in the determination of the actual net impact on fuel usage. On the negative side, AHS may have the impact of leading to an increase in driving, which on an aggregate basis of total fuel consumed would mean a fuel usage increase. Moreover, fuel economy is a function of speed, and increases in speed associated with AHS could lead to increases in fuel consumption. Also, if vehicles accessing or egressing the AHS develop into lengthy queues, vehicle fuel consumption would be affected.

However, AHS research and development will proceed along with other technological advances, in particular in areas such as vehicle fuel economy in addition to possible increases in the national corporate average fuel economy (CAFE) standards that would have the effect of reducing fuel consumption on a per mile basis. In addition, AHS have the potential for smoothing out the flow of traffic, as previously described, which could reduce vehicle fuel usage. As in the case for vehicle emissions, in the context of automated vehicles traveling with much smaller headways than presently possible, preliminary research has indicated there are fuel efficiency increases for all vehicles.[3] The net effect on fuel usage is unknown at the present time. Another travel-related issue resulting from possible increases in traffic volume and travel speeds would be an increase in *noise levels* at and adjacent to the AHS facility. With a potential increase in hourly lane capacity for automated lanes by a factor of possibly two or more over conventional nonautomated lanes, actual noise increases are a possibility.

3.2. Recommendations for Resolution of Travel-Related Issues

Recommendations to help alleviate the travel-related issues consist primarily of further research studies into the following areas that incorporate tools such as analytical techniques, simulation modeling, and empirical data:

- Develop modeling tools and the supporting data to more accurately represent the automated highway driving mode so that high-fidelity estimates of the impacts of AHS may be produced, in the areas of travel volume changes, mobility, land use impacts, emissions, energy consumption, and noise levels.
- Develop applications of AHS in the context of multimodal transportation systems, such as public transit and commercial vehicle operations.
- Identify AHS-related technology applications with potential environmental benefits, such as diagnostics on AHS check-in to determine vehicle emissions and fuel economy profile.
- Identify advanced technology research areas with potential AHS applications and investigate methods of developing beneficial linkages and synergistic effects, such as through emission control technologies, electric vehicles, and clean fuels.

- Investigate the feasibility of linking more closely with AHS such measures as travel demand management, land use management, congestion pricing, and parking pricing.

The NAHSC began its work after the conclusion of the PSA studies. An important component of this work is in the area of model and tool development for AHS concept design and evaluation purposes. Such modeling tools will eventually be used to investigate the impacts of AHS once specific AHS concepts, designs, and operational scenarios have been produced. The impact evaluation will be conducted relative to numerous measures of performance, including mobility, safety, land use, emissions, fuel usage, and noise levels. As the NAHSC has been in existence for a little more than a year and a half, development of these modeling tools has not been completed.

The NAHSC has developed specific system objectives and characteristics with stakeholder input.(4) Included in these objectives and characteristics is the specific application of AHS to (1) support automated transit operations, (2) enhance operations for freight carriers, and (3) support travel demand management and sustainable transportation policies. The current state of this work is the implementation of work plans with which to achieve these objectives.

3.3. *Infrastructure and Urban Form Issues*

There is the possibility that in certain very dense urban areas where AHS may be deployed, there could be additional infrastructure requirements either with the construction of new AHS facilities in entirely new rights-of-way (ROW), or within the same ROW a lateral expansion to accommodate additional lanes, median space, or shoulder space, or the construction of a new elevated facility if there is no space to build "out," only to build "up." This ROW expansion requires infrastructure changes that could have negative *visual impacts* as well as possible impacts on the *neighborhood cohesiveness*.

With the ability of an AHS facility to accommodate a sizable increase in the hourly lane capacity on the roadway, the concern is that there will be a negative impact on the local neighborhood, in the immediate vicinity of the AHS facility and entry and exit points. This could result from trying to accommodate the increase in vehicle access and egress in terms of the possible need for additional parking facilities, the capacity of nonautomated roadways adjacent to the AHS facility, and especially during an incident when traffic is diverted off the roadway onto surface streets.

Another potential concern, not unique to AHS, but possibly exacerbated with AHS are the *seismic safety issues* related to new construction of elevated structures for AHS use.

Another concern expressed with the deployment and operation of AHS combines environmental with equity issues. Areas in the neighboring vicinity of freeways are generally in less-well-to-do sections of a city relative to other areas. Freeway infrastructure modifications for AHS, additions in particular, could further exacerbate already existing negative impacts of living near a freeway for the residents in such areas. These people are probably less likely to be able to afford AHS-equipped automobiles than other more affluent individuals, and so the negative impacts of freeway modifications add to the following frequently stated AHS equity issue: Residents of poorer areas cannot afford to own such a vehicle, thus do not receive benefits since they would not be users of the system, and hence they must also absorb the consequences of infrastructure changes in their neighborhood. While the infrastructure modification issues are of concern and need to be addressed, the equity issue must be more thoroughly analyzed. The underlying assumptions of these criticisms and concerns are that AHS is primarily for SOVs and that only users of the AHS would derive benefits from it. While much more work is required to quantify the net impacts of AHS, early research results have shown that system nonusers can also derive benefits from AHS, i.e., travel-time benefits for users of non-AHS freeway lanes as well as travelers on arterials.[(5)] The modeling tools under development will be able to evaluate the impacts of AHS operations on non-AHS system users.

3.4. *Recommendations for Resolution of Infrastructure/Urban Form Issues*

Infrastructure modifications and impacts on the urban form of local neighborhoods resulting from the deployment and operation of AHS would have similarities to the implementation of highway projects today. Any project must successfully satisfy a multitude of requirements, including, in particular, a lengthy environmental impacts review process. This process should be reviewed to determine its applicability to the AHS case and plans made to develop modifications where needed. Measures to mitigate potentially negative consequences of AHS such as visual impacts or excessive traffic egress onto roadways adjacent to the AHS facility would be developed at the local level. For example, it may be mandated that lighter-weight material be used or methods be found to reduce the spatial requirements of elevated structures, including entry and exit areas. To accommodate increases in traffic volume resulting from access to and egress from an AHS facility onto nonautomated roadways could require street widenings, modifications to signal timing plans, or other measures.

Based on the development of AHS concepts, AHS designs, application scenarios, modeling tools, and performance measures, evaluations must be performed to ascertain the true local impacts, in and around AHS entry/exit points as well as in the surrounding vicinity of the AHS facility itself.

3.5. *Institutional Issues*

Institutional issues are primarily concerned with the relationships among the stakeholders in AHS work (research, development, deployment, and operation), the level of accurate information stakeholders have about each other, and barriers to making progress in the work. Numerous groups from a wide range of professional disciplines are involved in AHS work. This involvement varies from the level of spectator to more active participant. Even with the wide range of professional groups involved, there appears to be a division into two major groupings, based on differences in educational background, professional training, work experience, one's view of the world, and approaches to problem-solving. The barriers, unfortunately, help to maintain an "us versus them" outlook. One group consists primarily of the engineering fields, while the other group consists mainly of planners (urban, transportation, land use), economists, and members of the environmental community, who actually draw members from all such disciplines, including the engineering fields. Each group holds views spanning the range from extreme to moderate. There are areas of common interests that provide opportunities for bridging the gaps and achieving consensus. It is not the case that all engineers do not care about environmental issues or that the environmental community does not see the benefits of at least portions of AHS operation, in particular, the safety benefits. While the existence of barriers is real, the perception of the extent of the differences are quite large and make the barriers harder to resolve.

Another issue is the *lack of complete and accurate information and existence of misinformation among the stakeholders*. Some members of the nonengineering fields do not yet have a full and accurate appreciation of the AHS program from the microscopic view, either in terms of the set of individual technologies under investigation and their potential beneficial environmental impacts when placed in the context of a particular AHS design and application scenario or in the context of individual user services,[(6)] as well as from the more macroscopic view of user services' interrelationships. This lack of knowledge and complete information was made very evident during the information-gathering phase of this work as was previously discussed relative to travel-related issues. It has been difficult for some individuals to grasp the global picture of AHS research, and more generally Intelligent Transportation Systems research. To some individuals it is still an "alphabet soup" without a unifying purpose and objectives, while to others it does have an objective that simply runs contrary to their interests.

Moreover, unfamiliarity and misconceptions on both sides prevent a full and accurate appreciation of each other's standard and traditional methods of analysis and problem-solving, as well as a tendency to view such methods and practices as not completely effective and compatible with solving the other side's problems. This situation results in further distancing among the stakeholders. For example, one nonengineering view is that solutions to engineering problems have not had to account for the human element—uncertainty, randomness, much greater degree of unpredictability—on the scale present in AHS-related problems. Numerous pieces of legislation beyond the Intermodal Surface Transportation Efficiency Act (ISTEA), such as the Americans with Disability Act (ADA), the National Environmental Policy Act (NEPA), and the Clean Air Act and its Amendments (CAAA), have relevance and applicability to the deployment and operation of AHS and are unfamiliar to at least portions of the engineering community. Moreover, some members of the engineering community with a purely technical background do not appreciate the significance of these nontechnical issues and the challenges posed by them and view AHS research from a totally technical perspective.

An institutional issue related to the view that the AHS program is too "highway" oriented, even in its name, is the *need for much more active and central involvement in the program from the nonhighway agencies,* in particular, at the federal level the Environmental Protection Agency (EPA) and the Federal Transit Administration (FTA) who continue to have relatively minor roles in the program, as well as the MPOs at the regional and local level.

3.6. Recommendations for Resolution of Institutional Issues

The following three words are key ingredients to resolving the institutional issues: *education, communication, participation*. To break down barriers and correct misinformation among the major participants in order to forge a more common vision of a future transportation system with AHS as an integral component, the following suggested action items may serve as a first step toward achieving this goal and are offered for consideration:

- Perform systematic review and analysis of recent national legislation to find relevant and applicable pieces of legislation to the AHS program, of which ISTEA, CAAA, NEPA, and ADA are important examples.
- Stakeholder groups should review the full set of final PSA research results that will help alleviate at least some of the concerns previously raised.
- Consider instituting on a regular basis research/policy forums focused on AHS analogous to the following such conferences on ITS and the environment: IVHS Policy: A Workshop on Institutional and Environmental Issues (Monterey, California, 1992), National IVHS and Air Quality Workshop (Diamond Bar, California, 1993), and National Policy Conference on Intelligent Transportation Systems and the Environment (Arlington, Virginia, 1994).
- The current and second phase of AHS research and development should include an active dialogue with all stakeholder groups, especially those with opposing views to develop strategies to help bridge the gaps.
- AHS community should learn of and participate in conferences, forums, committees, special projects, and other related meetings sponsored by planning (transportation, urban, and land use), public transit, and environment-related communities, for example, meetings of Transportation Research Board committees.
- FHWA has begun a set of training courses with the purpose of educating regional planning professionals in the rudiments of the ITS planning process relative to standard regional planning methods.[7] In concert with evaluating the effectiveness of these training courses geared toward ITS in general, the feasibility of tailoring the courses to the needs of AHS should be made and if found to be of potential value, preparations should be made for their implementation.
- Stronger linkages with other transportation-related federal agencies, e.g., FTA and USEPA, need to be made.

As previously indicated, the NAHSC began its work after the conclusion of the PSA studies. The NAHSC has entered into a dialogue with members of several stakeholder groups, including those organizations with opposing views. The purpose of these discussions is to open lines of communication between all interested parties to develop strategies to help bridge the gaps. Moreover, a systematic review and analysis of national legislation to determine linkages with the AHS program has also begun.

4. CONCLUSIONS

An analysis of environmental issues associated with AHS was made to develop as comprehensive an examination as possible. The principal sources of information used in the analysis, individual interviews, and focus group participants in the engineering, planning, economics, and environmental areas allowed for a deep probe into views that might otherwise not come to light.

Environmental issues associated with AHS fell into three major categories: travel-related, infrastructure and urban form, and institutional. Travel-related issues arose from concerns over the consequences of AHS implementation and operation on how much additional travel will be made, by what means, and its secondary impacts on vehicle emissions and fuel usage. The major infrastructure and urban form issues relate to impacts from infrastructure changes resulting from AHS such as visual impacts and seismic safety concerns, as well as the impact on the local neighborhood as a result of potentially substantial increases in vehicle access and egress to and from nonautomated roadways. The institutional issues are centered around the relationships among the participants in AHS research and development. Examples of such issues are the barriers that exist between the two major groups of participants in this research, as well as the lack of complete and accurate information and attitudes that each group believes about the other group.

Primary suggestions for resolving these issues include (1) further research into developing modeling tools to more accurately represent the automated highway driving mode to produce reliable estimates of the impacts in areas of travel volume changes, mobility, land use, emissions, and energy consumption, (2) investigation of current methods for environmental impact review processes for applicability to the AHS case, determining and making necessary modifications, and (3) incorporating an aggressive process of education, communication, and participation to help dissolve the barriers and help forge a more common vision of a future transportation system with AHS as an integral component.

ACKNOWLEDGMENTS. This research was supported by the FHWA of the U. S. Department of Transportation under Contract DTFH61-93-C-00194, and by Delco Systems Operations under Purchase Order 67204 to the University of California at Berkeley. The author would like to especially thank Tom Horan of The Institute of Public Policy of George Mason University for his major contributions to planning and facilitating the focus group that served as a principal source of information for this research. The author would also like to expressly thank the focus group participants and the individual interviewees who freely gave their time and provided very valuable information to furthering AHS research. The author would also like to thank J. Richard Bishop of FHWA, William Stevens, Michelle Brahaney, and Rodney Lay of The MITRE Corporation, Herb Hall and Loren Bondersen of Delco Systems Operations, and Steven Shladover of the California PATH Program for their valuable inputs to this research.

REFERENCES

1. Federal Highway Administration, Precursor Systems Analyses of Automated Highway Systems, Broad Agency Announcement Request For Proposal Number DTFH61-93-R-00047 (Nov. 1992).
2. M. BARTH, Evaluating the Impact of IVHS Technologies on Vehicle Emissions using a Modal Emission Model, *National Policy Conference on Intelligent Transportation Systems and the Environment—Conference Papers,* Hubert H. Humphrey Institute of Public Affairs, University of Minnesota, June 1994.
3. M. ZABAT, Drag Measurements on a Platoon of Vehicles, California PATH Program Research Report UCB-ITS-PRR-93-27, University of California, Berkeley (1994).
4. National Automated Highway System Consortium, Automated Highway System (AHS) System Objectives and Characteristics Final Draft (Oct. 5, 1995).
5. Southern California Association of Governments, California PATH Program, Highway Electrification and Automation Technologies—Regional Impacts Analysis Project: Phase III: Impacts Analysis Results, California PATH Program Research Report UCB-ITS-PRR-93-21 (1993).
6. Intelligent Vehicle-Highway Systems Society of America, National Program Plan for Intelligent Vehicle-Highway Systems (IVHS) (Draft) (May 1994).
7. Federal Highway Administration, Three Day Training Course—Intelligent Vehicle Highway System (IVHS) Planning and Functional Requirements—An Overview, July 1994.

CHAPTER 16

AUTOMATED HIGHWAY SYSTEM DEPLOYMENT

A PRELIMINARY ASSESSMENT OF UNCERTAINTIES

RANDOLPH W. HALL AND H.-S. JACOB TSAO

1. INTRODUCTION

Automated Highway Systems (AHS) have the potential for a large highway capacity gain without requiring significant right-of-way acquisition. However, it is the most technology-intensive component of ITS (Intelligent Transportation Systems), and its deployment could be costly. In this high-yield but high-risk environment, it is particularly important to identify the critical issues, technical or not, that need to be resolved to ensure timely and efficient deployment of AHS.

As one step toward this task, AHS experts were interviewed to elicit their visions for AHS, and to capture their specialized expertise with respect to the most significant risks to AHS deployment. These interviews were supplemented by a series of meetings among systems researchers from PATH (Partners for Advanced Transit and Highways), California Department of Transportation (Caltrans), and

RANDOLPH W. HALL • Department of Industrial and Systems Engineering, University of Southern California, Los Angeles, California 90089-0193. H.-S. JACOB TSAO • California PATH Program, University of California, Berkeley, California 94804.

Automated Highway Systems, edited by Ioannou. Plenum Press, New York, 1997.

Lawrence Livermore National Laboratory, aimed at synthesizing a wide range of issues (as opposed to the interviews which, in each case, concentrated on a few issues in depth). This chapter documents the findings of these meetings.

The list of AHS deployment risks served as an input to a study funded by the Federal Highway Administration on AHS costs and benefits.[1,2] One aim of this study was to assess potential risks associated with AHS deployment, and to identify ways of overcoming these risks. We hope that an improved understanding of these risks will be valuable in prioritizing research efforts in AHS, to ensure that the major uncertainties are resolved as early as possible.

2. AN INFLUENCE DIAGRAM FOR AHS FEASIBILITY

The influence diagram depicted in Figure 1 illustrates relationships between factors that may dictate the "feasibility" of AHS. By feasibility, we do not require that the AHS produce a net societal benefit. Instead, feasibility is defined as: "people use AHS" and "people maintain their automated vehicles (AV)." Without these two events, the AHS could not exist (hence, it would be "infeasible").

For feasibility to occur, other events must occur first: (1) manufacturers must sell vehicles equipped for automation, (2) government (or perhaps the private

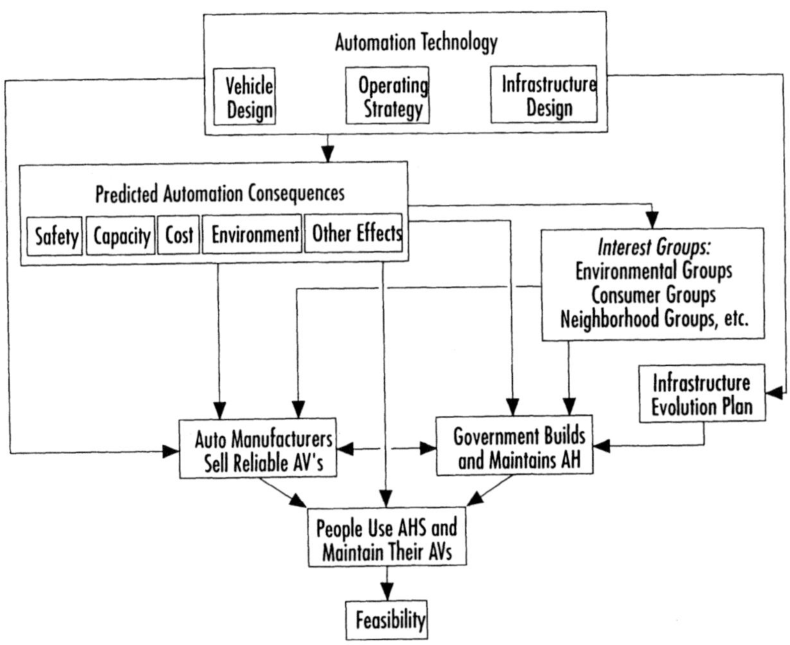

FIGURE 1. AHS feasibility influence diagram.

sector) must build the AHS infrastructure, and (3) users must perceive a net personal benefit from purchasing equipped vehicles. These events, in turn, depend on other factors. Moving up the diagram, government will only build and maintain an AHS if: there is sufficient support from interest groups; automation benefits are sufficient to justify the investment; and there exists a feasible plan for introducing AHS over time. Likewise, manufacturers will not sell equipped vehicles unless various antecedents are satisfied, with respect to consumer demand, interest group opposition, and so on. At the top of the diagram are the enabling technologies, with respect to vehicle technology, infrastructure technology, and operating strategy, which must also be developed to ensure feasibility. The significance of the diagram is that it merges the technical concerns with the nontechnical concerns, illustrating that a serious breakdown at either level may make AHS infeasible.

Based on this diagram, we organized the feasibility issues into seven categories, as follows.

2.1. People Use AHS

Given reliable vehicles and highway infrastructure, individuals must choose to purchase AHS-equipped vehicles, and use these vehicles on AHS equipped roadways. People must also maintain their vehicles properly to ensure continued usage.

2.2. Auto Makers Manufacture Equipped Vehicles

Given a feasible vehicular technology, auto manufacturers must choose to sell equipped vehicles, motivated by adequate return on investment, acceptable liability risk, adequate government investment in infrastructure, and a favorable regulatory environment.

2.3. Government Builds AHS Roadways

Given a feasible infrastructure technology, government must choose to build AHS, based on public and industry support, favorable cost/benefit ratio, acceptable liability risk, and so forth.

2.4. Highway Can Evolve

Given a feasible technology for constructing AHS, it must be possible to build the AHS within existing land uses and rights-of-way, at reasonable cost, and without unacceptable traffic disruption during construction.

2.5. Interest Groups Do Not Obstruct

Given a feasible technology, interest groups must not obstruct AHS, out of fear that it is not a wise investment, or fear that it will harm their constituents.

TABLE 1
Category 1: People Use AHS

Given reliable vehicles and infrastructure, individuals and fleet operators choose to use the AHS (or purchase equipped vehicles), and they choose to maintain their vehicles properly.

Issue No.	Issue	Evaluation 1	2	3
1.1	AHS is perceived as safe under actual operating conditions, accounting for intervehicle spacings and presence of lane barriers (if any).			
1.2	Public does not overly fear/distrust technology, due to • history or media representation of AHS safety • history of reliability of precursor technologies			
1.3	Attitudes toward relinquishing control of vehicle must not be an obstacle.			
1.4	Vehicle purchase and maintenance cost, and any use fees, must be acceptable relative to benefits.			
1.5	Time savings on fully automated AHS should be significant.			
1.6	Maintenance, entrance inspection, transition, etc. should be convenient.			
1.7	Liability and litigation concerns must not be an obstacle.			
1.8	AHS must be attractive relative to conventional modes, both automobile and transit.			
1.9	Ride comfort and vehicle performance must equal or exceed conventional vehicles.			

TABLE 2
Category 2: Auto Makers Manufacture Equipped Vehicles, and Participate in AHS

Given a feasible vehicular technology, auto makers choose to manufacture AHS equipped vehicles.

Issue No.	Issue	Evaluation 1	2	3
2.1	Manufacturing tooling and production costs must be acceptable.			
2.2	Ease and cost of proper maintenance must be acceptable.			
2.3	Market potential, accounting for varying climatic conditions, traffic problems, land availability, etc., must be adequate based on expected benefits, revenues, and costs.			
2.4	Component compatibility must not be a problem, due to lack of standards, or costs of imposing standards on all vehicles.			
2.5	Product liability and litigation must not be an obstacle.			

2.6. *Performance Is Adequate*

Given a feasible technology, the AHS must provide sufficient benefits to justify the costs and impacts, in terms of congestion reduction, safety improvement, and performance improvement, without major negative consequences.

TABLE 3
Category 3: Government Builds AHS Roadways

Given a feasible infrastructure technology, government must choose to build AHS roadways.

Issue No.	Issue	Evaluation 1	2	3
3.1	Construction/conversion technology is feasible for high-priority sites.			
3.2	AHS construction is cost-competitive relative to conventional roadways.			
3.3	Funds are available to extend system at acceptable pace.			
3.4	Liability and litigation are not obstacles.			
3.5	AHS is perceived as fair, relative to economic classes of drivers, relative to affected communities, and relative to regions of country.			
3.6	Enabling laws/traffic code are enacted on time.			
3.7	Roadway maintenance burden is manageable given costs and skill mix.			
3.8	Government expects vehicle industry to provide vehicles.			
3.9	Government perceives that societal benefits exceed costs.			

TABLE 4
Category 4: Highway Can Evolve

Given a feasible technology for constructing AHS, the AHS can be built within existing land uses and right-of-way.

Issue No.	Issue	Evaluation 1	2	3
4.1	Conventional lanes can be retrofitted to automated lanes when needed.			
4.2	New right-of-way can be acquired when needed, either for entry/exit, new routes or lanes.			
4.3	Initial AHS implementations, on small scale, can build momentum for later expansions.			
4.4	Benefits occur initially, even with low usage, and safety concerns do not prevent limited mixing of equipped and unequipped vehicles.			
4.5	Vehicles can be inspected, if necessary, without consuming excessive space or time.			
4.6	Interfaces with conventional roadways do not result in major bottlenecks.			
4.7	Conventional roadways provide sufficient capacity to support access and egress from the AHS.			
4.8	Where necessary, parking at destinations can support traffic volumes.			
4.9	Standards are enacted on time.			

TABLE 5
Category 5: Interest Groups do not Obstruct

Given a feasible technology, interest groups do not oppose AHS because it is not perceived as harmful to their constituents.

Issue No.	Issue	Evaluation		
		1	2	3
5.1	Environmental impacts (if any), with respect to land use, pollution, noise, and aesthetics, are acceptable relative to alternatives.			
5.2	Concerns that AHS might induce new traffic are mitigated.			
5.3	AHS is not perceived as inequitable, with respect to economic classes, neighborhoods, regions of the country, etc.			
5.4	AHS is supported by affected industries, such as aerospace, vehicle manufacturing, construction.			
5.5	AHS is not opposed due to general opposition to "big government," or opposed due to taxes needed to support public infrastructure.			

2.7. Technology Is Feasible

Vehicle/highway automation must be adequate to meet the rigors of full-scale deployment under a full range of operating conditions, including reliability, and maintainability.

3. FEASIBILITY ISSUES

To gain a better understanding of feasibility issues, we interviewed a wide range of specialists, reflecting expertise in many aspects of AHS. A goal of the interviews was to assess which of the barriers to AHS may prove to be the most difficult to surmount, and how they might be approached. The interviews were loosely structured to allow exploration of each specialist's expertise in depth. In these interviews, we emphasized the types of barriers within (6) and (7), because these were their primary areas of expertise. To follow up the interviews, a series of meetings was held among systems researchers from PATH, Caltrans, and Lawrence Livermore National Laboratory, especially within areas (1)–(5).

The interviews and meetings resulted in the list of issues in Tables 1–7, which correspond to the seven categories introduced in the previous section. The issues presented are not of equal priority. Evaluations 1–3 in the tables can represent a degree of urgency, roughly as follows: (1) issues needing attention as part of initial research efforts, (2) issues needing attention as part of development efforts, and (3) issues needing attention as part of deployment efforts. While it might be said that all issues need attention now, clearly some demand earlier attention than others, either because they serve as input to other areas of work, or because their resolution is essential to eventual deployment (i.e., unless a solution is found, AHS cannot

TABLE 6
Category 6: Performance is Adequate

Given a feasible technology, the AHS provides sufficient benefits to justify the costs and any negative impacts.

Issue No.	Issue	Evaluation 1	2	3
Subcategory A: Capacity gain				
6.1	There must be sufficient need for AHS capacity gains, based on traffic generation rates, O-D patterns, and trip lengths (i.e., capacity gains must translate into travel time savings).			
6.2	Close headway driving must be achievable in field operating conditions, providing stable traffic flow in high volume.			
6.3	Vehicle or system failures must not disrupt traffic and make travel times unreliable to an unacceptable extent.			
6.4	Collisions must not result in unacceptable delays, and the system must have the means for quickly removing accidents.			
6.5	Automated highway-to-highway interchanges should not prove to be excessively expensive, in which case conventional highway-to-highway interchanges could become bottlenecks.			
6.6	Origin–destination patterns and trip length distributions should be consistent with AHS design, if exits and entrances need to be spaced far apart.			
6.7	Vehicles must be able to exit at desired locations with high probability.			
6.8	In mixed conventional/automated highways, traffic must be capable of accessing the automated lanes through the manual traffic, and transition lanes, in sufficient volume to support the capacity.			
6.9	Access roadways and exit/entrance ramps must have sufficient capacity to support the AHS.			
6.10	Parking must be available to support AHS traffic volumes.			
6.11	Capacity gains in automated lanes must significantly exceed capacity losses (if any) in nonautomated lanes.			
6.12	If barriers are needed to separate lanes, they must not reduce capacity excessively, due to reducing capacity for lane changes, or due to obstructing accident clearance.			
Subcategory B: Comfort				
6.13	AHS provides a smooth ride in a range of operating conditions (e.g., no jerk, fast deceleration, etc.).			
6.14	People find the experience of automated driving comforTABLE and relaxing, including the transitions between manual and automated driving.			
6.15	Driving in AHS must give the appearance of safety, by providing adequate cues to travelers, and providing a pleasurable driving experience.			

(*continued*)

TABLE 6 (*continued*)

Issue No.	Issue	Evaluation		
		1	2	3
Subcategory C1: Safety (vehicles)				
6.16	Vehicle must be sufficiently reliable and fault tolerant that hazardous failures occur rarely, if ever.			
6.17	There must be assurance that safety critical aspects of vehicles are adequately maintained.			
6.18	Control software must be reliable under all conceivable operating conditions.			
Subcategory C2: Safety (roadside components)				
6.19	Roadside components must be inspectable, especially in remote locations and under adverse conditions.			
6.20	Roadside components must be sufficiently reliable, maintainable, and fault tolerant that hazardous failures occur rarely, if ever.			
Subcategory D1: Safety (human factors)				
6.21	Unequipped vehicles must be prevented from entering automated lanes (intentional or not) and causing collisions.			
6.22	The system must be secure from sabotage, both physically (e.g., vandalism on the roadway), and in its control systems (e.g., software viruses).			
6.23	Drivers must be capable of safely and reliably resuming manual control on exiting the AHS.			
Subcategory D2: Safety (after a collision)				
6.24	Collisions, if they occur, must not lead to catastrophic chain reactions.			
6.25	Emergency equipment must be capable of reaching the scene of an accident, even if lanes are separated by barriers. People must not be endangered in the aftermath of a collision, due to inability to exit to a safe location.			

exist). Through future research, it is our goal to examine the significance of each issue with respect to the above scheme, and to identify ways for resolving the issues.

The process was interesting in several respects. Perhaps the most striking aspect was that most researchers felt that the technological challenges within their own domain would not be insurmountable. The most critical issues, in their views, typically fell outside their domain, or were otherwise related to overall system design and institutional issues. In addition, it was apparent that the concept of the overall system varied from person to person. All of this points to a need for a more integrated vision of AHS, within which research efforts could be coordinated.

TABLE 7
Category 7: Technology is Feasible

Vehicle/highway automation is adequate to meet the rigors of full-scale deployment.

Issue No.	Issue	Evaluation		
		1	2	3
Subcategory A1: Control algorithms and actuators				
7.1	Control algorithms must be responsive to variations in external state (grade, curve, pavement friction, road wear/conditions, climate), especially with respect to uneven pavement friction and wind gusts.			
7.2	Control algorithms must be responsive to variations in own state (tires, brake wear, weight, engine performance).			
7.3	Control algorithms must be responsive to fault conditions, especially tire blow outs, and intrusions of objects or people into path of travel.			
7.4	Brake, engine and steering actuators must be reliable and safe under a wide range of operating conditions.			
Subcategory A2: Communication systems				
7.5	Assurance that interference does not destroy, corrupt or unduly delay messages.			
7.6	Accurate identification of surrounding vehicles.			
7.7	Assurance that messages are sent to/from the correct vehicles.			
7.8	Adequate speed to support control algorithms.			
7.9	Sufficient spectrum to support data traffic/safety critical messages.			
Subcategory B1: Sensors must provide				
7.10	Ability to determine vehicle's own position.			
7.11	Ability to see sufficiently far ahead within a lane.			
7.12	Ability to see sufficiently far ahead, to the side and behind in adjacent lanes (as well as an ability to distinguish lanes from each other).			
7.13	Ability to sense a vehicle's own state, and an ability to sense the external state.			
Subcategory B2: Others				
7.14	Inspection must be adequate and feasible, given space, capacity and cost constraints.			
7.15	Failures of safety critical systems must be detectable instantly (if not before-hand).			
7.16	System must be capable of responding to failures of safety critical system.			

4. CONCLUSIONS

To date, much of the research on AHS has centered on developing the enabling technologies for AHS components, and to some degree developing operating concepts and assessing impacts. These emphases are clearly appropriate for the initial stages of AHS development because they provide the foundation from which AHS feasibility will eventually be determined.

Quite independent of technological feasibility, critical issues lie ahead. It may be possible to determine in the near future whether AHS can be built on existing rights-of-way. If not, this may mean construction of facilities dedicated to automated vehicles. This may in turn affect the total design of the AHS, down to the level of sensors, actuators, and communication. Overall, this study was intended to identify these issues, and provide direction toward overcoming potential barriers to AHS.

ACKNOWLEDGMENTS. We would like to thank Rokaya Al-Ayat, Mohamed Al-Kadri, Tony Hitchcock, Bruce Hongola, Hong Lo, Mark Miller, Bin Ran, and Bobby Rao for their active participation in the systems discussions. Kwang-Soo Chang, Michael Fehling, Karl Hedrick, Tony Hitchcock, Petros Ioannou, Robert Johnston, Gordon Newell, Charles Pearson, Thomas Pollock, Masayoshi Tomizuka, Pravin Varaiya, Richard White, and Wei-Bin Zhang participated in the interviews. Our appreciation goes to Steven Shladover for his comments on this chapter.

This work was performed as part of the California PATH (Partners for Advanced Transit and Highways) Program of the University of California, in cooperation with the State of California, Business, Transportation, and Housing Agency, Department of Transportation, and the U. S. Department of Transportation, Federal Highway Administration.

REFERENCES

1. R. W. HALL, Cost/Benefit Analysis of Automated Highway Systems, Prepared for Federal Highway Administration, Contract DTFH61-93-C-001-99 (1995).
2. Lawrence Livermore National Laboratory, Analysis of Automated Highway System Risks and Uncertainties, Prepared for Federal Highway Administration, Contract DTFH61-93-C-001-99 (1995).

CHAPTER 17

SOCIETAL AND INSTITUTIONAL ASPECTS OF AHS DEPLOYMENT

WILLIAM B. STEVENS

1. COMMUNITY CHOICE

State and regional transportation planners are faced with an array of transportation problems to solve. After considering "nontransportation" solutions such as telecommuting* and demand management, the planners may decide that the surface transportation system requires expansion and/or modification to (1) reduce congestion, (2) handle new, projected growth; or (3) provide new connections to enhance mobility. An AHS will be one of many options that the planners will have to consider in meeting the transportation needs. Other options may include new or expanded nonautomated roadway; transit buses, light rail or heavy rail. At the local level, an AHS would be selected because it offers a cost-effective solution to the area's transportation needs. Local planners will be able to tailor the AHS to be compatible with the community's transportation plans for moving more people and goods. The AHS will do this by increasing the efficiency of an existing highway while significantly increasing system safety.

*Telecommuting—working at home or at an outlying suboffice, often with computer support, as an alternative to commuting daily to an office.

WILLIAM B. STEVENS • National AHS Consortium, Bethesda, Maryland 20817.

Automated Highway Systems, edited by Ioannou. Plenum Press, New York, 1997.

2. LAND USE

Any urban area is concerned with land use. Land use planners may seek to change the existing urban form—the overall social, business, and economic profile of the area. Plans for industrial areas, residential communities, parks, and revitalization of blighted areas are developed to improve the overall quality of life and economic health of the community.

Traffic demands in a community derive from its urban form. Transportation planners, when considering changes to a transportation system, are expected to respond to the demands of the existing urban form as well as any longer-range plans for its change; AHS will be no exception.

A significant change to the surface transportation system may change a community's land use—either deliberately or inadvertently. For example, a commuter rail corridor may induce the growth of new businesses and apartments near its stations; if the rail is elevated through an area, it could contribute to decay of the neighborhood if not properly planned. A new highway that extends from the central business district to beyond the existing suburbs to interconnect with another highway may well encourage additional vehicular commuters if not properly planned.

As transportation planners integrate their transportation plans with the community's land use plans, the AHS will be viewed as one of the approaches to be considered. That is, when AHS deployment is considered, it will be as part of the integrated planning for the community. The unique characteristics of AHS, however, will need to be carefully considered in this integration and planning. Some of these unique aspects are described below.

2.1. *Tailoring to Community Needs*

One of the major findings of the AHS Precursor Systems Analyses (PSA; 15 studies of AHS issues and risks sponsored by the Federal Highway Administration in 1993) was that the AHS can have many forms; the technology—even when allowing for national compatibility standards—can be tailored to meet a wide variety of transportation services.(1) The AHS will support passenger cars, trucks, and buses in a varied array of highway configurations; an AHS system could be designed to support all four-wheeled (or greater) vehicles, either intermixed or on exclusive lanes. Below are some examples:

- *Exclusive transit vehicle lanes*—separate lanes could be set up for transit vehicles on certain highways; for example, a reversible express bus lane could be established on a major artery in a large urban area. AHS would allow the vehicles to operate more efficiently and safely, and with greater trip predictability. The exits could correspond to parking lots and/or to terminal points for local transit vehicles. "Dual-mode" transit vehicles will enable the route flexibility that riders demand.

- *High-occupancy vehicles (HOV) only in rush hour*—because of limited parking and street space in many central business districts, transportation planners could decide that only vehicles with multiple passengers—transit vehicles, vanpools, car-pool vehicles—would be able to use the AHS lane(s) in rush hour; this policy could be tied to demand management policies regarding parking spaces.
- *Exclusive commercial vehicle lanes*—in areas of high truck traffic, such as between major east coast cities, separate lanes could be established for heavy vehicles. As with the transit vehicle lane, the AHS technology would ensure safe, efficient movement of goods with far greater trip predictability. The lanes could be extended into nearby docking facilities.
- *Sparse rural areas*—AHS-equipped vehicles on rural roadways could be designed to operate intermixed with non-AHS vehicles to help prevent two major causes of rural crashes, namely, crashes with animals and road departure. The AHS equipment would maintain a safe distance from the vehicle in front and help detect obstacles, including large animals, in the roadway. It would also help the driver keep the equipped vehicle in its lane and avoid road departure crashes. The driver could choose to turn these services on or off as desired.
- *Roadway-powered electric vehicles (RPEV)**—a region could stimulate the electric vehicle market by supplying power from the roadway to keep the vehicle moving without draining the vehicle's batteries. The vehicle would continue off the highway via battery power. Since there is possible correlation between the hardware necessary for RPEV and AHS, the incorporation of RPEV into an AHS system seems to offer substantial benefits for an RPEV system.

2.2. AHS Impact on Land Use

Highway construction has an impact on a community; a new highway right-of-way may cut through a neighborhood, disrupting traffic flow, emergency vehicle access, and the sense of the neighborhood. Adding additional lanes to an existing highway is less disruptive, if the existing highway right-of-way is adequate.

One AHS lane can provide throughput capacity equivalent to a two- or three-lane highway, but its construction should have far less negative impact than building a new two- or three-lane highway. An AHS lane can be added as a supplement to the existing highway network by conversion of an existing lane or addition of a new lane. Disruption to the local communities should be small relative to construction of a new highway, rail or light-rail facility. However, there will be some disruption; access ramps to the AHS lane will need to be added, the AHS lane will need an

*Roadway-powered electric vehicles—electric-powered vehicles that are able to draw some or all of their power from the roadway.

occasional or continuous breakdown lane, and in some locations where the existing right-of-way is insufficient to support the AHS conversion and/or addition, the right-of-way will need to be widened. Alternative approaches in high-density areas may be taken; for example, the AHS lanes could be elevated. This approach is being used in some communities today for the addition of new HOV lanes to existing freeways.[2,3]

The AHS could have a significant impact on a community's urban form. As noted above, the AHS lane, once constructed, would be used in accordance with the community's policies (e.g., bus and/or HOV only in rush hours, light vehicle only, trucks only); the AHS impact on urban form will be very dependent on the local policies.

The AHS may have characteristics similar to light rail in some areas since the AHS will add significant capacity in relatively confined spaces. If a manual HOV lane is converted to an HOV-only AHS lane, this could add a rush-hour people-movement capacity of an additional 4000 (two-person HOV) to 6000 (three-person HOV) passengers per hour in the same corridor. If the AHS lane is a bus-only lane, the peak-hour buses operating at close headways could approach the passenger capacity of a rail system.

One of the reasons for adding capacity to a highway system is to reduce congestion and thus reduce travel time. However, in the absence of policies to discourage it, increased highway capacity can result in induced demand that may offset some or all of the increased capacity. For example, it is possible that by the time new highway lanes open, the number of travelers on the highway has increased to the point that there is still congestion, and the number of vehicle miles traveled (VMT)* has increased. This could happen with AHS. Opening a new AHS, with its increased capacity, and in the absence of policies that reflect the community's desires, may result in increased VMT. And there may still be congestion; however, the congestion would be on the manual lanes—particularly at the AHS entry and exit points—not on the AHS lane(s). This is because the AHS will most likely be self-regulating; that is, it will accept vehicles up to, but not beyond, its optimum capacity; so by definition, the AHS lanes themselves will not become congested. The net result for AHS travelers, however, could be the same—waiting in long lines to enter AHS, and waiting in congestion after leaving AHS, will cause the AHS traveler's elapsed trip time to be unreliable and possibly not consistently shorter.

In summary, the goodness or badness of the AHS impact on a community will reflect (1) the extent of planning in preparing for the AHS use, (2) the extent to which the AHS is an integral part of the community's overall transportation planning, and (3) the coupling of the AHS with policies that reflect the community's desires. Each community will need to ask: will greatly increasing the passenger and vehicle capacity of an existing highway right-of-way create new demands and new land use that is a desirable part of the community's plan (e.g., enhance mobility and job access); or undesired? And how can land use and demand management policies help control or eliminate any undesirable results?

*Vehicle miles traveled—total miles traveled by a vehicle or vehicles on a trip or over a period of time.

Finally, planners will need to determine if AHS will contribute to the sustainability of the transportation system; that is, will AHS help meet the transportation needs of this generation without compromising the ability of future generations to meet their needs?

3. USER ACCEPTANCE

The potential users of AHS will primarily be those who use today's freeways. In addition, it is believed that some people who are concerned about driving freeways may feel comfortable traveling on the AHS. Highway users will choose to use the AHS if the perceived benefits of the system outweigh the costs. Different groups of users will have different perceptions of AHS costs and benefits; some of these are described below.

3.1. General Public

The general public will use AHS if they feel that it is a better, cost-effective way of getting to work, to the shopping center, to events, to vacation. As part of the PSA studies, four citizen focus groups were conducted.[(4)] After the system use was explained, there were discussions in which both positive and negative feelings about AHS were expressed. The groups seemed to have relatively little concern about the AHS technology and computer control, perhaps because it was a discussion of a hypothetical system, or perhaps because the public is exposed to high technology in virtually every aspect of life today and the public has generally become accustomed to it (for example, there were several references to the use of in-car cellular telephones, televisions, portable computers, and fax machines).

The concept of being able to increase the number of vehicles traveling on a highway using the existing roadway seemed attractive to the focus groups when translated into less congestion with shorter, more dependable trip times without having to build new freeways. Many felt that the ability to have free time for office work or for leisure while commuting or while on vacation, would also be a major advantage as would travel without the strain (or risk) of freeway driving. Some of the more enthusiastic said they would be the first to sign up for AHS.

One of the major concerns was cost—how much and who pays; even the enthusiastic said the cost would have to be reasonable. There were concerns about whether the AHS costs might exclude some groups of people. Also, there was concern about increasing pollution because of more cars on the road. Safety was felt to be important but was assumed to be a given; that is, for the system to be introduced, it would need to be very safe.

3.2. Trucking

Studies of the trucking industry were conducted as part of the PSA.[(2,5,6)] It was felt that trucking firms might be the first to sign up for AHS use if, in fact, the benefits were there. The trucking firms would be attracted to AHS by the promise of dependable and reliable trip times, and fewer accidents. It was also noted that the in-vehicle AHS equipment cost would be a much smaller percentage of a $100,000 truck cost than a $20,000 automobile cost.

Some potential shipping market segments were identified as potentially having an interest in AHS.[2] The long-haul tractor–trailer operators would be attracted to AHS as a means of reducing and/or eliminating run-off-the-road and exit crashes involving excessive speed. The larger problem for long-haul shippers, however, is the driver—the shortage of good drivers, driver fatigue, and cost. Several approaches for how AHS would help in these problems were suggested. At the least, it was felt that drivers would become less fatigued on an AHS; in fact, some suggested that drivers could perhaps doze on the long intercity segments; this would allow more miles to be covered by a driver in a 24-hour period. It was also pointed out that this approach could face problems from the unions—and from the public if the trucks were intermixed with lighter vehicles. Also, it was pointed out that there are other reasons for the truck driver, as a professional, to stay awake since he has responsibility for ensuring that the load and the trailer are okay, and must be prepared to respond if there are malfunctions. In theory, these conditions could be monitored automatically, but given the variety of load types, this would need some careful analysis. Carrying this further, it was suggested that if trucking lanes were separate, some trucks could eventually travel without drivers; one example given was truck-only lanes among intermodal distribution points. Daimler Benz[6] addressed the problem in their research by assuming caravans of trucks; the first truck would have a driver, and the second or third following trucks would be driverless and electronically linked to the first truck.

The Calspan study by Princeton University pointed out that many shipping companies are now using rail for trips of over 1000 miles because it is less expensive. The AHS potential for long-haul truck shipments would be limited by that trend unless driver costs can be reduced. Another possible limiting factor is the large number of independent owner-operator truckers on the road. Whereas the large trucking firms might see the benefit of an AHS and invest in it, the independent truckers are operating on a narrow margin; an added investment might be too much for some of them.

Delivery vans and light trucks in urban areas would have the same benefits as the other vehicles in the community; but truckers, in particular, would value the dependable trip times of an AHS. In some areas, the sizes of the vehicles using AHS could conceivably be limited by communities seeking to reduce heavy truck traffic in central business areas; also, a community might choose to limit truck access to AHS by time of day. For example, an AHS lane that is used only for buses and HOV vehicles in peak periods, could be used for light delivery trucks and buses in off-peak periods.

3.3. *Transit*

Transit use of AHS was analyzed by three of the PSA contractors, namely, Calspan, Raytheon, and BDM. Bus use of AHS lanes would have the same advantages as truck use: dependable, shorter trip times with higher safety. Once on the AHS, trip times would be very predictable.

The real power of coupling AHS with transit buses, however, comes if there are AHS lanes dedicated to buses in dense urban areas. In this case, the AHS buses take on operational characteristics that are very similar to a rail system. Passenger capacity of such a dedicated lane in peak periods would be limited by the number of buses and the availability of terminals in the central business district to handle the very high passenger flow. One PSA study was of the reverse bus lane in the Lincoln Tunnel between New Jersey and Manhattan.[2] It was shown that the increase in the number of passengers possible with AHS would be equivalent to three HOV lanes (with three-occupant restriction). This increase would be possible because AHS allows the buses to operate at closer headways. The tight lateral control tolerances of AHS would allow buses to stop very close to loading platforms for rapid loading and unloading, and for wheelchair access. These platforms could be along the bus lane or in restricted terminal areas, similar to a rail system. Operation could include narrow at-grade lanes, elevated guideway types of lanes, or even tunnels.

Outside of the central business district, a dedicated AHS bus lane might correlate to a rail system's impact on land use. For example, the areas along the AHS transit/HOV corridor might attract new offices and residences (apartments, town homes, single-family homes). Bus terminals could be located along the AHS lanes for transfers to and from local bus systems. However, unlike a rail system, the HOV buses will be able to exit from the AHS corridor and continue traveling through local neighborhoods into the residential neighborhoods. For this reason, the convenience and timeliness of access to AHS by commuters and shoppers are better than with a rail system where the travelers may first need to take a bus or drive to the rail station. With AHS, there would be less need to transfer or park at the suburban terminal; the transit vehicle or HOV vehicle in the neighborhood is the same vehicle that travels to the central business district or to the shopping center. Because of this, it could be argued that the degree of AHS land use impact along the travel corridor would be less than with a rail system; the argument by real estate developers along the travel corridor for greater ease of access would be less compelling with AHS.

4. EMISSIONS AND FUEL CONSUMPTION

Society is concerned about the quality of the nation's air and the increased consumption of the world's liquid fuel resources. A critical aspect of any transportation project is its net effect on the environment. AHS-related air quality issues are of particular importance because of the mandatory nature and strict standards of the Clean Air Act as amended.[2] Society generally wants "green" transportation systems.

In the 1991 Intermodal Surface Transportation Efficiency Act (ISTEA), VMT was identified as a primary control parameter. PSA researchers[2,3] suggested that AHS will help lower emissions per VMT by smoothing out traffic flow. Also, it was shown that platooned vehicles will reduce air resistance and therefore reduce emissions per VMT.[3] The AHS will also enable reduced emissions through support of transit, HOV traffic, congestion pricing, and the introduction of alternative propulsion (low and zero emission) vehicles. As mentioned above, Calspan showed that AHS technology could help establish the feasibility of an RPEV system.[2]

The concern is that an AHS might encourage/induce more internal combustion engine VMT; if so, then overall emissions and fuel consumption may increase even though emissions are reduced on a per-vehicle-mile basis. One PSA study showed that travelers, when given added highway capacity, will tend to increase their VMT until they use up a given amount of travel time.[4] By coupling AHS with demand management and land use planning, planners may be able to both meet the expanding transportation needs *and* reduce emissions. There was some question, however, about the reasonableness of using VMT as a clean air metric given the continued improvements in tail pipe emissions and the additional benefits of smooth operation offered by AHS; the particulate metrics of the Clean Air Act Amended may provide a more balanced view.

In short, AHS research to date is incomplete; there is a need for additional accurate and conclusive research to demonstrate the system's net effect on the environment.

5. LEGAL AND LEGISLATIVE ASPECTS

Just as the AHS must fit into a community, it must also be consistent with the legal and legislative structures that have been established by our society. The introduction and operation of an AHS, however, will present some new situations that may require new interpretations as AHS is fit into our legal context. Some of the major areas are discussed below.

5.1. *Tort Liability*

On today's highways, drivers can turn on the cruise control feature, and the system will automatically control the throttle of the vehicle and keep the vehicle at a constant speed. This is an early form of automated vehicle control. The driver can choose to turn the cruise control system on or off at any time, and the driver is expected to maintain awareness of the driving situation; if the driver crashes into the rear of another vehicle, the driver is responsible, not the cruise control unless the cruise control malfunctioned. This is a "driver-in-the-loop" system because the driver is responsible for the vehicle's operation even though he or she may temporarily delegate control to the cruise control. Other forms of automated vehicle control will be introduced to the public over the next few years; they will be driver-in-the-loop systems because the driver will maintain primary responsibility for the vehicle operation.

An AHS will be different. The major safety and throughput benefits of AHS come from the fact that manual drivers are not allowed on the dedicated AHS lanes, and drivers will not be able to assume control until the system allows it. The AHS is mostly a "driver-out-of-the-loop" system. When the AHS assumes control of the vehicle, "the system" also assumes a major level of liability for the consequences of any malfunction. Responsibility on an AHS will be shared by the instrumented AHS highway lanes (including the systemwide traffic management operations), the AHS instrumentation of the vehicles traveling on the AHS lanes, and the driver who will be responsible for maintenance and condition of the vehicle and its equipment.

There are several kinds of events that could occur on an AHS; each might have different sets of responsibilities. Below, some of the events are described and possible responsibilities are postulated:

- *Malfunction of roadside-based AHS*—the AHS owner/operator would probably be responsible for any malfunctions of roadside AHS components that might occur; the owner/operator may have recourse to the manufacturer.
- *Malfunction of vehicle-based AHS*—depending on the malfunction, this could be the vehicle owner (e.g., he or she is responsible for its proper maintenance); the system operator (e.g., it is the operator's responsibility to not allow anyone on the AHS lane whose AHS equipment is not under warranty or who has not had an authorized vehicle inspection in the past 6 months); the AHS equipment manufacturer (it failed during warranty); or the state and/or Federal government that established guidelines and procedures to ensure each vehicle's safe operation; or some or all of the above.
- *Malfunction of the vehicle*—again, depending on the malfunction, this could be the vehicle owner (e.g., responsible for its proper maintenance); the system operator (e.g., allowed an unauthorized vehicle on the lane); or the vehicle assembler or component manufacturer (e.g., a tire blowout occurred because of a manufacturer's defect); or some or all of the above.
- *Other*—e.g., earthquake, plane crashes in the road, truck falls off of an overpass onto AHS. Presumably there will be some kind of liability insurance that the owner/operator will need to carry to cover these types of incidents; negligence would also be a factor.

Appropriate and acceptable models for risk sharing among vehicles, infrastructure, and operators will need to be established.

Tort liability was not seen as a "showstopper" during the PSA studies if costs are controlled and safety is secure.[7] The ongoing ITS program will provide some basis for predetermining the conditions for AHS. But these questions remain[1]:

- Should federal legislative protection be sought to limit the amount of punitive damages that can be awarded?
- Should the user be expected to accept limited liability through a "user agreement" (perhaps as part of an AHS licensing procedure)? Are there driver and vehicle performance indicators that would serve as probable cause for police intervention?

- Can or should a mediation process be established to avoid nuisance lawsuits?

Efforts at tort reform, unrelated to ITS/AHS, are under way including changes to state/federal law, creative regulatory approaches, and insurance industry involvement.[2] The higher the level of command and control exercised and accepted, the more complex the resolution may need to become. Legal opinion[7] noted that the positive control of AHS should reduce the potential for serious crashes, thus reducing the total liability cost per vehicle mile of travel; this savings could be used to help compensate stakeholders who must assume shifts in liability. The most effective (and proactive) method to control costs related to liability is through careful product design and careful system operation. This may best accomplished by a fair and open regulatory framework.

5.2. *Privacy/Enforcement Issues*

The ITS program recognizes that society is concerned about the potential for invasion of a person's privacy; this concern is linked to living in the information era. The PSA research suggested that appropriate conditions can be designed into an AHS to ensure privacy; nevertheless it is one more area in which individuals who choose to use modern services will be electronically identified, and potentially tracked—hence the concern. An AHS may introduce a level of complexity over current ITS technologies which will require specific design attention to better ensure the public that information is not being misused.[6,7]

5.3. *Compliance with Current Regulations*

The AHS will be subject to the legislation that currently affects today's highway projects. It is also likely that the AHS will be heavily affected by regulations and legislation in other areas, such as Federal Communications Commission mandates. These could be regulations that have already been or will be enacted for other valid reasons such as regulation and operations of commercial vehicles and hazardous materials, or protection of the public right-of-way. As the AHS program progresses and the requirements of an AHS are determined, it will be necessary to stay up-to-date on the impacts of current and pending legislation, and to identify legislation that may be necessary for AHS deployment.[2]

5.4. *Special Licensing*

States may decide that special licenses should be considered for drivers using the AHS service—or potentially the partially automated vehicle control services as well. This would allow the states to ensure that the drivers understand and agree to their responsibilities on AHS; for example, drivers could be responsible for not overloading their vehicle, keeping their vehicle in good operating condition, not carrying oversized loads, not throwing debris on the roadway, getting state inspections periodically, and not tampering with the AHS equipment. Although some of these things could be verified at vehicle check-in, the state may choose to emphasize that if an incident occurs because of the driver's actions (e.g., an oversized load comes loose on the ramp approaching the check-in station), he or she is responsible.

As part of ensuring consistency and safety among the states implementing AHS, there will need to be some standardization at the national level. The PSA research suggests that the FAA regulatory model may supply an insight for important aspects of implementing a federal inspection program[(7)]; the model itself is probably far too complex, but might give some guidance if it were determined that some federal licensing guidelines were necessary and desirable.

6. EQUITY

The cost of the AHS and how it is paid for will be of increasing concern. In particular, equity in the use of the system will need to be considered. For example, should the system be available to the entire public or just for users who can afford the tolls and the options on their vehicles? This is an issue generic to public services in general, and to the broad ITS program of transportation improvements—and it may be an issue that will be resolved in different ways by different transportation planning organizations. A restrictive deployment could be subject to criticism even though AHS is expected to reduce congestion on both AHS and non-AHS roads. On the other hand, an AHS could be implemented in an area—particularly in an initial AHS deployment—in a manner so that only a few of the general public would directly benefit from the service (e.g., a truck-only lane). Covering the cost of such a system through a broad-based taxation plan could meet with resistance. Each region will need to consider the demographic and economic impacts of its AHS installations and pricing to avoid mistakes. A few of the issues are as follows:

- What contributions, if any, should be made by the federal government, state government, local communities, regional taxing authorities, and system users (truckers, transit authorities, and personal vehicle drivers)?
- Should the state and/or federal government provide incentives and/or help to individuals to equip their vehicles with AHS instrumentation (particularly on initial or test systems)?
- If a totally toll-financed system reduces taxing concerns, will everyone be able to afford access, or will it raise concerns about discrimination for people with lower incomes? Would the public utility type of AHS management and ownership reduce this concern?

7. AHS OPERATION AND MAINTENANCE

Improvements to the highway system have traditionally been handled by state and local transportation agencies. At some point, these same agencies will be considering AHS.

The AHS will introduce a new, high-technology level of complexity to those organizations that are responsible for highway functions and services. In some concepts, the AHS lane instrumentation would include advanced electronic sensors, computers and software, and integrated communications systems. Installation and maintenance of these systems may present a significant challenge to the operators.[1] Currently, state and local transportation authorities lack the manpower and technical expertise to operate and maintain an advanced AHS system with a lot of control in the infrastructure.[8] For example, maintenance of roadside electronics may involve relatively frequent circuit and/or software testing, component replacement, and system integration testing as the replacement components are brought on-line. An advanced AHS may employ traffic management functions that may involve real-time system monitoring; the operators for such a system will need special training.

Planners who recommend AHS must also realize that the funds for systems operations and maintenance must be adequate and consistent; the funds must be included in the operating budget as a nonnegotiable item.[8]

Some of the major operations and maintenance considerations in planning an AHS include the following:

- State transportation organizations are evolving. As planning for AHS begins, funds to build up and evolve the state's transportation departments will need to be made available so that technical staff can be hired and trained. Career paths will need to be established, job descriptions created, and so forth. This front-end cost will increase state DOT costs long before the AHS becomes operational.
- Facilities management firms can be hired. Full-service management of the AHS infrastructure could be privately provided. However, this could introduce questions regarding the liability of these firms when incidents occur.
- Facility ownership could be private, such as a private toll road.
- A separate public utility type of organization could be established to fund, build, and maintain AHS. Would an AHS commission need to be established?
- Insurance companies and insurance regulators will need to assess the impact of AHS operation on rates.
- Programs for inspection of AHS vehicles will need to be established.

Transportation agencies are beginning to realize that coordination and cooperation are essential to succeeding with current and future transportation improvements. However, many agencies experience a variety of organizational fragmentation both internally and across jurisdictions, especially with the transportation system growing out of a role that was strictly building and maintaining roadways to an expanded role that includes installing and upgrading electronic equipment. Many of the problems currently facing state and local transportation organizations have solutions that may rely on regional cooperation for optimal results.

Recently there have been indications of a gradual shift to deal with transportation issues on a regional basis erasing traditional local boundaries. Transportation agencies and urban planning organizations are beginning to recognize the value of cooperative efforts for major transportation endeavors. The formation of regional transportation authorities allows increasingly difficult transportation problems to be addressed and allows a changing transportation paradigm, which includes more technology-driven applications, to be accommodated. An AHS is likely to require an even higher level of coordination because of the complexity of the operations.(8)

8. NATIONAL CERTIFICATION

AHS planning assumes that the AHS will be a national system; that is, vehicles from one state will be able to operate in any of the other states without modification or redesign. Since the AHS will be tailorable to meet each community's needs, the national compatibility will be subject to restrictions that may be imposed in local communities regarding vehicle types and sizes; for example, heavy vehicles may be excluded from some central business districts.

National compatibility implies that there will be rigorous performance and design standards established for the manufacturing, operations, and maintenance of AHS components, both for components located on the vehicles and for those located on the roadside or traffic management centers. There could also be standards for driver examinations as well. Certainly standards organizations and the National AHS Consortium will be involved in establishing the standards. Certification that newly introduced components meet the standards will need to be considered. This certification could range from national testing organizations that test new products to state-operated inspection stations that periodically certify that individual vehicles are capable of AHS operation.

ACKNOWLEDGMENT. This chapter has been coordinated with selected participants of the National AHS Consortium (NAHSC) and the Federal Highway Administration. The views expressed are the author's and are not necessarily those of the NAHSC, its participants, the Federal Highway Administration, or The MITRE Corporation.

REFERENCES

1. L. BONDERSON, Report Overview, Precursor Systems Analysis Studies, Delco/Hughes Electronics, Dec. 1994.
2. J. ELIAS, Overview of the Precursor Systems Analyses Studies, Calspan Corp., Dec. 1994.
3. M. MARTIN, Commercial Trucking and Transit Analysis of AHS, Precursor Systems Analysis Contract, BDM Corp., Dec. 1994.
4. J. PITTENGER, Final Report, Precursor Systems Analyses Studies, Battelle, Dec. 1994.
5. S. SHLADOVER, Final Report, Precursor Systems Analyses Studies, PATH, Dec. 1994.

6. W. STEVENS *et al.*, Summary and Assessment of Findings from the Precursor Systems Analyses of Automated Highway Systems, MITRE, July 1995.
7. C. VICK, Institutional and Societal Aspects of AHS, Precursor Systems Analysis Contract, SAIC, Dec. 1994.
8. T. ZIMMERMAN, Commercial Trucking and Transit Analysis of AHS, conducted by Daimler Benz for Raytheon Corporation's Precursor Systems Analysis Team, Dec. 1994.

INDEX

Printed in the United States
97155LV00003B/107/A